Dengue

Tropical Medicine: Science and Practice

Series Editors: Geoffrey Pasvol
Dept. of Infection & Tropical Medicine
Imperial College London, UK

Stephen L Hoffman
Sanaria Inc., USA

Published

Vol. 1 Lymphatic Filariasis
Edited by Thomas B Nutman

Vol. 2 Amebiasis
Edited by Jonathan I Ravdin

Vol. 3 Schistosomiasis
Edited by Adel AF Mahmoud

Vol. 4 Malaria: A Hematological Perspective
Edited by Saad H Abdalla & Geoffrey Pasvol

TROPICAL MEDICINE
Science and Practice
Volume 5

Dengue

edited by
Scott B. Halstead
International Vaccine Institute, Korea

Series editors:
Geoffrey Pasvol and Stephen L. Hoffman

ICP
Imperial College Press

Published by

Imperial College Press
57 Shelton Street
Covent Garden
London WC2H 9HE

Distributed by

World Scientific Publishing Co. Pte. Ltd.
5 Toh Tuck Link, Singapore 596224
USA office: 27 Warren Street, Suite 401-402, Hackensack, NJ 07601
UK office: 57 Shelton Street, Covent Garden, London WC2H 9HE

British Library Cataloguing-in-Publication Data
A catalogue record for this book is available from the British Library.

DENGUE
Tropical Medicine: Science and Practice — Vol. 5

ISBN-13 978-1-84816-228-0
ISBN-10 1-84816-228-6

Typeset by Stallion Press
Email: enquiries@stallionpress.com

Printed in Singapore by Mainland Press Pte Ltd

This book is dedicated to my loving wife, Tot,
whose support has made this book possible

Contents

Contributors

Alan D. T. Barrett, PhD
Department of Pathology
University of Texas Medical Branch
301 University Blvd.
Galveston, TX 77555-0609, USA

David W. C. Beasley, PhD
Department of Microbiology and Immunology
University of Texas Medical Branch
301 University Blvd.
Galveston, TX 77555-0609, USA

Timothy P. Endy, MD, MPH
Associate Professor of Medicine
Chief, Infectious Disease Division
Department of Medicine
SUNY Upstate Medical University
725 Irving Avenue, Suite 304
Syracuse, NY 13210, USA

Francis A. Ennis, MD
Center for Infectious Disease and Vaccine Research
University of Massachusetts Medical School
55 Lake Avenue North, Room S6-862
Worcester, MA 01655, USA

Jeremy Farrar, FRCP, FRCP (Ed), FmedSci, PhD, OBE
Director, Oxford University Clinical Research Unit
The Hospital for Tropical Diseases
190 Ben Ham Tu, Ho Chi Minh City, Viet Nam
Professor of Tropical Medicine
Oxford University
Professor of International Health
London School of Hygiene and Tropical Medicine

Norman G. Gratz
Formerly, Chief, Vector Borne Diseases Division
World Health Organization (WHO)
Avenue Appia 20, CH-1211
Geneva 27, Switzerland

Scott B. Halstead, MD
Director, Pediatric Dengue Vaccine Initiative (PDVI)
Supportive Research and Development Program
5824 Edson Lane, N. Bethesda, MD 20852, USA

Brian H. Kay, AM, PhD, FAA, FACTM
Deputy Director, Australian Centre for International
and Tropical Health and Nutrition (ACITHN)
Professor in Tropical Health
Queensland Institute of Medical Research
Post Office Royal Brisbane Hospital
Brisbane, QLD 4029, Australia

Elli Leontsini, MD, MPH
Associate, Social and Behavioral Interventions Program
Department of International Health
Johns Hopkins Bloomberg School of Public Health
615 North Wolfe Street, Room E5034
Baltimore, MD 21205-2103, USA

Linda S. Lloyd, DrPH
Public Health Consultant
3443 Whittier St.
San Diego, CA 92106, USA

Ananda Nisalak, MD
Department of Virology
Armed Forces Research Institute of Medical Sciences
USAMC-AFRIMS, APO, AP 96546, USA

Alan L. Rothman, MD
Center for Infectious Disease and Vaccine Research
University of Massachusetts Medical School
55 Lake Avenue North, Room S6-862
Worcester, MA 01655, USA

LTC John M. Scherer, PhD, MT (ASCP)
United States Army Research Institute of Infectious Diseases
1425 Porter Street
Fort Detrick, MD 21702, USA

Wellington Sun, MD
Chief, Dengue Branch
Division of Vector-Borne Infectious Diseases
Centers for Disease Control and Prevention
1324 Calle Canada, San Juan, Puerto Rico 00920

David W. Vaughn, MD, MPH, FAAP
Director, Global Clinical Research and Development
GlaxoSmithKline Biologicals, North America
2301 Renaissance Boulevard, Mail Code RN0220
King of Prussia, PA 19406-2772, USA

Bridget Wills, MRCP, DM
Senior Clinical Research Fellow
Oxford University Clinical Research Unit
The Hospital for Tropical Diseases
190 Ben Ham Tu, Ho Chi Minh City, Viet Nam
Honorary Senior Lecturer/Consultant in Paediatric Infectious Diseases, St. Mary's Hospital
Imperial College Healthcare NHS Trust, London, UK

Peter Winch, MD, MPH
Associate Professor
Director, Social and Behavioral Interventions Program
Associate Chair, Department of International Health
Johns Hopkins Bloomberg School of Public Health
615 North Wolfe Street, Room E5030
Baltimore, MD 21205-2103, USA

Preface

Dengue is an enormous problem. It is both local and global. In its behavior, dengue most nearly resembles the intimate and ubiquitous respiratory diseases that make life miserable for human beings on an equal opportunity basis and whose treatments fill the pages of standard medical textbooks. In hundreds of millions of households on earth there lives a stealthy insect. It is happy in its domesticity, rearing its generations on human blood. Dengue, the commensal virus that this insect harbors, is perfectly adapted to a generational cycle — man to mosquito to man. It evolved from a common ancestor virus, no doubt one harbored by subhuman primates and transmitted by jungle mosquitoes. A period of isolation and the accumulation of a succession of small mistakes in RNA transcription resulted in the evolution of four distinct serotypes. Serial infection with one or more types is now the fate of perhaps 2.5 billion people who live in tropical and subtropical countries. This book describes aspects of the dengue story in a format designed for students at various levels. It is not a comprehensive presentation of contemporary dengue research, but rather a compact, accessible and authoritative description of major biological features in the complex story of dengue as a human disease.

Dengue: Overview and History. This chapter, written by the Editor, reviews the history of the term dengue, the viruses associated with the dengue syndrome and the history of dengue hemorrhagic fever. Statistics are provided showing the impact of dengue in terms of morbidity, mortality and cost during the past 40 years.

The Infections Agent. This chapter, written by highly productive arbovirologists Drs. David Beasley and Alan Barrett, University of Texas Medical Branch, Galveston, describes the genetic and amino acid sequences of the four dengue virus types and their genotypes, the structure of the entire virus and of key surface proteins. Descriptions are provided of the structure of gene products along with a summary of the current understanding of mechanisms of entry of dengue into cells, uncoating, transcription of RNA, translation of RNA, the composition and function of specified proteins, and the respective contribution of viral and host enzymes and replicative machinery to viral replication.

Epidemiology. In this authoritative chapter, the Editor describes the bionomics of dengue mosquito vectors together with many of the elements that contribute to influencing the transmission of dengue viruses, including the mosquito, factors in the environment and the complex controls contributed by the host.

Resistance to Infection. US Army physicians, led by Dr. David Vaughn, who spent more than two decades conducting research on dengue vaccines, have written this chapter. They summarize knowledge of the antibody responses to dengue infection and the role of antibodies in protection against dengue infections. The current status of dengue vaccine development is reviewed.

Clinical Features of Dengue. Written by Dr. Jeremy Farrar, Director, Oxford University Clinical Research Unit, Ho Chi Minh City, Vietnam, this chapter describes the signs and symptoms of mild and severe dengue diseases, including unusual manifestations.

Treatment. Principles and details of treating children and adults with mild and severe dengue syndromes are given by Dr. Bridget Wills, an acknowledged world authority who conducts clinical research on dengue illnesses at the Oxford Clinical Research Unit, Ho Chi Minh City, Vietnam.

Pathogenesis. This is the beginning of the story of severe dengue disease. The Editor describes the role that dengue antibodies play in governing the outcome of infection by neutralizing or enhancing dengue virus infections, providing controls for the major afferent mechanism that governs the intensity of dengue infections and the severity of dengue diseases.

Elimination of Infection. This chapter, by Dr. Alan Rothman, Professor of Medicine, University of Massachusetts Medical Center, a world authority on immune responses to dengue infection, summarizes the rich contemporary research experience on the cellular immune response to dengue virus infection. Emphasis is on the kinetics and specificity of the cell-mediated responses to dengue infection and the role these play in elimination of dengue-virus-infected cells and in producing the pathophysiological changes of dengue diseases.

Pathophysiology. The Editor describes how and when immunological responses to dengue infections generate chemokine and cytokine responses. Evidence concerning the role that cytokines play in the pathophysiological responses to dengue infection is presented and discussed. New findings that provide amazing insights into the compelling story of antibody-dependent enhancement of dengue infection are presented.

Diagnosis. Diagnosis of recent primary or secondary dengue and other flavivirus infections is presented. Included are serological methods, virus isolation, detection of viral antigen by various methods, and the use of PCR and molecular probes to detect and quantitate viral nucleic acid in blood or infected tissues. This chapter has been written by Prof. Timothy Endy along with dengue virologists who served as virologists at the Armed Forces Research Institute of the Medical Sciences in Bangkok, Thailand, and at the Walter Reed Army Institute of Research, Washington, D.C.

Mosquito Control. Control of mosquito vectors is covered in three chapters. Chapter 11 provides a seminar on classical methods of the control of dengue vector mosquitoes. It has been written by an authority with a specialist experience or perspective — the late Dr. Normal Gratz, Chief of Vector Control, World Health Organization. This chapter assesses the strengths and weaknesses of contemporary vector control, including the proper and improper use of insecticides and other mechanical or chemical antilarval methods. In Chap. 12, Dr. Brian Kay, Director, Entomology, Queensland Institute of Medical Research, Brisbane, Australia, a world authority, discusses biological control of mosquito larvae. And in Chap. 13, Drs. Peter Winch and Elie Leontsini, Associate Professors, Department of International Health, Johns Hopkins School of Hygiene and Public Health, review methods for community-based programs to effect the control of dengue mosquito vectors.

Controversies. Unique to this series are discussions on major controversies in dengue pathogenesis and diagnosis. Dr. Jose Rigau-Perez, formerly with the CDC Dengue Laboratory in Puerto Rico, critiques the usefulness of the current WHO case definition of dengue hemorrhagic fever as an epidemiological and clinical management tool, with rebuttal by the Editor. David Morens, Assistant to the Director, NIAID, NIH, and the Editor present and discuss the pros and cons, respectively, of four major controversies relating to the causation of severe dengue disease.

Acknowledgements. Many thanks to Lizzie Bennett of Imperial College Press, London and Joy Quek of World Scientific Publishing Company, Singapore for their assistance in preparing, revising and tracking the submission of manuscripts. The Editor wishes to thank the contributors to this book, who remained faithful despite an editing process that consumed much of a decade. The months that the Editor has spent traveling the world to learn about dengue and the lonely hours spent writing and rewriting the dengue story have exacted a price in abandonment and missed social occasions, the brunt being borne by his lovely and tolerant wife. He owes her debts far beyond repayment.

1

Dengue: Overview and History

Scott B. Halstead

Dengue fever is a benign syndrome caused by several arthropod-borne viruses and characterized by biphasic fever, myalgia or arthralgia, rash, leukopenia and lymphadenopathy. Dengue hemorrhagic fever (DHF) is a severe, often fatal, febrile disease caused by dengue viruses characterized by abnormalities of hemostasis and capillary permeability that leads, in severe cases, to a protein-losing shock syndrome (dengue shock syndrome, DSS). It is thought to have an immunopathological basis. There are no synonyms for dengue viruses in current use in the English language. There is general agreement that there are four antigenically distinct members of the dengue subgroup of the genus Flavivirus.[1,2] These are named dengue types 1 through 4. Historical synonyms for the dengue fever syndrome are bilious remitting fever, la dengue, dandy fever, breakbone fever and breakheart fever. Synonyms for DHF include infectious thrombocytopenic purpura, dengue shock syndrome, and Philippine, Thai or Singapore hemorrhagic fever.

Host Range

Inoculation of dengue strains of known human pathogenicity does not produce demonstrable infection in adult chickens, lizards, guinea pigs, rabbits, hamsters or cotton rats.[3,4] Many genera of subhuman primates are susceptible to infection by dengue viruses. Species belonging to *Macacus*, *Pongidae*, *Certhopicicus*, *Cercocebus*, *Papio*, *Hylobates* and *Pan* can be infected by bites of virus-infected mosquitoes or by injection of infectious virus preparations.[4,5] Infection is essentially asymptomatic but viremia occurs at levels sufficient to infect mosquitoes. Simmons and colleagues[5] were the first to note that wild-caught *Macaca philippinensis* resisted dengue infection, whereas *Macaca fuscatus* (Japanese macaque) were susceptible. Work in Malaysia has revealed an extensive jungle dengue transmission cycle involving canopy-feeding monkeys and *Aedes niveus*, a species that feeds on both monkeys and humans.[6] In the early 1980's an extensive epizootic of dengue virus type 2 involving subhuman primates was recognized over a wide area of West Africa.[7] During successive epizootics, dengue 2 strains have been recovered from humans after being transmitted by mosquitoes that also feed on subhuman primates.[8] Sylvatic dengue 2 strains were poorly transmitted by the urban vectors *Aedes aegypti* and *Aedes albopictus*.[9] By contrast, *Ae leutocephalus* and *Ae furcifer*, probable sylvatic vectors in Africa, readily transmitted urban dengue 2 viruses.[10] As yet, urban dengue 2 viruses have not been isolated from infected subhuman primates in Africa. From genetic and epidemiological studies, it has been concluded that urban human dengue and jungle monkey dengue are relatively compartmentalized.[11] The full geographic range of the zoonotic cycle, the range of mosquito vectors and the composition of the subhuman primate zoonotic reservoir are not known.

In the urban cycle, dengue is vectored by anthropophilic mosquitoes that breed in and around human habitations. The virus travels along routes of transportation. Although there are important genetic differences between human and monkey dengue viruses and it is surmised that zoonotic viruses do not readily enter the urban cycle,[11,12] the question whether dengue viruses can be reintroduced from the sylvatic cycle may become important if intensive vaccination should eradicate one or more types from a major region. If human herd immunity lessens and if populations of *Aedes aegypti* are left undisturbed, zoonotic dengue viruses might reseed into the urban cycle.

Geographical Distribution

Dengue fever outbreaks have been documented on every continent except Antarctica.[13] During the 18th and 19th centuries, epidemics occurred in newly colonized lands, largely because of the necessity for domestic storage of water in frontier areas. Shipboard or garrison outbreaks often confined to nonindigenous settlers or visitors were reported in Africa, the Indian subcontinent and Southeast Asia.[3,14–16] It is not known how many dengue viruses have been introduced into the Western Hemisphere since international commerce began after 1492. The American genotype of dengue 2, first isolated in Trinidad in 1951, may have been transmitted in the American tropics for much of the 20th century and is likely to have been the cause of the outbreaks in Panama and Cuba prior to World War II.[17,18] During WWII, dengue virus infections were common among combatants in the Pacific, spreading to staging areas such as Japan, Hawaii and Polynesia.[19] In the 50 years since WWII, following the introduction of all four serotypes from Southeast Asia, major outbreaks of DHF have occurred in the Pacific islands, tropical America and South Asia.[13,16,20–22]

Dengue 1 of Asian origin appeared in the Caribbean in 1977, and spread to and remained endemic in coastal Central and South America, Brazil and Mexico.[16,20,23] This virus briefly visited the United States along the Texas–Mexico border. A sharp outbreak of Southeast Asian genotype dengue 2 hospitalized 116,000 persons on Cuba in 1981 — the hemisphere's first DHF/DSS epidemic.[24] In 1986–90, dengue 1 spread throughout most of coastal Brazil, and from there to Paraguay and then to Peru and Ecuador.[16] In 1990, more than 9000 dengue cases were reported from Venezuela; 2600 of them were classified as DHF, with 74 deaths reported. Dengue types 1, 2 and 4 viruses were isolated from this outbreak. Since then, DHF/DSS has spread to Colombia, French Guiana, Guyana, Brazil and Nicaragua, and to a smaller extent Puerto Rico. An interesting outbreak of American genotype DEN 2 resulted in thousands of secondary infections in Iquitos, Peru, but did not result in any DHF/DSS cases.[25] In 1995, dengue virus type 3 was introduced into the region.[16]

Dengue 1 and 2 viruses were recovered from humans with mild clinical illnesses in Nigeria in the absence of epidemic disease.[26] In 1983, dengue virus type 3 was recovered from Mozambique.[16]

DHF-like disease was described clinically in Thailand from 1950 and in the Philippines from 1953. DHF was first described in Singapore and Malaysia in 1962, Vietnam in 1963, India in 1963, Ceylon (Sri Lanka) in 1965, Indonesia in 1969, Burma in 1970, China in 1985, and Kampuchea and Laos from about 1985.[16,27,28] Large outbreaks of DHF were reported from Sri Lanka and India since 1988, French Polynesia since 1990, Pakistan in 1992 and Bangladesh in 2000.[16,29] These outbreaks of severe disease had been preceded by periods of silent dengue transmission.[30]

Disease Burden and Cost

Disease burden

For more than 40 years, countries endemic for DHF have reported hospitalized cases and deaths to regional WHO offices. Samples from reported cases, in some instances, have been serologically verified. While there may be underreporting, it is unlikely to be of an order of magnitude. For the period 1956–2004, Table 1 lists the geographic distribution of 4,975,807 DHF cases (an unknown but small number are dengue fever) and 68,977 deaths (1.4% C.F) reported to national health authorities. Clearly, DHF is a huge problem in Southeast Asia. Note that the 67,295 deaths over a 40-year period in Asia means an average of 1682 per year. The highest number of DHF deaths reported to WHO has never exceeded 2500.

Dengue fever cases are reported most notably from the American hemisphere and the Pacific Islands, with 6.9 million case reports over the past four decades (Table 2). The numbers have escalated decade by decade.

Table 1. Dengue hemorrhagic fever cases and deaths reported to WHO regional offices, 1956–2004.

Western Pacific		Southeast Asia		Eastern Mediterranean		Americas	
C	D	C	D	C	D	C	D
2,822,263	35,632	2,028,058	31,663			125,486	1682

Table 2. Reports of dengue fever cases by region, 1960–2004.

Years	Americas*	Pacific Islands†
1960–1969	88,219	
1970–1979	774,797	
1980–1989	1,030,723	11,175
1990–1999	2,708,631	43,550
2000–2004	2,200,603	50,085
Total	6,802,973	104,810

*Reports to PAHO. †Reports to WPRO.

The reliability and completeness of reports of dengue fever cases have not been subjected to research scrutiny.

When encountered by susceptible individuals, strains of some serotypes may result in inapparent human infections. For example, when a Southeast Asian genotype dengue 2 virus circulated in Santiago de Cuba in 1997, over 95% of primary infections in adults were silent.[31] But, in this same outbreak, virtually all adults who were immune to a different dengue serotype (dengue 1 in this instance) became ill when infected with dengue 2, with illnesses varying from classical dengue fever to DHF. Other types are more virulent. Virgin soil outbreaks of dengue 1 usually result in febrile disease, often with high attack rates.[32] Simmons *et al.*,[5] who infected susceptible adult volunteers, using mosquitoes serially transferred a single strain of dengue 1 and observed virtually 100% overt disease. In 1995, an American genotype dengue 2 was introduced into Iquitos, Peru. Only a small number of mild febrile illnesses were reported to health authorities, none with DHF.[25] It was surprising, then, to find that more than 80% of the population had been infected with dengue 2 while at the same time 60% experienced secondary dengue infections.

These studies are important, because despite the large number of cases of DHF and dengue fever reported throughout the world it can be anticipated that vastly more people are inapparently infected by one or more dengue viruses. Dengue antibody prevalence studies have not been reported from the Americas, making it impossible to estimate annual dengue infection rates for these areas. In some Southeast Asian countries annual dengue infection rates have been conservatively estimated at 10%. If this rate is applied to the at-risk population of children less than 15 years old in South and Southeast Asia, approximately 732.1 million[a], then as many as 73 million children may experience at least one dengue infection each year. A substantial but unknown portion of adults is susceptible to dengue. If Asian dengue infection rates are applied to the under-15-year-old populations of all South and Central American countries, 166.6 million (UNICEF Annual Report, 1995), then another 17 million dengue infections occur annually. This estimate does not allow for homotypic or heterotypic immunity, nor are dengue infections in adults estimated. Thus, 50–100 million dengue virus infections occur annually, and not 50–100 million dengue fever cases as often claimed.[33]

Dengue is rarely reported from Africa, although infections occur there.[8,15,16,34] As an explanation of the absence of disease in indigenous Africans, a 1997 seroepidemiological study of children of African ancestry living in Haiti showed high prevalence of dengue antibodies to all four types of prevalence without any recognized dengue disease.[35] These results suggest that blacks are genetically resistant and seldom express severe dengue illness and may only experience milder syndromes when infected.

Cost

Only a few local or country-level estimates have been made of the cost incurred by dengue, principally DHF. Estimates in Puerto Rico,[36] placed losses at US$8 million for a population of 2.5 million; in Cuba the 1981 DHF epidemic cost US$103 million,[24] including the cost of control measures (US$43 million), medical care (US$41 million), costs of lost productivity (US$14 million), and loss of salaries of adult patients (US$5 million) for 344,203 patients. Two outbreaks in Thailand, in 1976–77 and

[a] Data from 1993 global population estimates, UNICEF Annual Report, 1995. Included are the under-16-year-old populations of China (only 30% living in the southern provinces), Vietnam, the Philippines, Cambodia, Laos, Thailand, Malaysia, Singapore, Indonesia, Myanmar, Sri Lanka, Bangladesh (dengue endemicity not well documented), India (dengue endemicity not known in detail) and Pakistan (dengue endemicity not well known).

1994, cost US$7 million[23] and US$51 million,[37] respectively. In Vietnam, during the 1998 epidemic (234,866 cases; 383 deaths), the total direct patient costs exceeded US$2 million (US$9 per case). The government spent another US$1 million on a national antidengue mosquito control program. To obtain some perspective, the US$3 million allocated for curing or preventing DHF/DSS is borne in a country in which the per capita gross national product is US$365.[38]

The most comprehensive method for estimating costs associated with dengue illnesses of all degrees of severity employs disability-adjusted life years (DALY's).[39] DALY's were used to assess the burden of dengue in Myanmar, resulting in an estimated 83.8 DALY's lost due to dengue per year per million population from 1970 to 1997.[40] In Puerto Rico from 1984 to 1994, the economic burden due to dengue, estimated to be 658 DALY's per year per million population, was found to be as important as that for tuberculosis, sexually transmitted diseases (except HIV/AIDS), and the cluster of tropical diseases or intestinal helminths.[41] A household- and population-based study in Thailand found that out-of-pocket costs to families in rural Central Thailand were US$24, while 426.9 DALY's per year per million population were lost due to DF or DHF.[42] Other studies have estimated the cost of Thai government subsidies to hospitals at US$38.65–US$54.60 per DHF case.[37] A 2001 World Health Report attributed 241.6 DALY's per year per million population to dengue for all SE Asian countries.[43] All DALY calculations use a multiplication factor of at least 10 on the assumption of underreporting. For severe disease in countries with strong epidemiological services, this degree of underreporting is likely to be off the mark. Milder disease certainly is underreported. The conclusion from all these studies is that costs attributed to dengue illnesses are in the range of the tropical diseases cluster (schistosomiasis, leishmaniasis, trypanosomiasis, onchocerciasis and filariasis).

In order to estimate the full economic impact of dengue illness, in 2005–2006, a single protocol study was conducted in five countries in the Americas (Brazil, El Salvador, Guatemala, Panama, and Venezuela) and in three Asian countries (Cambodia, Malaysia, and Thailand) (personal communication, Dr. Jose Suaya). Participants were patients with suspected or confirmed dengue treated at hospitals or ambulatory facilities, each of whom were interviewed on one to two occasions, medical records were abstracted, and medical cost data were obtained from facility budget, reimbursements, and utilization statistics. The study estimated the cost per treated case of dengue and made extrapolations based on the 2001–2005 dengue cases and deaths reported to the World Health Organization. Cost calculations consisted of direct medical costs (public and private sector ambulatory and inpatient care), non-medical costs (e.g. transportation, extra food expense), and indirect costs (e.g. days lost by patient and other household members from school, work, or other activities). Costs were expressed in 2005 international dollars (I$) to adjust for purchasing power parity across countries. Overall mean costs were I$514 and I$1394 per ambulatory and hospitalized case, respectively. Mean ambulatory costs varied by country ranging from I$158 (Guatemala) to I$699 (Brazil) and hospital costs from I$752 (Guatemala) to I$2182 (Thailand). Shares of direct medical, direct non-medical, and indirect costs were 23%, 5%, and 72%, for ambulatory patients and 68%, 9%, and 23% for hospitalized patients, respectively. The distribution of shared costs varied between countries and age groups. Costs for unconfirmed and confirmed cases did not differ significantly within countries or between adults (> 15 years) and children (< 15 years). For the period 2001–2005, the annual average cases and deaths reported in these eight countries were 574,000 and 399, respectively. The corresponding aggregate annual economic cost of these cases and deaths was estimated to be above I$587 million. If adjustment for under-reporting of dengue were included, this cost could increase to I$1.8 billion.

Another burden imposed by dengue is the nationwide attempts to control populations of the mosquito, *Aedes aegypti.* In the modern era only two countries, Singapore and Cuba — both islands — have successfully controlled this mosquito. The Ministry of the Environment (Singapore) reported evidence that the house index has been held below 1% for more than 30 years. This success has produced a generation of children with little or no exposure to dengue. But the program increased the political damage caused by continuing dengue cases. Infections among Singaporeans are due to viruses imported from adjacent or trading countries, all of which have hyperendemic dengue.

WHO and the Pan American Health Organization reported that in most tropical countries, *Aedes aegypti* control programs grind on endlessly, seemingly without any benefit. In Brazil, annual expenditures on dengue control have exceeded US$600 million, constituting 1.6% of the total health budget. In Thailand, despite annual expenditures reported by the Ministry of Health on the order of US$11 million, in 2001 the country suffered its largest-ever DHF epidemic.

The DALY method does not describe the fear, disruption and political turmoil caused by epidemics of an urban-based capricious fatal infectious disease. Dengue epidemics have caused considerable political upheaval. An appropriate measurement tool must reflect social disruption costs of dengue.

History

Evolution of dengue viruses

Evidence suggests that the human dengue virus evolved as a parasite of subhuman primates.[12] Animals infected with the ancestral dengue virus must have become separated for prolonged periods, permitting the evolution of viruses whose envelope proteins differed sufficiently to escape cross-neutralization — hence the four virus types. The probable spread from Africa during historical times of *Aedes aegypti* throughout the world created an ecologic niche permitting an urban transmission cycle.[44,45] Genetic evidence from the few strains studied suggests that the four sylvatic virus types were independently imported into the urban cycle within the past 1000 years,[12] and from ecological evidence this may have occurred in tropical Asia.

Dengue fever

Only after the isolation and characterization of dengue viruses during WWII was it possible to attribute past outbreaks to dengue viruses. While dengue-like disease may have been described in accounts written in China in 992 and the West Indies in the 1600's, without a detailed clinical picture it is hazardous to attribute these outbreaks to dengue.[16] Yellow fever and quite possibly dengue viruses were imported into the New World in the 1600's and the chikunguyna virus in 1827–8 as a consequence of the

African slave trade. The first account of a well-characterized clinical syndrome that bears the hallmarks of dengue is that of Benjamin Rush,[46] who cared for patients during a 1780 Philadelphia outbreak. He noted that the August–September febrile exanthem was confined to persons residing along the Delaware River waterfront. Rush reported a sudden onset with "pains in the head…sometimes in the…eyeballs…. A few complained of the flesh being sore to the touch…nausea universally, and…vomiting, accompanied by a disagreeable taste in the mouth...a rash often appeared on the third and fourth days…. Most of those who recovered, complained of...a total want of appetite…. But the most remarkable symptom (post-illness) was an uncommon dejection of the spirits…one (patient) aptly proposed…to change the name of the disorder to *Break-heart fever.*" There is only one etiology known to produce this constellation of acute phase and convalescent symptoms — the dengue virus.

Rigau-Perez found that the term "breakbone fever" (*quebranta huesos*) was in use in Spain at about the same time as the Philadelphia outbreak.[47] Outbreaks surmised to be of dengue etiology based upon similar clinical and epidemiological features were common in North America during the 18th and 19th centuries along the Atlantic coast, on the Caribbean islands and in the Mississippi basin.[3] Dengue viruses were almost certainly the cause of the five- and seven-day fevers that afflicted European colonists in tropical Asia during the same period.[26] Similar diseases occurred among settlers in tropical Australia.[14]

Observations that culminated in the recognition of dengue as an arthropod-borne agent were initiated in Lebanon in 1902.[48] *Aedes aegypti* mosquitoes were first identified as dengue vectors by Bancroft.[49] This was confirmed and substantiated by the classical studies of Cleland, Bradley and McDonald,[50] Chandler and Rice,[51] Siler *et al.*[3] and Simmons *et al.*[5] Ashburn and Craig,[52] using human volunteers, demonstrated that the etiological agent of dengue was present in the blood of patients and that it passed through a Lilliput diatomaceous earth filter. Working in 1922 and 1929, respectively, Siler *et al.*[3] and Simmons *et al.*[5] serially transferred dengue viruses to US Army volunteers by the bite of *Aedes aegypti.* For each of these strains, they identified an intrinsic incubation period in human beings of 3–8 days and an extrinsic incubation period in mosquitoes of 8–11 days. They also documented postinfection immunity in people and monkeys, and the unsusceptibility to dengue infection of most domestic animals. From neutralizing antibody studies of sera from these volunteers bled in 1973, viruses were identified as dengue types 4 and 1, respectively.[53]

The modern era of dengue research began in 1943–44, when dengue viruses were recovered in the laboratory following the intracerebral inoculation of suckling mice.[54,55] Failure of strains to cross-protect in human volunteers led to the designation of dengue viruses types 1 and 2.[56,57]

Other viruses

Using historical accounts, the etiology of febrile exanthems can easily be confused. The most dramatic mix-up seems to have been caused by a virus known now as chikungunya, an alphavirus in the *Togaviridae* genus. Chikungunya derives its name from the Makonde word meaning "that which bends up," referring to the characteristic symptom, arthralgia.[58] An account cited for many years as the initial description of dengue fever was, in fact, probably on chikungunya. David Bylon,[59,60] "*stads cirurgyn*" of Batavia (Jakarta), described an epidemic of a febrile disease with an acute onset and joint involvement in 1779. Dr. Bylon, who himself contracted the illness, wrote that "it was last May 25, in the afternoon at 5:00 when I noted while talking with two good friends of mine, a growing pain in my right hand, and the joints of the lower arm, which step by step proceeded upward to the shoulder and then continued onto all my limbs; so much so that at 9:00 that same evening I was already in my bed with high fever…." "It's now been three weeks since I…was stricken by the illness, and because of that had to stay home for 5 days; but even until today I have continuous pain and stiffness in the joints of both feet, with swelling of both ankles; so much so that when I get up in the morning, or have sat up for a while and start to move again, I cannot do so very well and going up and down stairs is very painful."…"Natives, Chinese, slaves; no race escaped, as both sexes, children, adults and old people were all affected equally; not only in this city of Batavia, but also in the surrounding area." Remarkably, in 1779 in Cairo and Alexandria, another outbreak occurred of a disease that bears close resemblance to chikungunya fever.[61] Pandemics of a chikungunya-like syndrome were reported from India in 1824–25,

1871–72, 1923, 1964–65 and 2005–07. The latter two epidemics have been virologically identified as being due to the chikungunya virus.[62,63]

Introduction of the term "dengue"

An interesting pandemic of "dengue" (likely chikungunya) occurred in the years 1870–73, appearing first on the coast of East Africa, then on the Arabian coast and at Port Said.[3,64,65] From there the disease was carried by emigrant steamers to Bombay, Calcutta, and to Java. The 1870 outbreak led to the discovery that the Swahili word for the disease was "*ki-dinga pepo*".[64] "*Pepo*" is derived from the Swahili — "to sway, reel, stagger or totter".[66] "*Ki*" is a diminutive. Modern usage of "*dinga*" or "*denga*" in Swahili does not explain the term "*ki-dinga*." Christie[64] noted that the term "*denga*", or "*dyenga*," had been used to designate the disease in East Africa in an earlier outbreak in 1823 and in the similar disease occurring in the West Indies in 1827–28.[65] It must be assumed that the chikungunya virus spread to the Caribbean with the African slave trade carrying along its Swahili name and was known locally by various homonyms. It was only after the 1828 outbreak in Cuba that the Spanish word "*dengue*" came into general use in the medical literature, continuing to this day.[15] The term "dandy fever" — a homonym of "*dyenga*" with an apt meaning — was used in English colonies, from where it entered the English medical literature.[67] There is no evidence that chikungunya was ever imported into the American hemisphere after 1827–28. Over time, the syndromic term originally applied to chikungunya illnesses came to be associated with those caused by the dengue viruses. There are written records of the use of "dengue" to describe an illness occurring among members of the Spanish Court in 1800.[47] This early usage might still have an African origin, as there were extensive Spanish connections with Africa over many centuries of trade and exploration.

Meaning of "dengue"

When Sabin[19] inquired into the etymology of "dengue," the standard dictionary meaning of the word was "affectation." Dengue researchers were unable to make a connection between this term and characteristic symptoms and signs of dengue. However, an interesting connection does exist — but with chikungunya, not dengue. Contemporary observers were struck by the postillness arthralgia and disability caused by "dengue." Dumaresq[68] noted: "A person on the disappearance of this fever would attempt to rise from the bed, feeling not much loss of strength, and a consciousness of being able to move about and attend to a little business; but how egregiously would he be mistaken when he assumed the upright posture! The joints felt as if fettered or ankylosed…the appearance of persons in the streets and everywhere else must have been truly pitiable…here one would be seen dragging his legs after him, supported on crutches; and there another with limping gait and various contortions of countenance." Stedman[67] noted an even more extreme manifestation of dengue "It is even said that when the disease first appeared in St. Thomas, several negroes, who, being all at once attacked with pain in the knees, had fallen down, were actually apprehended by the police for drunkenness." Lehman,[69] Lazaretto physician to the port of Philadelphia, interviewed a ship captain from Cuba who averred "It [dengue] is a vulgar phrase, and implies a 'staggering weakness,' and is somewhat similar in its import to our term of 'corned' [drunk] as applied to a man reeling about from intoxication." A modern Spanish dictionary defines the colloquial meaning of "dengue" as "strut, swagger."[66] While the African etymological origin of "*dinga*" is still not understood, the original meaning of "*ki-dinga pepo*" was consistently maintained from Swahili to Spanish and seems an apt name for a disease that produces postillness "staggers."

Dengue hemorrhagic fever

From 1897 to 1902 in Australia, in 1928 in Greece, and in 1931 in Taiwan, a severe syndrome characterized by shock, hemorrhagic manifestations and death was described during dengue epidemics.[14,70–72] These early outbreaks suggest that DHF/DSS is not a new phenomenon, contradicting the hypothesis that DHF is caused by newly acquired genetic and phenotypic changes.[73]

Australia, 1897

During the summer of 1897, the inhabitants of the coastal towns of North Queensland suffered their fourth consecutive annual visitation of dengue fever. The unusual virulence of that epidemic engaged the attention of Dr. F. E. Hare of Charters Towers, who undertook to write a general description of the outbreak by polling physicians practicing in the region.[70] From 19 correspondents Hare obtained records of 60 fatalities, 30 of these in children. The severe illness in children was most characteristic. Hare remarked that these cases were "amongst the most startling that occur in medical practice. In nearly all of these [children] death must be, I think, attributed to the intensity of uncomplicated disease.... All...were previously healthy children; their ages varied between 3 and 14 years...the manner of death was in the majority almost identical, very rapid heart failure and collapse [which] occurred at the crisis on the fifth day of fever, and death ensued from two to 48 hours later.... The patient exhibits all the signs of acute hemorrhage, jactitation (i.e., a frightful restlessness), extreme irritability of temper... terminating (sometimes) in a state exactly resembling the (shock) stage of cholera." This is a remarkably accurate description of dengue shock syndrome (Chap. 5). It is notable that a small number of DSS-like deaths occurred in an outbreak in which thousands of adults and children experienced classical dengue fever syndrome. Hare himself noted that "large numbers [of persons] who had had the disease two years previously were again attacked equally as badly or even more severely." It should be noted that the youngest death in this series was a three-year-old child, old enough to have experienced an infection during the 1895 dengue epidemic.

Greece

During 1927, in refugee-swollen Athens and Piraeus, there was a barely perceived mild dengue fever outbreak. This was succeeded in 1928 by the most explosive and virulent dengue epidemic ever recorded.[74] In August and September alone, there were over 650,000 cases, with 1060 deaths.[71] Retrospective serological studies undertaken using the relatively nonspecific hemagglutination-inhibition test suggested that only a single dengue virus, type 1, caused the 1927–28 dengue epidemics. Papaevangelou and Halstead[75] found dengue type 1 and type 2 monospecific neutralizing antibodies and dengue 1 plus 2 antibodies in the majority of individuals who had lived in Athens during the outbreak. A subsequent study of 111 residents of Athens born in 1927, 1928 or in 1931–35 demonstrated predominantly monotypic DEN 1 and DEN 2 antibodies in individuals who were children during the 1927–28 epidemic, but no dengue antibodies in persons born after 1930.[76]

The epidemiological circumstances of dengue transmission in Greece were uniquely favorable for a retrospective serological study. From all available evidence, dengue transmission stopped abruptly after 1928.[77,78] The epidemics of 1927 and 1928 can be attributed largely to a huge increase in breeding habitats for *Aedes aegypti.* These were created by the need to store water because of a severe water shortage in Athens and Piraeus. This was exacerbated by the large number of temporary shelters erected to house many of the 1.5 million refugees repatriated from Turkey following the Greco-Turkish war of 1922 and the 1923 Treaty of Lausanne.[79] After 1928, a sustained period of refugee resettlement, the opening in Athens of a municipal water system with a capacity of 40–50 gallons per capita per day[79] and mosquito abatement ended dengue transmission permanently. Since 1935, there have been few records of *Aedes aegypti* in the Eastern Mediterranean.[80]

Taiwan

Japanese physicians in Formosa described a few fatal cases among children hospitalized in 1931 whose signs and clinical course suggest DSS.[72]

Post–World War II

Modern DHF pandemic

A comparative lull in reports of dengue activity followed the withdrawal of major foreign forces from tropical Asia at the end of the War. This lull was broken in the mid-1950's by the unexpected recovery of dengue viruses from a "hemorrhagic fever" of children.[81] Hammon, who was in the Philippines to study poliomyelitis, isolated two dengue virus types new to science, calling these dengue 3 and 4.[82] Two years later, Hammon and coworkers again recovered dengue viruses from similar cases among children in Bangkok, Thailand, labeling these dengue 5 and 6.[83] The "new" syndrome described in Manila, the Philippines, in 1954 was called Philippine hemorrhagic fever, because of clinical similarities to epidemic hemorrhagic fever on the Korean peninsula.[84] Retrospectively, records at Bangkok's Siriraj Hospital documented children with DSS-like syndromes during every rainy season from 1950.[85] Often, these cases received the final diagnosis: "thrombocytopenic purpura with cardiovascular collapse." When the same reviewers examined 572 pediatric charts from Siriraj Hospital for July and August, 1932–1942, no DHF/DSS-like cases were found.

Causal hypotheses

First impressions, 1958–62

In Thailand, there was a complication. A significant fraction of all hospitalized cases were caused by chikungunya, an alphavirus. Many patients with "Thai hemorrhagic fever" had simultaneous serological responses to dengue and chikungunya viruses. The immediate question was "Why were dengue and chikungunya viruses suddenly causing a severe and fatal disease?"

Early observations resulted in four hypotheses the causation of hemorrhagic fever:

(1) "*Hemorrhagic variants*," specifically dengue types 3–6, were responsible;
(2) *Role of chikungunya*. In Thailand, as opposed to the Philippines, chikungunya, a nondengue virus, seemed to be causing up to 20% of cases. One idea was that simultaneous infections with dengue and chikungunya might account for severe disease.[86] It seemed possible that the chikungunya virus might have gained virulence since freshly isolated strains produced hemorrhagic enteritis in suckling rodents.[87]
(3) *Immune response*. The very first serological studies produced evidence that many patients with Thai and Philippine hemorrhagic fever (THF and PHF) experienced extremely high antibody responses to dengue viruses. This suggested that patients had been infected previously with

an antigenically related virus. Because not all viruses circulating in these countries were known, initial earlier infection could only be guessed at.

(4) *Human genetic factor*. During the 1962 epidemic in Thailand, predominantly Caucasian expatriates, both children and adults, suffered dengue fever, but not THF.[88] It seemed possible that Caucasians were genetically resistant to severe dengue disease.

Confusion, 1963–65

Observations made in 1963 and 1964 were helpful, but sometimes contradictory. Dasaneyavaja and coworkers[89] reported that the chikungunya virus had not been isolated from shock or fatal THF cases. And, because THF and PHF were clinically similar and the chikungunya virus did not occur in the Philippines, it seemed unlikely that chikungunya was a necessary component of the etiology of hemorrhagic fever. As to genetic factors in the host, Halstead and Yamarat[85] called attention to earlier episodes of severe and fatal hemorrhagic fever associated with dengue fever outbreaks among Caucasians in Australia and Greece. They concluded that Caucasians were not genetically resistant to hemorrhagic fever. They reasoned that failure of Thai dengue strains to cause hemorrhagic fever in Caucasians infected with dengue viruses could only be interpreted to mean that dengue viruses were not inherently virulent. A factor "somehow acquired through continuous exposure to environmental or immunologic conditions of Bangkok"[90] seemed more plausible. Halstead and Yamarat[85] called attention to a small THF outbreak in 1964 in which primary-type antibody responses seemed to predominate. At the 1964 WHO conference, Hammon formulated an "immunological response" hypothesis about THF but, after deliberation, discarded it in favor of the virus virulence hypothesis.[91]

Increasing clarity, 1966–67

The WHO Seminar on Mosquito-borne Haemorrhagic Fevers held in Bangkok, 19–24 October 1964, was notable for two events — the introduction of the term "dengue hemorrhagic fever (DHF)"[92] and an agreement that better case definition would improve etiological classification.

This was soon accomplished. In 1966, Cohen and Halstead[93] published the classical description of the dengue shock syndrome, describing fully the central underlying pathophysiology as being due to the leakage of fluid and protein through damaged capillaries. This and other work in Bangkok led to the introduction of logical principles of resuscitation. Focussing on dengue shock syndrome (DSS) cases led to a breakthrough on etiology. Halstead *et al.*[94] found a strong correlation between secondary-type dengue antibody responses and cases with DSS. Analysis was strengthened by new methods in immunology that made it possible to reliably distinguish primary and secondary dengue antibody responses by their association with different immunoglobulin types.[95]

Sequential infections

When multiple dengue virus serotypes are in circulation, solid evidence, for an etiologic role of infection sequence can be obtained by comparing the prevalence of secondary-type antibody responses among DHF/DSS cases with that in milder dengue illnesses. Two important conditions apply: cases must (1) demonstrate clinically significant vascular permeabilty and (2) be one year of age or older.[96] The special category of infants DHF/DSS is discussed at length in Chap. 7.

Analysis of antibody responses during DHF/DSS cannot directly answer the question whether severe disease was associated with a second, third or fourth dengue infection. Very few DHF/DSS cases were hospitalized early enough during illness for acute phase sera to retain preillness attributes. An attempt to solve this problem was made by comparing Bangkok age-specific DHF/DSS hospitalization rates with age-specific rates of second, third and fourth dengue infections generated by a mathematical model.[97] Only second dengue virus infection curves fit DHF/DSS hospitalization data.

It was obvious that only if children were followed from their first through successive dengue infections could it be determined if a second, third or fourth infection resulted in DHF/DSS. Pioneer studies were conducted on Koh Samui island, Thailand, in 1966 and 1967;[98–100] DHF/DSS cases were shown to occur only in children who circulated dengue HI antibodies prior to onset of severe disease. The HI test method used could not accurately answer the question whether any of these children had experienced one or more prior dengue infections. The answer required the neutralization test. Many years were to pass before direct evidence for the etiological importance of a single dengue infection was gathered. These data are summarized in Chap. 7.

Immune enhancement, 1968–77

Immune enhancement of dengue virus replication was described *in vitro* and *in vivo* in 1973. Increased growth of dengue 2 virus was observed in cultures of peripheral blood leukocytes (PBL's) obtained from dengue-immune rhesus monkeys.[101] Enhanced levels of viremia were observed in rhesus monkeys during second infections as compared with initial dengue 2 infections.[102] It was not immediately evident how "immune enhancement of dengue virus infection" worked. It seemed possible that viruses might replicate in memory T lymphocytes transformed by dengue antigen into lymphoblasts. At the time, it was known that several viruses could grow in phytohemagglutinin-transformed T lymphoblasts.[103]

Antibody-dependent enhancement, 1977–present

Epidemiological evidence had identified two groups at risk of DHF/DSS: children experiencing second dengue infections and infants born to dengue-immune mothers who developed DHF during their first dengue infection.[96] The only plausible immunological factor that could tie together these observations was antibody. Dengue antibody somehow modulated subsequent infection. Very quickly, it was found that dengue antibody at nonneutralizing concentrations enhanced dengue viral growth in cultures of human and rhesus peripheral blood leukocytes.[104] Mononuclear phagocytes specifically supported dengue virus replication *in vitro* and permitted enhanced infections in the presence of infectious immune complexes.[105] Optimal conditions for *in vitro* infection enhancement were described.[106] Finally, enhancement of dengue 2 viremia was achieved in rhesus monkeys that had been inoculated with small concentrations of passively transferred human dengue antibody.[107] Other laboratories began to study "antibody-dependent enhancement (ADE)".[108]

Confirmation in humans

In a small study, somewhat higher viremias had been demonstrated in a few humans during secondary compared with primary dengue 3 infections.[109] More recent formal studies have demonstrated that viremia titers correlate with disease severity during secondary dengue infections. Peak viremia titers correlated with subsequent disease severity in Thai children experiencing a secondary dengue 2 infection.[110] This study is unique, in that viremia titers were measured in successive bleedings starting early in the course of infections. These same children circulated nonstructural protein 1 (NS-1) at levels that correlated with and predicted disease severity.[111] NS-1 is released from dengue-infected cells and its blood concentration is thought to reflect total cellular dengue infection. A correlation between viremia and disease severity was also found in Thai children experiencing dengue 3 infections.[112] Finally, it was observed in Taiwanese patients that dengue RNA titers even after defervescence correlated with disease severity.[113] The contemporary ADE hypothesis of DHF/DSS pathogenesis assigns preillness dengue antibodies an enhancing or neutralizing (afferent) role that up- or down-regulates dengue infection of mononuclear phagocytes. An efferent role (elimination of infected cells) is played by T-cell-mediated immunity. This process produces cytokines that mediate both vascular permeability and abnormal hemostasis by mechanisms discussed in subsequent chapters.[13,114]

References

1. DeMadrid AT, Porterfield JS. The flaviviruses (group B arboviruses): a cross-neutralization study. *J Gen Virol* 1974;23:91–96.
2. Karabatsos N (ed.) *International Catalog of Arboviruses, 1985*. American Society of Tropical Medicine and Hygiene, Ft. Collins, Colorado, 1985.
3. Siler JF, Hall MW, Hitchens AP. Dengue: its history, epidemiology, mechanisms of transmission, etiology, clinical manifestations, immunity and prevention. *Philippine J Sci* 1926;29:1–304.
4. Schlesinger RW (ed.) *The Togaviruses: Biology, Structure, Replication.* Academic Press, New York, 1980.

5. Simmons JS, St John JH, Reynolds FHK. Experimental studies of dengue. *Philippine J Sci* 1931;44:1–247.
6. Rudnick A. Lim TW. Dengue fever studies in Malaysia. *Inst Med Res Malaysia Bull.* 1986;23:127–197.
7. Roche JC *et al.* Isolement de 96 souches de virus dengue 2 partir de mosquiques captures en Cote-D'Ivoire et Haute-Volta. *Ann Virol (Inst Pasteur)* 1983;134:233–244.
8. Traore-Laminzana M *et al.* Dengue-2 outbreak in southeastern Senegal during 1990: virus isolations from mosquitoes (*Diptera: Culicidae*). *J Med Entomol* 1994;31:623–627.
9. Moncayo AC *et al.* Dengue emergence and adaptation to peridomestic mosquitoes. *Emerg Inf Dis* 2004;10:1790–1796.
10. Diallo M *et al.* Potential role of sylvatic and domestic Afrinca mosquito species in dengue emergence. *Am J Trop Med Hyg* 2005;73:445–449.
11. Rico-Hesse R. Molecular evolution and distribution of dengue viruses type 1 and 2 in nature. *Virology* 1990;174:479–493.
12. Wang E *et al.* Evolutionary relationships of endemic/epidemic and sylvatic dengue viruses. *J Virol* 2000;74:3227–3234.
13. Halstead SB. Immunological parameters of togavirus disease syndromes. In: Schlesinger RW (ed.) *Togaviruses: Biology, Structure, Replication.* Academic, New York, 1980, pp. 107–173.
14. Lumley GF, Taylor FH. *Dengue*, School of Public Health and Tropical Medicine Service Publication No. 3. Glebe, NSW, Australia, 1943.
15. Carey DE *et al.* Dengue viruses from febrile patients in Nigeria, 1964–68. *Lancet* 1971;1:105–106.
16. Gubler DJ. Dengue and dengue hemorrhagic fever: its history and resurgence as a global public health problem. In: Gubler DJ, Kuno G (eds.) *Dengue and Dengue Hemorrhagic Fever.* CAB International, Wallingford, UK, 1997, pp. 1–22.
17. Rosen L. Observations on the epidemiology of dengue in Panama. *Am J Hyg* 1958;68:45–58.
18. Guzman MG *et al.* Dengue hemorrhagic fever in Cuba, 1981: a retrospective seroepidemiologic study. *Am J Trop Med Hyg* 1990;42:179–184.
19. Sabin AB. Dengue. In: Rivers TM (ed.) *Viral and Rickettsial Infections of Man.* Lippincott, Philadelphia, 1952a, pp. 556–568.

20. PAHO. *Conference Proceedings: Dengue in the Caribbean, 1977,* Scientific Publication 375. PAHO, Washington, DC, 1979.
21. Halstead SB. Epidemiology of dengue and dengue hemorrhagic fever. In: Gubler DJ, Kuno G (eds.) *Dengue and Dengue Hemorrhagic Fever.* CAB, Wallingford, UK, 1997, pp. 23–44.
22. Monath TP. Dengue: the risk to developed and developing countries. *Proc Natl Acad Sci USA* 1994;91:2395–2400.
23. Halstead SB. Selective primary health care: strategies for control of disease in the developing world. XI. Dengue. *Rev Infect Dis* 1984;6: 251–264.
24. Kouri G *et al.* Dengue haemorrhagic fever / dengue shock syndrome: lessons from the Cuban epidemic, 1981. *Bull World Health Organ* 1989; 67:375–380.
25. Watts DM *et al.* Failure of secondary infections with American genotype dengue 2 viruses to cause dengue haemorrhagic fever. *Lancet* 1999;354:1431–1434.
26. Carey DE. Chikungunya and dengue: a case of mistaken identity? *J Hist Med* 1971;26:243–262.
27. Halstead SB. Mosquito-borne hemorrhagic fevers of South and Southeast Asia. *Bull World Health Organ* 1966;35:3–15.
28. Halstead SB. Pathogenesis of dengue: challenge to molecular biology. *Science* 1988;239:476–481.
29. Rahman M *et al.* First outbreak of dengue hemorrhagic fever — Bangladesh. *Emerg Inf Dis* 2002;8:738–740.
30. Hossain MA *et al.* Serologic evidence of dengue infection before onset of epidemic — Bangladesh. *Emerg Inf Dis* 2003;9:1411–1414.
31. Guzman MG *et al.* Epidemiological studies on dengue in Santiago de Cuba, 1997. *Am J Epidemiol* 2000;152:801–808.
32. Ko YC. Epidemiology of dengue fever in Taiwan. *Kao Hsuing I Hsueh Ko Hsueh Tsa Chi* 1989;5:1–11 (in Chinese).
33. Rigau-Perez J *et al.* Dengue and dengue haemorrhagic fever. *Lancet* 1998a;352:971–977.
34. Saluzzo JF. Isolation of dengue 2 and dengue 4 viruses from patients in Senegal. *Trans R Soc Trop Med Hyg* 1986;80:5.
35. Halstead SB *et al.* Haiti: absence of dengue hemorrhagic fever despite hyperendemic dengue virus transmission. *Am J Trop Med Hyg* 2001;65:180–183.

36. Von Allmen SD *et al.* Epidemic dengue fever in Puerto Rico, 1977: a cost analysis. *Am J Trop Med Hyg* 1979;28:1040–1044.
37. Okanurak K, Sornmani S, Indaratna K. The cost of dengue hemorrhagic fever in Thailand. *Southeast Asian J Trop Med Public Health* 1997b;28:711–717.
38. Shepard DS *et al.* Cost-effectiveness of a pediatric dengue vaccine. *Vaccine* 2004;22:1275–1280.
39. Murray CJL. Quantifying the burden of disease: the technical basis for disability-adjusted life years. *Bull World Health Organ* 1994;72: 429–455.
40. Naing CM. Assessment of dengue hemorrhagic fever in Myanmar. *Southeast Asian J Trop Med Public Health* 2000;31:636–641.
41. Meltzer MI *et al.* Using disability-adjusted life years to assess the economic impact of dengue in Puerto Rico: 1984–1994. *Am J Trop Med Hyg* 1998;59:265–271.
42. Clark DV *et al.* Economic impact of dengue fever / dengue hemorrhagic fever in Thailand at the family and population levels 2005;72: 786–791.
43. World Health Organization. *The World Health Report, 2001: Statistical Annex Table 3.* World Health Organization, Geneva, 2001.
44. Mattingly PF. Genetical aspects of the *Aedes aegypti* problem. I. Taxonomy and bionomics. *Ann Trop Med Parositol* 1957;51:392–408.
45. Mattingly PF. II. Ecological aspects of the evolution of mosquito-borne virus diseases. *Trans R Soc Trop Med Hyg* 1960;54:97–112.
46. Rush B. An account of the bilious remitting fever, as it appeared in Philadelphia in the summer and autumn of the year 1780. *Med Inq Obs Philadelphia* 1789;1:104–117.
47. Rigau-Perez J *et al.* The early use of break-bone fever (*Quebranta huesos*, 1771) and dengue (1801) in Spain. *Am J Trop Med Hyg* 1998b;59:272–274.
48. Graham H. The dengue: a study of its pathology and mode of propagation. *J Trop Med* 1903;6:209–214.
49. Bancroft TL. On the etiology of dengue fever. *Australas Med Gaz* 1906;25:17–18.
50. Cleland JB, Bradley B, McDonald W. On the transmission of Australian dengue by the mosquito *Stegomyia fasciata*. *Med J Aust* 1916;2: 179–184.

51. Chandler AC, Rice L. Observations on the etiology of dengue fever. *Am J Trop Med* 1923;3:233–262.
52. Ashburn PM, Craig CF. Experimental investigations regarding the etiology of dengue fever. *J Infect Dis* 1907;4:440–475.
53. Halstead SB. Etiologies of the experimental dengues of Siler and Simmons. *Am J Trop Med Hyg* 1974;23:974–982.
54. Kimura R, Hotta S. On the inoculation of dengue virus into mice. *Nippon Igaku* 1944;3379:629–633.
55. Sabin AB, Schlesinger RW. Production of immunity to dengue with virus modified by propagation in mice. *Science* 1945;101:640–642.
56. Sabin AB. The dengue group of viruses and its family relationships. *Bacteriol Rev* 1950;14:225–232.
57. Sabin AB. Research on dengue during World War II. *Am J Trop Med Hyg* 1952b;1:30–50.
58. Robinson MC. An epidemic of virus disease in Southern Province, Tanganyika Territory, in 1952–53. *S Afr Med J* 1955;49:28–32.
59. Bylon D. Korte aatekening, wegens eene algemeene ziekte, dorgans genamd de knokkel-koorts. *Verandelingen van het Bataviaasch Genootschap van Kunsten Wetenschappen* 1780;2:17–30.
60. Pepper OHP. A note on David Bylon and dengue. *Ann Med His* 1941;3:363–368.
61. Hirsch A. Dengue, a comparatively new disease: its symptoms — geographic distribution — characteristics of dengue as an epidemic disease of the tropics. In: Hirsch A (ed.) *Handbook of Geographical and Historical Pathology,* Vol. I. New Sydenham Society, London, 1883, pp. 466–520.
62. Sarkar JK *et al*. Chikungunya virus infection with haemorrhagic manifestations. *Indian J Med Res* 1965;53:921–925.
63. Mavalankar D, Shastri P, Raman P. Chikungunya epidemic in India: a major public-health disaster. *Lancet Infect Dis* 2007;7:306–307.
64. Christie J. Remarks on "Kidinga Pepo": a peculiar form of exanthematous disease. *Br Med J* 1872;1:577–579.
65. Christie J. On epidemics of dengue fever: their diffusion and etiology. *Glasgow Med J* 1881;3:161–176.
66. Johnson F. *A Standard Swahili–English Dictionary.* Oxford University Press, Oxford 1939.

67. Stedman GW. Some account of an anomalous disease which raged in the islands of St. Thomas and Santa Cruz in the West Indies, during the months of September, October, November, December, and January 1827–28. *Edinburgh Med Surg J* 1828;30:227–248.
68. Dumaresq PJ. An account of dengue, danga, or dandy fever, as it occurred in New Orleans and in the person of the writer, communicated in a letter to one of the editors. *Boston Med Surg J* 1828;1:497–502.
69. Lehman GF. Account of the disease called dengue, which has prevailed so extensively at Havanna. *Am J Med Sci* 1828;2:477–480.
70. Hare RE. The 1897 epidemic of dengue in North Queensland. *Australas Med Gaz* 1898;17:98–107.
71. Copanaris P. L'epidemie de dengue en Grece au cours de l'ete 1928. *Office Int Hyg Publique Bull* 1928;20:1590–1601.
72. Nomura S, Akashi K. Uber todesfalle infolge von haemorrhagie im verlauf des denguefiebers. *Taiwan Agakkai Zasshi* 1931;30:1154–1156.
73. Zanotto PM de A *et al.* Population dynamics of flaviviruses revealed by molecular phylogenies. *Proc Natl Acad Sci* 1996;93:548–553.
74. Anonymous. The dengue epidemic in Greece. *League Nations Mon Epidemiol Rep* 1928;7:334–335.
75. Papaevangelou G, Halstead SB. Infections with two dengue viruses in Greece in the 20th century: did dengue hemorrhagic fever occur during the 1928 epidemic? *J Trop Med Hyg* 1977;46:46–51.
76. Halstead SB, Papaevangelou G. Transmission of dengue 1 and 2 viruses in Greece in 1928. *Am J Trop Med Hyg* 1980;29:635–637.
77. Theiler M *et al.* Etiology of the 1927–28 epidemic of dengue in Greece. *Proc Soc Exp Biol Med* 1960;103:244–246.
78. Pavlatos M, Gordon-Smith CE. Antibodies to arthropod-borne viruses in Greece. *Trans R Soc Trop Med Hyg* 1964;58:422–424.
79. Mears EB. *Greece Today: The Aftermath of the Refugee Impact.* Stanford University Press, Palo Alto, 1929.
80. Curtin TJ. Studies of *Aedes aegypti* in the Eastern Mediterranean. *J Med Entomol* 1967;4:48–50.
81. Hammon W McD *et al.* Viruses associated with hemorrhagic fevers of the Philippines and Thailand. *Science* 1960a;131:1102–1103.
82. Hammon W McD *et al.* New hemorrhagic fevers of children in the Philippines and Thailand. *Trans Assoc Am Physicians* 1960b;73:140–155.

83. Hammon W McD, Sather GE. Virological findings in the 1960 hemorrhagic fever epidemic (dengue) in Thailand. *Am J Trop Med Hyg* 1964;13:629–641.
84. Quintos FN *et al.* Hemorrhagic fever observed among children in the Philippines. *Philipp J Pediatr* 1954;3:1–19.
85. Halstead SB, Yamarat C. Recent epidemics of hemorrhagic fever in Thailand: observations related to pathogenesis of a "new" dengue disease. *J Am Public Health Assoc* 1964;55:1386–1395.
86. Hammon W McD. Discussion. In *Symposium on Haemorrhagic Fever*, 10–11 August 1961. SEATO Medical Research Monograph No. 2, Bangkok, 1961, pp. 125–126.
87. Halstead SB, Buescher EL. Hemorrhagic disease in rodents infected with virus associated with Thai hemorrhagic fever. *Science* 1961;134:473–476.
88. Halstead SB *et al.* The Thai hemorrhagic fever epidemic of 1962: a preliminary report. *J Med Assoc Thai* 1963;46:449–465.
89. Dasaneyavaja A, *et al.* Laboratory observations related to prognosis in Thai haemorrhagic fever. *J Trop Med Hyg* 1963;66:35–41.
90. Halstead SB, Yamarat C. Hemorrhagic fevers of Southeast Asia. In: *Proceedings of the Seventh International Congress on Tropical Medicine and Malaria, Vol. III.* Rio de Janeiro, 1963, pp. 279–298.
91. Hammon W McD. Immunological response: possible role of human response as an etiologic factor. *Bull World Health Organ* 1966;35: 55–56.
92. Halstead SB. Mosquito borne haemorrhagic fevers of South and South-East Asia. *Bull World Health Organ* 1966;35:3–15.
93. Cohen SN, Halstead SB. Shock associated with dengue infection. I. The clinical and physiologic manifestations of dengue hemorrhagic fever in Thailand, 1964. *J Pediatr* 1966;68:448–456.
94. Halstead SB *et al.* Hemorrhagic fever in Thailand: newer knowledge regarding etiology. *Jpn J Med Sci Biol* 1967;20S:96–102.
95. Russell PK *et al.* Antibody response in dengue and dengue hemorrhagic fever. *Jpn J Med Sci Biol* 1967;20S:103–108.
96. Halstead SB *et al.* Observations related to pathogenesis of dengue hemorrhagic fever. IV. Relation of disease severity to antibody response and virus recovered. *Yale J Biol Med* 1970;42:311–328.

97. Fischer DB, Halstead SB. Observations related to pathogenesis of dengue hemorrhagic fever. V. Examination of age-specific sequential infection rates using a mathematical model. *Yale J Biol Med* 1970;42:329–349.
98. Winter PE *et al.* An insular outbreak of dengue hemorrhagic fever. I. Epidemiologic observations. *Am J Trop Med Hyg* 1968;7:590–599.
99. Russell PK *et al.* An insular outbreak of dengue hemorrhagic fever. II. Virologic and serologic studies. *Am J Trop Med Hyg* 1968;17:600–608.
100. Winter PE *et al.* Recurrence of epidemic dengue hemorrhagic fever in an insular setting. *Am J Trop Med Hyg* 1969;18:573–579.
101. Halstead SB *et al.* Immunologic enhancement of dengue virus replication. *Nat New Biol* 1973a;243:24–26.
102. Halstead SB *et al.* Studies on the pathogenesis of dengue infection in monkeys. II. Clinical laboratory responses to heterologous infection. *J Infect Dis* 1973b;128:15–22.
103. Wheelock EF, Edelman R. Specific role of each human leukocyte type in viral infections. 3. 17D yellow fever virus replication and interferon production in homogeneous leukocyte cultures treated with phytohemagglutinin. *J Immunol* 1969;103:429–436.
104. Halstead SB, O'Rourke EJ. Antibody enhanced dengue virus infection in primate leukocytes. *Nature* 1977a;265:739–741.
105. Halstead SB, O'Rourke EJ. Dengue viruses and mononuclear phagocytes. I. Infection enhancement by non-neutralizing antibody. *J Exp Med* 1977b;146:201–217.
106. Halstead SB, O'Rourke EJ. Dengue viruses and mononuclear phagocytes. II. Identity of blood and tissue leukocytes supporting *in vitro* infection. *J Exp Med* 1977c;146:218–229.
107. Halstead SB. *In vivo* enhancement of dengue infection with passively transferred antibody. *J Infect Dis* 1979;140:527–533.
108. Porterfield JS. Antibody-dependent enhancement of viral infectivity. *Adv Virus Res* 1986;31:335–355.
109. Gubler DJ *et al.* Epidemic dengue hemorrhagic fever in rural Indonesia. I. Virological and epidemiological studies. *Am J Trop Med Hyg* 1979;28: 701–710.
110. Vaughn DW *et al.* Dengue viremia titer, antibody response pattern, and virus serotype correlate with disease severity. *J Infect Dis* 2000;181:2–9.

111. Libraty *et al.* High circulating levels of the dengue virus nonstructural protein NS1 early in dengue illness correlate with the development of dengue hemorrhagic fever. *J Infect Dis* 2002b;186:1165–1168.
112. Libraty *et al.* Differing influences of virus burden and immune activation on disease severity in secondary dengue-3 virus infections. *J Infect Dis* 2002a;185:1213–1221.
113. Wang WK *et al.* High levels of plasma dengue viral load during defervescence in patients with dengue hemorrhagic fever: implications for pathogenesis. *Virology* 2003;305:330–338.
114. Kurane I, Ennis FA. Cytokines in dengue virus infections: role of cytokines in the pathogenesis of dengue hemorrhagic fever. *Sem Virol* 1991;5:443–448.

2

The Infectious Agent

David W. C. Beasley and Alan D. T. Barrett

Classification

The family *Flaviviridae* contains three genera: *Flavivirus*, *Pestivirus* and *Hepacivirus*. The names of these genera are derived from the Latin for "yellow" (*flavus*), "plague" (*pestis*), and the Greek for "liver" (*hepatos*), respectively. These three genera are included in the family *Flaviviridae* on the basis of similar virion morphology and genome organization. There are no serological cross-reactions between the genera. The *Flavivirus* genus contains 67 human and animal viruses. The term "flaviviruses" is generally considered to be restricted to the *Flavivirus* genus. Yellow fever virus (YFV) is the prototype member of the *Flavivirus* genus.

On the basis of their ecology, flaviviruses have been termed "arboviruses" or arthropod-borne viruses to denote the fact that many are transmitted between vertebrate hosts by mosquitoes or ticks. It should be noted that other families of viruses are also termed "arboviruses," including members of *Togaviridae*, *Bunyaviridae*, *Reoviridae*, *Orthomyxoviridae* and *Rhabdoviridae*. Although arboviruses share ecological similarities, they have different virion morphology and genome organization.

Serological relationships between flaviviruses have been determined by antibodies (monoclonal or polyclonal) associated with the biological activities of hemagglutination-inhibition (HAI) and neutralization, which are both encoded by the envelope protein. A third method of identification based on complement-fixation activity has been associated with the NS1 protein. Neutralization tests have been used to identify eight serological subgroups together with 17 viruses that were not assigned to any serogroups due to poor cross-reactivity with other members of the *Flavivirus* genus.[1] One of these serogroups is the dengue serogroup, which contains four viruses, termed dengue 1, dengue 2, dengue 3 and dengue 4 (DENV-1, -2, -3, -4).

Nucleotide sequencing of regions of the NS5, NS3 or E genes of most flaviviruses has resulted in identification of genetic relationships that closely follow those of serological relationships.[2–4] However, the genetic analysis has allowed more detailed relationships to be established than was possible by serological studies, and current classification of the genus *Flavivirus* is based on genetic groupings.[5] In addition, genotypes have been identified for each flavivirus that has been examined. These genotypes tend to relate to the geographic origin of the virus strain. Although genotypes have been reported, each flavivirus exists as a single serotype. Nucleotide sequences for portions of most flaviviruses can be found in Genbank.

Dengue serogroup

These viruses are often described as four serotypes of the dengue virus. However, this nomenclature is somewhat misleading, as the four dengue viruses are antigenically and genetically distinct rather than being four serotypes of the same virus as exemplified by the three serotypes of the poliovirus. It is more accurate to consider the dengue viruses as four related viruses that cause very similar disease in humans. The complete nucleotide sequences of the genome of approximately 20 flaviviruses have been determined. This includes all four dengue viruses, members of the Japanese encephalitis and tick-borne encephalitis serogroups plus the non-vector-borne viruses Rio Bravo, Montana myotis leukoencephalitis, Modoc, Tamana bat and Apoi (Fig. 1). The Japanese encephalitis serogroup is similar to the dengue serogroup in that it contains viruses that cause very similar disease in humans (in this case encephalitis) but are found mainly in different geographic locations, e.g. Murray Valley encephalitis virus (MVEV) in Australasia, St. Louis encephalitis virus (SLEV) in the Americas, Japanese encephalitis virus (JEV) in Asia, and West Nile virus (WNV) in Africa, Europe, the Middle East and the Americas. As can be seen in Fig. 1, the members of the Japanese encephalitis serocomplex show a similar level of identity as the four dengue viruses to one another.

Virion Structure and Morphology

Flavivirus particles are spherical in shape, with a lipid envelope; the particles are approximately 50 nm in diameter. The lipid envelope is derived from host cell membranes and constitutes 15–20% of the total weight of the virus particle. Carbohydrates represent 9–10% of the weight of virus particles and are found as glycolipids and glycoproteins. Virions contain three structural proteins. The small basic capsid (C) protein surrounds the genome of the virus while the envelope contains two proteins known as the envelope (E) and the membrane (M). Two types of virions are recognized: mature extracellular virions contain M protein, while immature intracellular virions contain precursor M (prM), which is proteolytically cleaved during maturation to yield M protein (Fig. 2).

Cryoelectron microscopy (cryo-EM) has been used to determine the three-dimensional structure of the DENV-2 virion.[6,7] These studies confirmed the general size and arrangement of the virion as determined by earlier electron-microscopic and biochemical analyses. Fitting of the tick-borne encephalitis virus (TBEV) E protein X-ray crystallographic structure[8] suggested that 90 E protein dimers are closely packed on the virion surface in a "herringbone" arrangement. The dimers lay parallel to the lipid membrane, as had been proposed by Rey *et al.*,[8] giving the virion a smooth surface. The capsid structure within the lipid envelope was not well resolved, suggesting that the core protein was poorly ordered or that the arrangement of the capsid structure itself was variable in relation to the external envelope structures. Some icosahedral arrangement of the viral RNA was observed.

In contrast, a cryo-EM structure of the immature (prM-containing) dengue virion revealed a dramatically different arrangement, with spikes

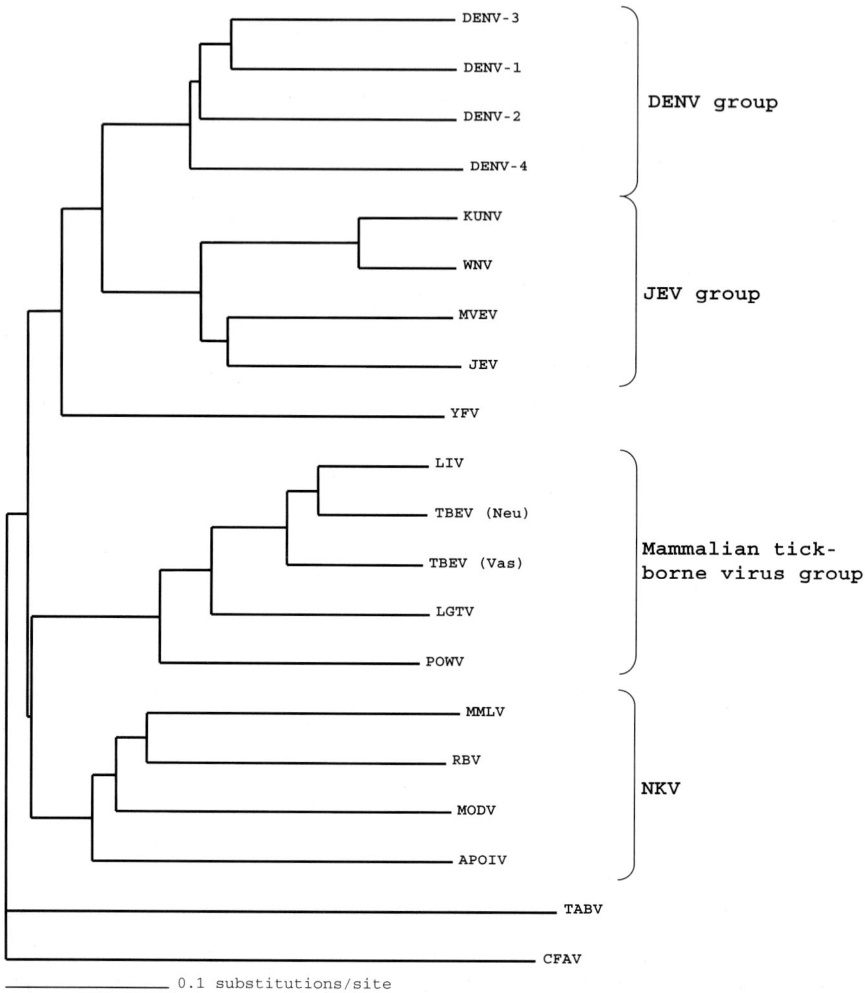

Fig. 1. Neighbor-joining tree compiled with whole genome sequences of flaviviruses available in the Genbank database using a Kimura two-parameter distance formula. Sequence alignments were performed using AlignX/VectorNTI (Informax Inc., MD) and phylogenetic analysis with PAUP. Abbreviations are as defined in the Seventh Report of the International Committee on Taxonomy of Viruses.[4] Viruses included (with accession numbers) are as follows: dengue 1–4 (DENV-1, -2, -3, -4) DVU88535, DEN2CGA, DENCME, M14931; Kunjin (KUNV) KUNCG; West Nile (WNV) AF196835; Murray Valley encephalitis (MVEV) NC_000943; Japanese encephalitis (JEV) JEU14163; yellow fever (YFV) YFU17066; Louping Ill (LIV) LIVGEN; tick-borne encephalitis (TBEV) strains Neudorlf (Neu) TEU27495 and Vasilchenko (Vas) AF069066;

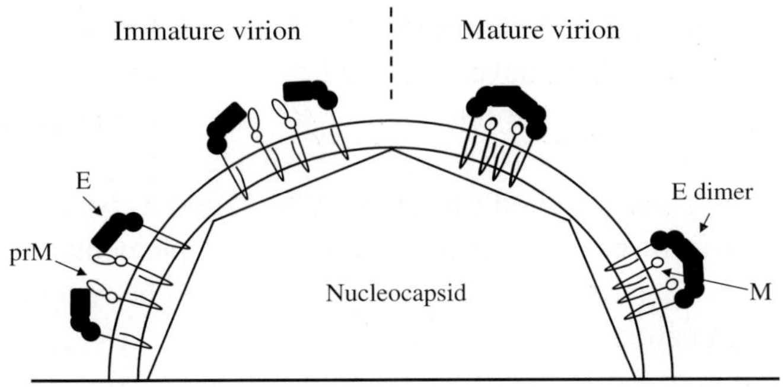

Fig. 2. Schematic diagram of the structure of immature (cell-associated) and mature flavivirus virions. Modified from Ref. 186.

composed of trimeric prM/E heterodimers projecting from the virion surface.[9] As a result of this arrangement, the immature virion was slightly larger (~ 60 nm diameter) than the mature particle.

The dengue virus genome is one single-stranded, positive-sense RNA molecule of approximately 11,000 nucleotides (i.e. it is infectious). At the 5′ terminus there is a type I cap (m^7GpppAmp) followed by the conserved dinucleotide AG, while the 3′ terminus lacks a poly-A tail. The genome encodes a single large open reading frame of approximately 10,200 nucleotides that encodes 10 proteins, 3 structural proteins encoded by the 5′ quarter of the genome and 7 nonstructural (NS) proteins encoded by the remainder of the coding region. The gene order is C-prM-E-NS1-NS2A-NS2B-NS3-NS4A-NS4B-NS5. The genome 5′ noncoding region (NCR) is very similar in size for most flaviviruses (approximately 100 nucleotides), while the 3′ NCR is variable in size (approximately 400–600 nucleotides).[10–12]

Several conserved sequences have been identified within the NCR's of mosquito-borne and tick-borne flaviviruses, and these may play some

---

Fig. 1. ***(Continued).*** Langat (LGTV) AF253419; Powassan (POWV) PWARPT; Montana myotis leukoencephalitis (MMLV) NC_004119; Rio Bravo (RBV) AF144692; Modoc (MODV) NC_003635; Apoi (APOIV) AF160193; Tamana bat (TABV) AF285080; cell-fusing agent (CFAV) NC_001564. NKV — no known arthropod vector. Bootstrap support for all branches (500 replicates) ≥ 91.

role in RNA replication and translation.[12] In addition, conserved 5′ and 3′ NCR stem-loop structures have been predicted for many flaviviruses, and these may play a role in the initiation of RNA replication by serving as attachment sites for a replicase complex consisting of viral (NS3, NS5 and possibly others) and host cell proteins.[13] The 5′ and 3′ NCR's of DENV and other flaviviruses also appear to encode complementary sequences that facilitate circularization of the genome that may be required for efficient viral replication.[14,15]

Translation of the genome results in formation of a single polyprotein which is subsequently cleaved by the action of both host cell enzymes and a virus-derived protease complex, consisting of the NS2B and NS3 proteins, to yield the individual proteins (Fig. 3).[11,12,16]

Immune Response Induced by Dengue Viruses

The major immunogen for inducing neutralizing antibodies is the E protein, although neutralizing epitopes have been identified on the M protein.[17] The NS1 protein has been shown to induce antibody-mediated cellular cytotoxicity, and antibodies against NS1 can mediate passive protection in mouse and primate models.[18] The major target of cell-mediated immunity is NS3 but T cell epitopes have also been identified on envelope, membrane, NS1 and NS2A proteins.

Proteins Encoded by the Virus

Flavivirus structural proteins

Capsid protein

The viral nucleocapsid consists of the genome surrounded by the 113-amino-acid (13–16 kDa) capsid (C) protein. Although amino acid sequence homology of C is low between the flaviviruses, C proteins share a similar distribution of basic amino acids and similar hydrophobicity profiles.[19] In particular, two highly basic domains at the N and C termini of C protein have been identified and shown to bind specifically to the 5′ and 3′ NCR's of the viral RNA genome, presumably via interactions with conserved stem loop structures.[19]

```
                1                                                                         75
DEN1      (1) MNNQRKKTGRPSFNMLKRARNRVSTVSQLAKRFSKGLLSGQGPMKLVMAFIAFLRFLAIPPTAGILARWGSFKKN
DEN2      (1) .......ARNTP......E.......Q..T....L.M.Q.R..L..F..LV......T........K...TI..S
DEN3      (1) .........K..I.....V.....G........R...N.........................V.....T...S
DEN4      (1) -M.....VV..P......E......PQG.V....T..F..K..LRM.L...T...V.S........K...QL...

              76                                         C    |# prM/M                   150
DEN1     (76) GAIKVLRGFKKEISNMLNIMNRRKRSVTMLLMLLPTALAFHLTTRGGEPHMIVSKQERGKSLLFKTSAGVNMCTL
DEN2     (76) K..N.....R...GR....L...R.TAG.II..I..VM.......N........R..K........ED.......
DEN3     (76) ......K..........S.I.K..KTSLC.M.M..AT.|....S.D...R...GKN..........AS.I.....
DEN4     (75) K...I.I..R...GR....L.G....TIT..C.I..VM.|S.S..D...L...A.H...RP.....TE.I.K...

              151                                                  pr    |↓M             225
DEN1    (151) IAMDLGELCEDTMTYKCPRITETEPDDVDCWCNATETWVTYGTCSQTGEHRRDKRSVALAPHVGLGLETRTETWM
DEN2    (151) M..........I.....FLKQN..E.I.....S.S........TT......E..|...V...M.........
DEN3    (151) .......M.D..V.....H...V..E.I.....L.S........N.A.........|........M..D...Q...
DEN4    (150) .......M....V.....LLVN...E.I.....L.S...M....T.S..R..E..|....T..S.M.....A....

              226                                                prM/M |# E               300
DEN1    (226) SSEGAWKQIQKVETWALRHPGFTVIALFLAHAIGTSITQKGIIFILLMLVTPSMAMRCVGIGNRDFVEGLSGATW
DEN2    (226) .......HA.RI...I.......IM.AI..YT...THF.RAL.....TA.A..|T...I..S.......V..GS.
DEN3    (226) .A....R.VE...........TIL......Y....L...VV...........|T.....V.............
DEN4    (225) .......HA.R..S.I..N...ALL.G.M.YM..QTGI.RTVF.V.M...A..|YG.....V........V..GA.

              301                                                                        375
DEN1    (301) VDVVLEHGSCVTTMAKDKPTLDIELLKTEVTNPAVLRKLCIEAKISNTTTDSRCPTQGEATLVEEQDTNFVCRRT
DEN2    (301) ..I.............N.....F..IE..AKQ..T...Y.....LT.............PS.N....KR...KHS
DEN3    (301) ........G.......N........Q...A.QL.T.......G..T.I............I.P....Q.Y..KH.
DEN4    (300) ..L.....G......QG.....F..T..TAKEVAL..TY....S...I..AT.......PY.K....QQYI...D

              376                                                                        450
DEN1    (376) FVDRGWGNGCGLFGKGSLITCAKFKCVTKLEGKIVQYENLKYSVIVTVHTGDQHQVGNETTEHGTTATITPQAPT
DEN2    (376) M...............GIV...M.T.KKNMK..V..P...E.TIVI.P.S.EE.A...D.GK..KEIK....SSI
DEN3    (376) Y.................V.....Q.LESI...V..H.....T..I..............--Q.V..E..S..S.
DEN4    (375) V...............GVV.....S.SG.IT.NL..I...E.T.V....N..T.A...D.SN..V..M...RS.S

              451                                                                        525
DEN1    (451) SEIQLTDYGALTLDCSPRTGLDFNEMVLLTMEKKSWLVHKQWFLDLPLPWTSGASTSQETWNRQDLLVTFKTAHA
DEN2    (451) T.AE..G..TV.ME...............Q..N.A...............LP..D.QGSN.IQKET.....NP..
DEN3    (449) A.AI.PE..T.G.E............I....KN.A.M..R...F..........T.KTP....KE......N...
DEN4    (450) V.VK.P...E.....E..S.I.....I.MK.K..T.................A..D..EVH..YKERM....VP..

              526                                                                        600
DEN1    (526) KKQEVVVLGSQEGAMHTALTGATEIQTSGTTTIFAGHLKCRLKMDKLTLKGMSYVMCTGSFKLEKEVAETQHGTV
DEN2    (526) ...D.......................M.SGNLL.T.......R....Q......S....KF.VV..I.......I
DEN3    (524) .............................G.S...............K......A..LNT.V.K...S......I
DEN4    (525) .R.D.T..........S..A....VDSGDGNHM.......KVR.E..RI.....T..S.K.SID..M.......T

              601                                                                        675
DEN1    (601) LVQVKYEGTDAPCKIPFSSQDEKGVTQNGRLITANPIVTDKEKPVNIEAEPPFGESYIVVGAGEKALKLSWFKKG
DEN2    (601) VIR.Q...DGS......EIM.LEKRHVL.....V.....E.DS...........D..III.VEPGQ...N.....
DEN3    (599) .IK.E.K.E.........TE.GQ.KAH.........V..K..E..............N..I.I.D....IN.YR..
DEN4    (600) V.K.....AG....V.IEIR.VNKEKVV..I.SST.LAENTNSVT...L.....D....I.V.NS..T.H..R..

              676                                                                        750
DEN1    (676) SSIGKMFEATARGARRMAILGDTAWDFGSIGGVFTSVGKLIHQIFGTAYGVLFSGVSWTMKIGIGILLTWLGLNS
DEN2    (676) ....Q.I.T.M...K...............L......I..AL..V..AI..AA......I...L..VII..I.M..
DEN3    (674) .............................V...LN.L..MV.....S..TA.......I......V....I....
DEN4    (675) ........S.Y...K......E.......V..L...L..AV..V..SVYTTM.G....MIR.L..F.VL.I.T..

              751                        E |#   NS1                                      825
DEN1    (751) RSTSLSMTCIAVGMVTLYLGVMVQADSGCVINWKGRELKCGSGIFVTNEVHTWTEQYKFQADSPKRLSAAIGKAW
DEN2    (751) ......VSLVL..V..........|.....VS..NK.........I.DN...........PE..SK.AS..Q..H
DEN3    (749) KN..M.FS...I.II......V...|.M........K..............................VAT..AG..
DEN4    (750) .N..MA.......GI..F..FT...|.M...AS.S.K..........VDN...........PE..A..AS..LN.H

              826                                                                        900
DEN1    (826) EEGVCGIRSATRLENIMWKQISNELNHILLENDMKFTVVVGDVSGILAQGKKMIRPQPMEHKYSWKSWGKAKIIG
DEN2    (826) ...I.....V.....L.....TP......S..EV.L.IMT..IKG.MQA..RSLQ...T.L.....T.....MLS
DEN3    (824) .N.......T..M..LL....A....Y..W...I.L......IT.V.E...RTLT.....L.....T..L...VT
DEN4    (825) KD.......T.....V.....T....YV.W.GGHDL...A...K.V.TK..RALT.PVSDL.....T......FT
```

Fig. 3. Alignment of polyprotein amino acid sequences of dengue viruses 1–4. Cleavage sites defining N termini of individual proteins are indicated by vertical lines (# — host cell signalase cleavage; Δ — viral NS2B/NS3 protease cleavage; ↓ — furin-like protease). Underlined shaded regions are potential glycosylation sites.

```
                901                                                                       975
DEN1     (901)  ADVQNTTFIIDGPNTPECPDNQRAWNIWEVEDYGFGIFTTNIWLKLRDSYTQVCDHRLMSAAIKDSKAVHADMGY
DEN2     (901)  TESH.Q..L....E.A...NTN....SL........V..........EKQDVF..SK........NR........
DEN3     (899)  .ET..SS......S.....SAS....V.........V..........EV...L.........V..ER........
DEN4     (900)  PEAR.S..L....D.S..PNER....SL........M......M.F.EGSSE.............Q.........

                976                                                                      1050
DEN1     (976)  WIESEKNETWKLARASFIEVKTCIWPKSHTLWSNGVLESEMIIPKIYGGPISQHNYRPGYFTQTAGPWHLGKLEL
DEN2     (976)  ....AL.D..IEK.......S.H.....................NFA..V.........H.............M
DEN3     (974)  ....Q..GS...EK..L......T...............D.....SLA.......H....H..............
DEN4     (975)  ....S..Q..QIEK..L......L...T...........Q.L...S.A..F......Q..A...V.........I

                1051                                                                     1125
DEN1    (1051)  DFDLCEGTTVVVDEHCGNRGPSLRTTTVTGKTIHEWCCRSCTLPPLRFKGEDGCWYGMEIRPVKEKEENLVKSMV
DEN2    (1051)  ...F........T.D............AS..L.T.............YR.............L........N.L.
DEN3    (1049)  ..NY.......IS.N..T..........S..L...............YM.............IN.....M...LA
DEN4    (1050)  ..GE.P....TIQ.D.DH.........AS..LVTQ........M....L.............LS.....M...Q.

           NS1 | Δ NS2A                                                                  1200
DEN1    (1126)  SAGSGEVDSFSLGLLCISIMIEEVMRSRWSRKMLMTGTLAVFLLLTMGQLTWNDLIRLCIMVGANASDKMGMGTT
DEN2    (1126)  T..HGQI.N....V.GMALFL..ML.T.VGT.HAILLVAVS.VT.ITGNMSFR..G.VMV....TMT.DIG..V.
DEN3    (1124)  .....K..N.TM.V..LA.LF.....GKFGK.HMIA.V.FT.V..LS..I..RGMAHTL..I.S....R....V.
DEN4    (1125)  T..Q.TSET..M....LTLFV..CL.R.VT..HMILVVVITLCAIIL.G...M..L.AL..L.DTM.GRI.G-QI

                1201                                                                     1275
DEN1    (1201)  YLALMATFRMRPMFAVGLLFRRLTSREVLLLTVGLSLVASVELPNSLEELGDGLAMGIMMLKLLTDFQSHQLWAT
DEN2    (1201)  ....L.A.KV..T..A...L.K...K.LMMT.I.IV.LSQSTI.ETIL..T.A..L.M.V..MVRKMEKY..AV.
DEN3    (1199)  ....I...KIQ.FL.L.FFL.K.....N...G...AMA.TLR..EDI.QMAN.I.L.L.A...I.Q.ETY...TA
DEN4    (1199)  H..I..V.K.S.GYVL.VFL.K.....TA.MVI.MAMTTVLSI.HD.M..I..ISL.LIL..IV.Q.DNT.VGTL

                1276                                                           NS2A | Δ NS2B
DEN1    (1276)  LLSLTFVKTTFSLHYAWKTMAMILSIVSLFPLCLSTTSQK-TTWLPVLLGSLGCKPLTMFLITENKIWGRKSWPL
DEN2    (1276)  IMAILC.PNAVI.QN...VSCT..AV..VS..F.TSSQ..-AD.I.LA.TIK.LN.TAI..T.LSRTNKKR....
DEN3    (1274)  .V...CSN.I.T.TV..R.ATL..AGI..L.V.Q.SSMR.-.D...MTVAAM.VP..PL.IFSLKDTLK.R....
DEN4    (1274)  A.....IRS.MP.VM..R.IMAV.FV.T.I...RTSCL..QSH.VEITALI..AQALPVY.M.LM.GAS.R....

                1351                                                                     1425
DEN1    (1350)  NEGIMAVGIVSILLSSLLKNDVPLAGPLIAGGMLIACYVISGSSADLSLEKAAEVSWEEEAEHSGASHNILVEVQ
DEN2    (1350)  ..A.....M....A.......I.MT...V...L.TV...LT.R....E..R..D.K..DQ..I..S.PILSITIS
DEN3    (1348)  ...V....L....A....R....M....V...L.......T.T....TV....D.T......QT.V...LMIT.D
DEN4    (1349)  .......GL..L.G.A...........MV...L.L.A..M.............N.Q.D.M.DIT.S.PI.E.KQD

                1426                                              NS2B | Δ NS3           1500
DEN1    (1425)  DDGTMKIKDEERDDTLTILLKATLLAISGVYPMSIPATLFVWYFWQKKKQRSGVLWDTPSPPEVERAVLDDGIYR
DEN2    (1425)  E..S.S..N..EEQ.....IRTG..V...LF.V...I.AAA..L.EV....A.....V....P.GK.E.E..A..
DEN3    (1423)  .....R...D.TENI..V...TA..IV..IF.Y.....ML..HT...QT........V.....TQK.E.EE.V..
DEN4    (1424)  E..SFS.R.V.ETNMI.L.V.LA.ITV..L..LA..V.MTL..M..V.T....A...V...AATKK.A.SE.V..

                1501                                                                     1575
DEN1    (1500)  ILQRGLLGRSQVGVGVFQEGVFHTMWHVTRGAVLMYQGKRLEPSWASVKKDLISYGGGWRFQGSWNAGEEVQVIA
DEN2    (1500)  .K.K.I..Y..I.A..YKE.T..............HK...I.....D............KLE.E.KE......L.
DEN3    (1498)  .K.Q.IF.KT..G...QK................THN......N................LSAQ.QK........
DEN4    (1499)  .M...F.KT.....IHM..............S.ICHETG.......D.RN.M........LGDK.DKE.D...L.

                1576                                                                     1650
DEN1    (1575)  VEPGKNPKNVQTAPGTFKTPEGEVGAIALDFKPGTSGSPIVNREGKIVGLYGNGVVTTSGTYVSAIAQAKASQEG
DEN2    (1575)  L.....RA...K..L..TNA.TI..VS...S.........IDKK..V..........R..A.......TEK.I.D
DEN3    (1573)  .........F..M..I.Q.TT..I................I.....V..........KN.G...G...TN.EPD.
DEN4    (1574)  I......H...K..L....LT..I..VT............I..K..VI.........K..D.....T..ERIG.P

                1651                                                                     1725
DEN1    (1650)  PLPEIEDEVFRKRNLTIMDLHPGSGKTRRYLPAIVREAIRRNVRTLVLAPTRVVASEMAEALKGMPIRYQTTAVK
DEN2    (1650)  -N....DI...K...........A..K............K.GL..I.........A..E..R.L.......P.IR
DEN3    (1648)  .T..L.E.M.K.................K..........K.RL..I.........A..E.M..L.........T.
DEN4    (1649)  -DY.VDEDI...KR.........A..K.I..S......LK.RL..I.........A..E..R.L.......P...
```

Fig. 3. (*Continued*).

```
                1726                                                                       1800
DEN1  (1725)  SEHTGKEIVDLMCHATFTMRLLSPVRVPNYNMIIMDEAHFTDPASIAARGYISTRVGMGEAAAIFMTATPPGSVE
DEN2  (1724)  A....R.........................L........................E.....G..........RD
DEN3  (1723)  .....R.........................L........................................TAD
DEN4  (1723)  .....R............T....ST......L.V.........S.V..........E...............ATD

                1801                                                                       1875
DEN1  (1800)  AFPQSNAVIQDEERDIPERSWNSGYDWITDFPGKTVWFVPSIKSGNDIANCLRKNGKRVVQLSRKTFDTEYQKTK
DEN2  (1799)  P......P.M....E......S..HE.V...K...........A.....A.......K.I........S..V..R
DEN3  (1798)  .......P................NE.....V...........A..V..........K.I...............
DEN4  (1798)  P.....SP.E.I..E.......T.F.....YQ...........A..........S..K.I...........P...

                1876                                                                       1950
DEN1  (1875)  NNDWDYVVTTDISEMGANFRADRVIDPRRCLKPVILKDGPERVILAGPMPVTVASAAQRRGRIGRNQNKEGDQYI
DEN2  (1874)  T....F.............K.E........M.....T..E............HS............PKN.N....
DEN3  (1873)  L....F.............I................T.........................V...PQ..N....
DEN4  (1873)  LT...F...............G..............P...........I...P.............PAQ.D..YV

                1951                                                                       2025
DEN1  (1950)  YMGQPLNNDEDHAHWTEAKMLLDNINTPEGIIPALFEPEREKSAAIDGEYRLRGEARKTFVELMRRGDLPVWLSY
DEN2  (1949)  ...E..E....C...K.................SM.......VD.................D...........A.
DEN3  (1948)  F......K............................................K..S.................AH
DEN4  (1948)  FS.D..K..................Y.......T..G.....TQ.....F.....Q...................

                2026                                                         NS3  |Δ NS4A
DEN1  (2025)  KVASEGFQYSDRRWCFDGERNNQVLEENMDVEIWTKEGERKKLRPRWLDARTYSDPLALREFKEFAAGR|RSVSGD
DEN2  (2024)  R..A..IN.A........IK...I....VE.............K......KI.......K.........|K.LTLN
DEN3  (2023)  ......IK.T..K..........I...............K...................K...D.....|K.IAL.
DEN4  (2023)  ....A.IS.K..E...T......I.....E.....R...K...........V.A..M..KD.....S..|K.ITL.

                2101                                                                       2175
DEN1  (2100)  LILEIGKLPQHLTQRAQNALDNLVMLHNSEQGGKAYRHAMEELPDTIETLMLLALIAVLTGGVTLFFLSGRGLGK
DEN2  (2099)  ..T.M.R..TFM..K.RD.....AV..TA.A..R..N..LS..PE.L...L..T.L.TV...IF..LM....I..
DEN3  (2098)  .VT...RV.S..AH.TR..........T..H..R.....V....E.M...L..G.MIL....AM..LI..K.I..
DEN4  (2098)  ILT..AS..TY.SS..KL....I....TT.R..R..Q..LN...ESL.....V..LGAM.A.IF...MQ.K.I..

                2176                                                        NS4A  |Δ NS4B
DEN1  (2175)  TSIGLLCVIASSALLWMASVEPHWIAASIILEFFLMVLLIPEPDRQRTPQDNQLAYVVIGLLFMILTAA|ANEMGL
DEN2  (2174)  MTL.MC.I.TA.I...Y.QIQ..............I.......EK.........T....AI.TVVAATM|....F
DEN3  (2173)  .....I......GM....D.PLQ...SA.V....M........EK...............I.TLAAIV.|......
DEN4  (2173)  L.M..ITIAVASG...V.EIQ.Q......I.............EK.........I..ILTI.TI.GLI.|......

                2251                                                                       2325
DEN1  (2250)  LETTKKDLGIGHAAAENHHHAAMLDVDLHPASAWTLYAVATTIITPMMRHTIENTTANISLTAIANQAAILMGLD
DEN2  (2249)  ..K......L.-SITTQQPESNI..I..R.............FV...L..S...SSV.V.........TV....G
DEN3  (2248)  .....R...MS-KEPGVVSPTSY...................V....L......S...V..A......VV.....
DEN4  (2248)  I.K..T.F.FY----QVKTETTI.....R..............L...L.......S..L..A.......V....G

                2326                                                                       2400
DEN1  (2325)  KGWPISKMDIGVPLLALGCYSQVNPLTLTAAVFMLVAHYAIIGPGLQAKATREAQKRTAAGIMKNPTVDGIVAID
DEN2  (2323)  ....L...........I..............L.L.......................A.............TV..
DEN3  (2322)  .........L...................I..LL..T...................................MT..
DEN4  (2319)  ....LHR..L......M........T....SLV..LV..................................TV..

                2401                                                                       2475
DEN1  (2400)  LDPVVYDAKFEKQLGQIMLLILCTSQILLMRTTWALCESITLATGPLTTLWEGSPGKFWNTTIAVSMANIFRGSY
DEN2  (2398)  ...IP..P........V...V..VT.V.M.........AL......IS.....N..R..................
DEN3  (2397)  ....I..S........V...V..AV.L.....S.....VL......I............................
DEN4  (2394)  .E.IS..P........V...V..AG.L........F..VL......IL.....N..R.........T........

                2476       NS4B  |Δ NS5                                                      2550
DEN1  (2475)  LAGAGLAFSLMKSLGGGR|RGTGAQGETLGEKWKRQLNQLSKSEFNTYKRSGIIEVDRSEAKEGLKRGEPTKHAVS
DEN2  (2473)  ......L..I..NTTNT.|...NI.......W.SR..A.G....QI..K...Q....TL....I....TDH....
DEN3  (2472)  .......L.I...V.T.K|...S..........KK.....RK..DL..K...T....T..........I.H....
DEN4  (2469)  ..........I.NAQTP.|...TT.............S.DRK..EE......L....T...SA..D.SKI.....

                2551                                                                       2625
DEN1  (2550)  RGTAKLRWFVERNLVKPEGKVIDLGCGRGGWSYYCAGLKKVTEVKGYTKGGPGHEEPIPMATYGWNLVKLYSGKD
DEN2  (2548)  ..S..........M.T.....V.............G...N.R....L.............S.......R.Q..V.
DEN3  (2547)  ..S...Q......M.I...R........................R............V..S.....I...M....
DEN4  (2544)  ..SS.I..I...GM...K...V............M.T..N..............................H..V.
```

Fig. 3. (***Continued***).

```
                2626                                                                      2700
DEN1   (2625) VFFTPPEKCDTLLCDIGESSPNPTIEEGRTLRVLKMVEPWLRG-NQFCIKILNPYMPSVVETLEQMQRKHGGMLV
DEN2   (2623) ........................V.A.......NL..N..NNNT.....V........I.KM.AL...Y..A..
DEN3   (2622) ..YL.................S..V..S..I..........KN-......V......T.I.H..RL.........
DEN4   (2619) ..YK.T.QV...........S....................SSKPE....V......T.I.E..KL......N..

                2701                                                                      2775
DEN1   (2699) RNPLSRNSTHEMYWVSCGTGNIVSAVNMTSRMLLNRFTMAHRKPTYERDVDLGAGTRHVAVEPEVANLDIIGQRI
DEN2   (2698) ..............L.NAS.....S...I....I.....R.K.A...P.....S...NIGI.S.IP......K..
DEN3   (2696) ..............I.N.......S...V..L.......T..R..I.K...........NA...TP.M.V..E..
DEN4   (2694) .C..............GAS.....S..T..K.......TR.......K.........S.ST.T.KPDMT...R.L

                2776                                                                      2850
DEN1   (2774) ENIKNGHKSTWHYDEDNPYKTWAYHGSYEVKPSGSASSMVNGVVRLLTKPWDVIPMVTQIAMTDTTPFGQQRVFK
DEN2   (2773) .K..QE.ETS....Q.H............T.QT......G.............V.....M...............
DEN3   (2771) KR..EE.S......DE...............AT......I....K........V.....M...............
DEN4   (2769) QRLQEE..E.....QE...R.........APST...........K..............L...............

                2851                                                                      2925
DEN1   (2849) EKVDTRTPKAKRGTAQIMEVTARWLWGFLSRNKKPRICTREEFTRKVRSNAAIGAVFVDENQWNSAKEAVEDERF
DEN2   (2848) .......QEP.E..KKL.KI..E...KE.GKK.T..M...............L..I.T...K.K..R.....S..
DEN3   (2846) ........RPMP..RKV..I..E...RT.G...R..L.......K...T...M....TE....D..RA.....E.
DEN4   (2844) ........QP.P..RMV.TT..N...AL.GKK.N..L......IS............QE.QG.T..S...N.S..

                2926                                                                      3000
DEN1   (2924) WDLVHRERELHKQGKCATCVYNMMGKREKKLGEFGKAKGSRAIWYMWLGARFLEFEALGFMNEDHWFSRENSLSG
DEN2   (2923) .E..DK..N..LE...E...........................................L..............
DEN3   (2921) .K..D.......L...GS.................................Y........L...........Y..
DEN4   (2919) .E..DK..A..QE...ES.................R........................L......G....W..

                3001                                                                      3075
DEN1   (2999) VEGEGLHKLGYILRDISKIPGGNMYADDTAGWDTRITEDDLQNEAKITDIMEPEHALLATSIFKLTYQNKVVRVQ
DEN2   (2998) ...............V..KE..A..............LE..K..EMVTNH..G..KK..EA..............
DEN3   (2996) ......................A..................H..E...QQ.D...RQ..NA...........K..
DEN4   (2994) .......R.....EE.D.KD.DL.....................EL..EQ.A.H.KI..KA...........K.L

                3076                                                                      3150
DEN1   (3074) RPAKNGTVMDVISRRDQRGSGQVGTYGLNTFTNMEAQLIRQMESEGIFSPSELETPN-LAERVLDWLKKHGTERL
DEN2   (3073) ..TPR.....I................................G..V.KSIQHL.VT-EEIA.QN..ARV.R...
DEN3   (3071) ..TPK.....I...K............................G..VL.KAD..N.HP.EKKITQ..ETK.V...
DEN4   (3069) ..TPR.A...I...K....................V.......A..VITQDDMQN.KG.K...EK...EC.VD..

                3151                                                                      3225
DEN1   (3148) KRMAISGDDCVVKPIDDRFATALTALNDMGKVRKDIPQWEPSKGWNDWQQVPFCSHHFHQLIMKDGREIVVPCRN
DEN2   (3147) S.............L.....S...............Q.....R.....T..........E.......VL......
DEN3   (3146) ....................N..L...............Q.....H.............E.......KL.....P
DEN4   (3144) ..............L.E..G.S.LF....................KN..E.........KIF.....SL......

                3226                                                                      3300
DEN1   (3223) QDELVGRARVSQGAGWSLRETACLGKSYAQMWQLMYFHRRDLRLAANAICSAVPVDWVPTSRTTWSIHAHHQWMT
DEN2   (3222) ....I....I......................S.....................SH.............K.E...
DEN3   (3221) ....I....I................A.....T............S.........H...................
DEN4   (3219) ....I....I................A.....S............SM.......TE.F.................

                3301                                                                      3375
DEN1   (3298) TEDMLSVWNRVWIEENPWMEDKTHVSSWEDVPYLGKREDRWCGSLIGLTARATWATNIQVAINQVRRLIGNENYL
DEN2   (3297) .....T.......Q.........P.E...EI........Q.........S.....K...T......S.....E.T
DEN3   (3296) .....T........D........P.TT............Q........TS.....Q..LT..Q...S.....EF.
DEN4   (3294) .....K........D..N.T...P.H....I........L........SS.....K..HT..T...N...K.E.V

                3376                3395
DEN1   (3373) DFMTSMKRFKNESDPEGALW
DEN2   (3372) .Y.P.....RK.EEEA.V..
DEN3   (3371) .Y.P.....RK.EES...I.
DEN4   (3369) .Y.PV...YSAP.ES..V.-
```

Fig. 3. (*Continued*).

C protein is found in two forms in infected cells. Cleavage by host cell signalases in the endoplasmic reticulum (ER) occurs after a stretch of hydrophobic residues at the C terminus and yields a membrane-anchored form of C protein (C_{anch}). The hydrophobic tail can also be cleaved by the viral NS2B/3 protease to release the mature virion-associated form (C_{vir}). C_{vir} remains integrated with the membranes of the ER via a conserved internal hydrophobic domain, which presumably acts similarly to a hydrophobic signal sequence and functions as a membrane anchor.[20] Hence, C_{vir} is thought to maintain a hairpin-loop orientation in the membrane of the ER, with the N and C terminal RNA-binding domains remaining in the cytoplasm of the cell.

A solution structure for recombinant DENV-2 capsid protein equivalent to C_{vir} has been determined by nuclear magnetic resonance (NMR) spectroscopy and revealed a homodimeric structure, with each C monomer comprising four alpha helices in slightly different spatial orientations.[21] Distribution of charges across the highly basic C protein structure was nonuniform and allowed the identification of a charged cleft that probably functions as the primary RNA-binding site.

C protein has been detected in the nucleus and nucleoli, as well as the cytoplasm, of DENV-infected cells.[22,23] Nuclear localization appears to be dependent on a bipartite nuclear localization sequence (NLS) between residues 85–100.[24] The DENV core protein has also been shown to interact with a cellular regulatory protein, heterogeneous nuclear ribonucleoprotein K (hnRNPk), suggesting that, in addition to its role in virion structure and assembly, C protein may also be involved in regulation of viral replication.[25]

Envelope proteins

The viral envelope consists of a lipid bilayer in which are embedded the envelope (E) and membrane (M) proteins (Fig. 2).[10] The lipid composition of the envelope is dictated by the composition of the host cell membrane from which the virus buds.[26]

M protein may be found in two forms, depending upon the maturity of the virus. In cell-associated (immature) virions a 165-amino-acid, glycosylated (at residue 69) pre-M (prM) protein is observed which forms a heterodimer with the E protein.[9,27] Studies using DENV and JEV have demonstrated that cosynthesis of prM and E proteins is essential for proper folding, membrane association and assembly of the flavivirus E protein, suggesting that prM may act as a chaperone for E protein folding in the ER.[28,29] Immature MVEV particles have been reported to be up to 400 times more acid-resistant than mature virions, indicating that the E–prM interaction also serves to protect the E protein from irreversible inactivation during transport to the cell surface in acidic post-Golgi vesicles.[30]

The prM protein contains six cysteine residues that form three disulphide bridges.[31] During viral release, prM is cleaved and a large amino-terminal portion is removed, leaving the disulphide-free, unglycosylated M protein.[32] This cleavage is apparently mediated within the trans-Golgi network and post-Golgi vesicles by a host-cell-derived furin-like protease and is dependent upon an irreversible, low-pH-induced conformational change within the prM protein.[33]

Immature TBEV virions derived by specific mutation of the furin cleavage site or growth in the presence of furin inhibitors have been shown to be incapable of fusing to target cells, suggesting that cleavage of prM is necessary for the generation of fusion activity of mature virions.[34,35] *In vitro* cleavage using recombinant furin was employed to confirm the above and converted immature prM-containing virions to their mature, M-containing functional form.[33] Similarly, cryo-EM analysis revealed that treatment of immature DENV virions with furin resulted in morphology consistent with that of the mature virion structure.[9] Small numbers of E/prM heterodimers remain uncleaved and can be found on extracellular flaviviruses.[27]

The E protein is a 495-amino-acid protein (except for DENV-3, which has 493 amino acids) which is glycosylated in dengue viruses (potential glycosylation sites at residues 69 and 155) but not all other flaviviruses, and is the major constituent of the virus envelope.[32,36] E has important functional roles in virus attachment to cells and fusion with membranes, and is the major target for neutralizing antibody.[37–39] In addition, mutations within E have been observed to significantly influence the function and virulence of dengue and other flaviviruses (for a review see Ref. 40).

E is the most highly conserved structural protein, and all flavivirus E proteins contain 12 conserved cysteine residues which form 6 intramolecular disulphide bridges. In addition, residues 98–111 are highly conserved in all flaviviruses,[41] and structural changes have been detected in this region following low pH treatment of the virus.[42] Given that low-pH-induced activation of fusion activity plays an important role in the infectivity of flaviviruses,[43] this region has been proposed to be the flavivirus "fusion domain," responsible for the fusion of flaviviruses to their target cells.[42] A later study using recombinant protein fragments showed an increased binding of fragments of this conserved region to target cells following acid treatment at pH 5.0, and specific mutation of a single residue (E107) of the TBEV E protein was shown to significantly impair fusion activity, confirming the importance of this domain.[44,45]

The greatest advance in the understanding of the physical structure of the E protein came with the crystallization of the truncated soluble ectodomain of this protein from Central European TBEV and its subsequent X-ray diffraction analysis to produce a biophysical model of the E protein.[8,46] This structural model revealed the E protein dimer to have dimensions of approximately 150 Å × 55 Å × 30 Å, with the long axis of the E protein monomer parallel to the virion membrane. Three distinct structural domains were identified: a central β barrel (domain I), an elongated dimerization domain (domain II) and a C terminal immunoglobulin-like domain (domain III). The three domains were identified as being homologous to domains C, A and B, respectively, of an earlier antigenic model.[41]

Subsequent crystal structures of recombinant DENV-2 and -3 E protein ectodomains retained the same dimeric structure and overall protein fold of the TBEV E protein structure, although several loops differed in their precise placement.[47–49] In particular, a four-residue loop in domain III (E382-385 in DENV-2) that is present in mosquito-borne but not tick-borne flaviviruses, and which has been proposed as a potential receptor binding and tropism determinant, formed a solvent-exposed bulge on the outside face of the domain (Fig. 4).

The E protein dimer is stabilized by polar and van der Waals interactions between the parallel segments of domain II in each monomer, and contact between the distal end of domain II of one monomer and the

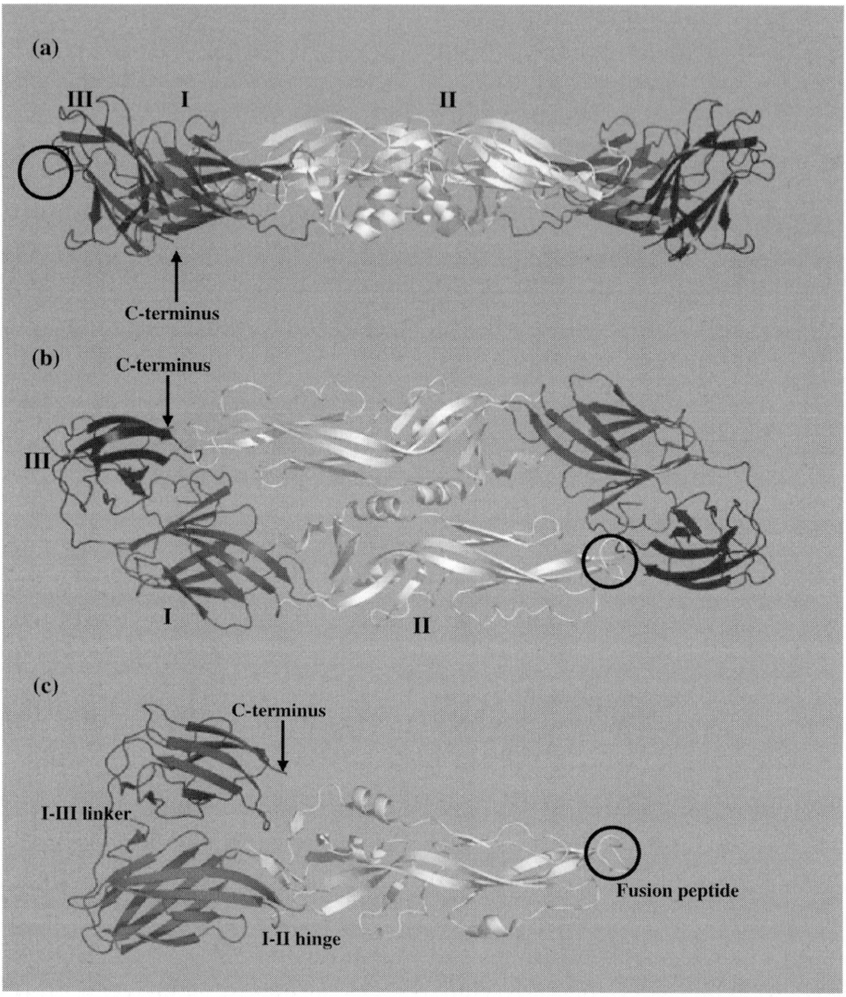

Fig. 4. X-ray crystallographic structure of the dengue virus type 2 E protein ectodomain dimer shown in **(a)** lateral and **(b)** overhead views. Domains I, II and III and the C terminus leading to the stem and transmembrane regions are indicated on one monomer. The mosquito-borne virus-specific loop formed by residues E382–E385 in domain III is circled in (a) and the fusion loop at the tip of domain II is circled in (b). **(c)** shows the structure of an E monomer in the post-fusion conformation (overhead view). Domain II is oriented similarly to (a) and (b). Note the relative rotation and reorientation of domains I and III (resulting in the C terminus moving closer to the fusion peptide) and the extension of the domain I-III linker required to accommodate these changes. Structural data obtained from PDB files 1OAN and 1OK8.

crevice between domains I and II of the other.[8] This second interaction is apparently stabilized by the carbohydrate moiety of domain I. These interactions leave two approximately 20 Å "holes" on either side of the central domain II α helices, and it had been speculated that prM may project through these holes until it is cleaved to M in the mature virion. However, as described above, a cryo-EM structure of the immature dengue virion has shown the arrangement of surface structures to be significantly different to those observed in the mature virion, with prM/E heterodimers projecting upward from the virion surface.[9] Consistent with other experimental observations, prM shielded the E protein fusion peptide located in domain II.

Located below the E ectodomain are two alpha-helical regions and the two hydrophobic membrane-spanning domains. Studies using full-length and truncated DENV-2 E protein showed that trypsin cleavage of the E protein ectodomains from the membrane-spanning carboxy terminal stem abolished the association between prM and E, suggesting that major prM binding sites are located in the stem anchor region of E protein.[50] However, other studies with YFV[51] and MVEV[30] suggest that prM interacts with domain III of the E protein. Analysis of a DENV-2 virion cryo-EM structure revealed that the two alpha helices located in the stem below domain III lie parallel to the virion surface and appear to be embedded within the upper surface of the lipid bilayer.[7]

Exposure to acidic pH induces a conformational change in the E protein accompanied by a transition from dimeric to trimeric forms[52] and is associated with dramatic conformational rearrangements of the E protein structural domains (Fig. 4c).[53] This rearrangement has been observed with detergent-solubilized virion-derived E protein dimers, but not with trypsin-cleaved dimer, suggesting involvement of the carboxy terminal membrane-spanning region or the wider E protein network on the virion surface.[8,54] Rearrangement to a trimeric form is critically dependent upon exposure to low pH, presumably due to a requirement for protonation of the dissociated monomeric E protein prior to trimerization.[55] The interface between domains I and II was shown to form a flexible hydrophobic pocket in DENV-2 E protein X-ray crystal structures.[47,48] The authors proposed that structural flexibility in this region may be a critical factor in the initiation of fusion, and might allow some vertical movement of domain II that facilitates more dramatic structural rearrangements leading to E protein trimerization and membrane fusion. The stem alpha helices in the upper surface of the lipid membrane may also contribute to the forces required for conformational rearrangements associated with fusion.[7]

Trimerization of E results in the formation of a hydrophobic pocket, made up of the conserved "fusion loop" at the distal end of each domain II, which embeds within the target membrane. Rearrangements of domains I and II with respect to domain III suggest that a folding back of the E protein trimer occurs which would bring the virion and target membranes into close proximity, leading to lipid mixing and membrane fusion.[53]

Nonstructural proteins

NS1

The NS1 protein is a variably glycosylated 44–49 kDa (353–354 amino acids) protein which exists primarily as a heat-labile dimer, but can be found in other oligomeric forms with varying functional activities. The DENV NS1 protein contains 2 conserved glycosylation sites (at residues 130 and 207) and 12 conserved cysteine residues.[56] Although not identified in DENV-infected cells, NS1 proteins of varying lengths can also be produced due to C terminal cleavages at alternative sites within the downstream NS2A protein.[57–59]

Immunolocalization of NS1 by both light and electron microscopy has shown it to be associated with double-stranded replicative form (RF) RNA, suggesting involvement in viral RNA replication.[60] Furthermore, mutations within NS1, including mutations resulting in loss of either or both glycosylation motifs, have been shown to significantly inhibit or abolish viral replication and RNA accumulation, and affect virulence in animal models.[61–64] Inhibition of replication is most prominent during the initial phases of replication, suggesting an important role for NS1 in early replication events.[63] NS1 may also play some role in virion assembly/maturation, as it has been found to be associated with immature E protein in the lumen of the ER.[65]

NS1 is cotranslationally inserted into the lumen of the endoplasmic reticulum, where it is glycosylated and forms dimers following cleavage from the N terminal E protein signal sequence and C terminal NS2A protein by host-derived proteases.[56,66] Some mutational analyses of NS1 suggested that monomeric forms of NS1 can retain functional activity within the host cell,[67] although dimerization has been reported to be essential for processing of NS1 through cellular export pathways.[68] Soluble NS1 secreted from dengue-virus-infected cells appears to be a hexamer of three dimeric subunits, while membrane-bound NS1 at the cell surface is dimeric.[69]

The processing of the N glycans during transport through exocytic pathways appears to be critical for efficient maturation and release of NS1 from DENV-infected cells. One of the two N glycans can be modified to a complex sugar chain, while the other is apparently protected from further modification during transport to the cell surface, possibly by the dimerization of NS1 molecules.[56,69,70] Formation of soluble hexameric NS1 is dependent on the processing of the first N glycan to its complex form.[69]

A portion of dimeric DENV-2 NS1 expressed in infected cells is posttranslationally modified by addition of a glycosyl-phosphatidylinositol (GPI) anchor which allows membrane anchoring of NS1 on the cell surface.[71] Binding of anti-NS1 antibodies to this membrane-anchored NS1 was also reported to induce signal transduction leading to protein tyrosine phosphorylation which might affect virus replication rates within infected cells. This was the first report of a GPI-anchored virus-derived protein.

NS3

The NS3 protein (67–70 kDa; 618–623 amino acids) is the second-largest protein encoded in the dengue virus genome and its primary amino acid sequence is the most conserved between the dengue viruses. Several functions for NS3 have been demonstrated: it is the viral protease involved in cleavage of the translated viral polyprotein, and plays roles in viral RNA replication through nucleotide triphosphatase (NTPase), RNA 5′ triphosphatase (RTPase) and helicase activities.

Protease activity of NS3 has been demonstrated for several flaviviruses, including dengue, using recombinant or cell-culture-derived NS3 protein, but requires the presence of the NS2B protein for efficient activity.[72–76] A trypsin-like serine protease domain was identified in the N terminal 180 amino acids of NS3,[77] and the protease activity of the DENV-2 NS3 protein has been precisely mapped to the N terminal 167 amino acids.[75] A conserved catalytic triad (His51-Asp75-Ser135) was identified in this region by sequence homology and mutational analyses,[78–80] and a crystal structure of DENV-2 NS3 protease confirmed the close spatial association of these residues.[81] The protease domain of flavivirus NS3 proteins contains several conserved sequences, and mutation of residues within these regions has also been shown to inhibit protease activity.[82] Structural and mutational studies of the NS2B–NS3 protease complex have determined that the NS2B cofactor is critical for protease activity and acts to both stabilize the NS3 structure and form part of the substrate binding site.[83,84]

NS3 has been identified in close association with viral RNA in infected cells,[85] and the C terminal portion of NS3 contains domains associated with RNA replication, including RNA binding, NTPase and helicase activities.[73,75,86–89] Several motifs critical for the NTPase activity of DENV-2 NS3 have been identified immediately downstream of the protease domain in the region of residues 184–201.[75] In particular, a small region of highly basic amino acids (residues 196–200) may be involved in RNA–protein interactions essential for NS3 NTPase activity. The same region also appears to function as a RTPase involved in capping of the viral RNA genome.[90] An X-ray crystallograpic structure of the C terminal region of DENV-2 NS3, including the helicase and NTPase domains, has been determined and should allow more detailed mutational studies to determine the specific enzymatic mechanisms for NS3 in viral RNA replication and facilitate the design of specific antiviral compounds.[91]

The NTPase activity of DENV-1 NS3 can be modulated *in vitro* by the NS5 protein, suggesting that interactions between these two proteins may induce conformational changes in NS3 that enhance its NTPase activity.[92] In line with this finding, NS3 of DENV and JEV have been shown to complex with NS5 *in vitro* and/or *in vivo*.[93,94] Similarly, the observed *in vitro* helicase activity of recombinant DENV-2 NS3 was significantly lower than that reported for related hepaci- and pestiviruses, which may indicate a requirement for interaction with other viral replication components.[75] However, a recombinant JEV NS3 protein displayed helicase activity more comparable to that reported for other members of *Flaviviridae*.[95] It was suggested that the increased activity of JEV NS3 may have been a result of differences in the protocols used to express and purify the dengue and Japanese encephalitis virus proteins.

Recombinant DENV-2 NS3 protein has been reported to be capable of converting replicative form RNA templates to replicative intermediates, suggesting that NS3 may also encode some RNA polymerase activity.[96] Consistent with this finding, these authors also identified motifs in dengue and other flavivirus NS3 proteins similar to sequences found in several animal and plant virus RNA-dependent RNA polymerases.

NS5 — the viral RNA-dependent RNA polymerase (RdRp)

The NS5 protein (104–106 kDa; 900–905 amino acids) is the largest protein encoded in the dengue virus genome and has two putative functional activities: the C terminal portion encodes an RdRp, while the N terminal region contains potential methyltransferase (MT) motifs.[97–99] A short central region of DENV-2 NS5 has also been reported to contain two functional nuclear localization sequences,[100,101] and the protein has been reported to have both a cytoplasmic and a nuclear distribution.[93]

The polymerase activity of NS5 has been demonstrated directly and indirectly in several *in vitro* assays. Monoclonal antibodies against NS5 have been used to block RNA replication in dengue-virus-infected cell lysates,[88] while recombinant DENV NS5 proteins have been used to replicate virus-derived and other RNA templates.[96,102] Engineered deletions in either the RdRp and/or MT motifs of NS5 abolished replication activity of Kunjin virus (KUNV) subviral replicons *in vitro*, although deletions in the C terminal RdRp region could be compensated for by functional NS5 provided in *trans*.[103,104] In contrast, larger deletions including parts of the N terminal domain, could not be complemented, suggesting that this region of the protein is a critical *cis*-acting element, possibly involved in protein–protein interactions essential to the formation of the viral replication complex.[104] X-ray crystallography and analysis of the enzymic activity of a bacterially expressed DENV-2 NS5 MT domain (residues 1–296) defined the cap GTP-binding function of this domain and identified novel structural features that might be targeted for design of antiviral inhibitors for DENV and other flaviviruses.[99]

Posttranslational modification of NS5 appears to be important in regulating both its function and its cellular distribution. NS5 of DENV-2 has been shown to exist in differentially phosphorylated states, with the degree of phosphorylation affecting both the interaction of NS5 with NS3, which is critical to formation of the viral replication complex, and the localization of NS5 within the infected cell.[93] In these studies, highly phosphorylated NS5 dissociated from NS3 in the cell cytoplasm and was transported to the nucleus. In contrast, phosphorylation of the central NLS has been reported to inhibit nuclear transport of DENV-2 NS5.[100] Regardless, a potential nuclear function for NS5 remains obscure.

Other nonstructural proteins

Flaviviruses also encode four small hydrophobic proteins — NS2A, NS2B, NS4A and NS4B — which are not well conserved in primary sequence between viruses, but share similar structural features, including internal hydrophobic domains which may act as membrane-spanning segments.[58] These proteins may also act as antagonists of host innate immune responses.[105]

NS2B is a cofactor required for the proteolytic activity of the NS3 protease[72–76] and has been demonstrated in close association with NS3 in KUNV-infected cells by electron microscopy.[85] The NS2B protein has a central hydrophilic domain bounded by hydrophobic segments, and the region required for NS3 protease activation has been localized within the central hydrophilic region of 40 amino acids.[74,106] Activity of the NS2B/NS3 protease of DENV-2 is dependent upon membrane association of the NS2B protein via its N and C terminal hydrophobic segments, possibly by allowing interaction between the hydrophilic domain of NS2B with NS3 through induced conformational change.[76,106,107] This requirement for membrane association of the protease complex has also been reported for WNV,[79,108] but was not necessary in the activity of proteases from YFV[78] or TBEV.[80]

Immunoelectron microscopy of KUNV-infected cells has located NS2A and NS4A proteins in close association with other molecules defining the viral replicase complex, including both NS3/NS5 and double-stranded "replicative form" viral RNA. In addition, these proteins have been shown to bind specifically to the 3′ noncoding region of viral template RNA and to the other replication-associated proteins, suggesting that NS2A may play a direct role in replication of viral RNA, and that NS4A may act in targeting or anchoring the replication complex.[109] Mutations at an internal cleavage site in the YFV NS2A protein have recently been reported to prevent generation of infectious viruses, despite normal RNA replication and protein synthesis, suggesting a role for this protein in virion assembly and/or release.[110]

NS4B has been identified in association with cytoplasmic membranes involved with viral replication and this appears to be mediated via a series of internal transmembrane domain sequences.[111,112] NS4B of DENV-2 also functions as an interferon antagonist via blocking of STAT1 activation.[113]

Deletions of these four small NS proteins during *trans*-complementation experiments using a Kunjin replicon suggested that all four must be translated in *cis* with other viral NS proteins for replication to occur, possibly because these proteins may not assume the correct membrane associations outside the context of the viral polyprotein.[114]

A potential role for NS2A, NS2B and NS4A as "viroporins" has also been proposed. Expression of these proteins from JEV in transfected mammalian cells led to altered membrane permeability and growth inhibition and it was suggested that, in the virus-infected cell, these proteins may form hydrophobic pores in cellular membranes leading to induction of cytopathic effect.[115]

Replication Strategy

Attachment/entry/fusion/uncoating

The attachment and entry of flaviviruses into cells involves an initial attachment to the plasma membrane of the target cell followed by uptake via receptor-mediated endocytosis and pH-dependent fusion with the endocytic vesicle membrane.[116,117] For DENV-2, domain III of the E protein has been shown to be involved in binding to mammalian and mosquito cells.[118] Exposure of the virus to acidic pH in endosomic vesicles following attachment and uptake into a target cell results in major structural modifications at the virion surface, including the trimerization of E protein and an associated rearrangement of domain II allowing that domain to project up from the virion surface, thereby exposing the flavivirus-conserved fusion domain.[8,52,55]

Direct fusion with the membranes of mosquito C6/36 cells has also been reported.[119] However, inhibition of SLEV infection of mammalian and mosquito cells by growth in the presence of acidotropic amines such as ammonium chloride has been reported, suggesting that the endosome plays an important role in the entry of flaviviruses into both mammalian and mosquito cells.[120]

Gollins and Porterfield[116] studied the kinetics of WNV entry into P388D1 cells and observed that, within one minute of binding, virus particles could be observed at invaginations of the cell membrane and were contained within small, coated vesicles. Within five minutes, viral antigens were no longer detected at the cell surface (leading the authors to suggest that direct fusion and uncoating at the cell membrane did not occur), and internalized virions were observed in prelysosomal vesicles where pH-dependent fusion of virion and cell membranes presumably occurred. In contrast, more recent experiments examining DENV uptake into BHK cells suggested a slower rate of entry, with only 50% of the adsorbed virus entering the cell within 25 minutes.[121]

Potential receptor molecules

In vitro attachment of dengue and other flaviviruses to cell lines of human and other mammalian and mosquito origin has been reported to be dependent on the presence of glycosoaminoglycans (GAG's), in particular heparan sulfate, on the surfaces of those cells.[121–125] In each case, binding of viruses was reported to be specifically inhibited by pre- or coincubation with heparin and/or pretreatment with heparinase. Binding of recombinant DENV-2 E protein domain III to mammalian cells also appears to occur via interaction with GAG molecules, although this is not the case for binding to mosquito cell lines.[118] However, the consensus of these and other authors[126–129] is that GAG binding most probably represents an interaction preliminary to binding to a second, possibly multi-component, receptor or receptor complex. At least to some extent, this property may be an artifact of the tissue culture passage history of the virus strains examined.

Specific dengue virus binding to proteins from several cellular origins — including human lymphocytic[126,127] and hepatoma cells,[123,125] as well as other mammalian[130–132] and mosquito cells[133,134] — has been investigated in attempts to identify specific protein receptor molecules. In almost all cases, more than one cellular protein has been reported to bind virus antigen, with molecular weights of these proteins ranging from approximately 30 to 80 kDa. Limited biochemical characterization of two virus-binding proteins from Vero cells suggested that they are probably glycoproteins.[132] Subsequent studies have identified two heat-shock proteins, HSP90 and HSP70, as potential receptor molecules for DENV attachment to mammalian cell lines, including monocyte/macrophage cells.[135]

Bacterial lipopolysaccharide (LPS), which is believed to bind to a receptor complex consisting of CD14 and an unidentified transmembrane protein, was also able to block attachment of DENV-2 to human monocyte/macrophage cells.[136] However, preincubation of cells with anti-CD14 antibodies did not inhibit dengue virus infection of those cells, suggesting that CD14 was not the specific dengue receptor. Rather, it has been proposed that a CD14-associated molecule may be utilized by dengue viruses for binding to monocytic cells. Dengue virus entry into human monocytic cells had previously been reported to be dependent on a trypsin-sensitive receptor, although this protein was not identified.[130] Subsequently, LPS has also been shown to inhibit dengue virus binding to some cell lines of lymphocytic origin, although the degree of inhibition varied significantly with different virus strains and cell types.[127] More recently, DC-SIGN (CD209, a c-type lectin) has been shown to be a cellular receptor for all four dengue viruses on dendritic cells.[137–139] A cryo-EM structure of DENV-2 in complex with a recombinant carbohydrate recognition domain (CRD) of DC-SIGN showed that the interaction between the virion and the putative receptor molecular occurs via carbohydrate at residue 67 of the E protein, one of the two E protein glycosylation sites in DENV.[140] Each CRD interacted with the two E67 carbohydrates located within an

E dimer, leaving a significant exposed surface area for additional or subsequent interactions with other receptor molecules.

An alternative mechanism of dengue virus attachment and entry into cells of monocyte/macrophage origin is believed to occur during anamnestic infections. Complexes formed between virus and pre-existing cross-reactive antibody bind to F_c or complement C3 receptors on the surfaces of those cells and are taken into the cell by receptor-mediated endocytosis.[141–144] This antibody-mediated infection (also known as antibody-dependent enhancement) is 5–6 times more efficient than that of free virus binding to the monocyte DENV receptor[145] and is believed to contribute to the development of hemorrhagic symptoms (dengue hemorrhagic fever and dengue shock syndrome) by amplification of the viremia.[146]

Translation and polyprotein processing

Viruses with positive-sense RNA genomes do not contain a polymerase in virus particles; rather, the input RNA is translated to express the polymerase. Following virus entry and uncoating, the positive-sense RNA genome is translated by host cell machinery to yield a single large polypeptide that is co- and posttranslationally cleaved by host-cell- and virus-derived proteases to yield the individual structural and nonstructural proteins.

The efficiency of genome translation has been reported to be a critical determinant of infectivity differences between DENV-2 strains in primary cell cultures.[147] As mentioned previously, flavivirus genomes are capped RNA's, but studies with DENV-2 suggest that they can also initiate translation in a cap-independent fashion via interaction of the 5′ and 3′ NCR's.[148] This redundancy in translation mechanisms is hypothesized to allow the infecting virus to compete with capped cellular RNA's to effectively recruit cellular proteins required for translation, and the transition between cap-dependent and -independent translation may be triggered by the available quantities of a cap recognition protein, eukaryotic initiation factor 4E (eIF4E). Elements within the DENV 3′ NCR also act to enhance translation of genomic and other mRNA's without the 5′ NCR.[149] Flavivirus genomes include several potential translation initiation codons near the

5′ end, and translation of DENV-2 RNA is initiated under the direction of a conserved hairpin structure within the capsid protein coding region.[150]

The viral NS2B/NS3 protease cleaves at a dibasic cleavage site: usually Lys–Arg, Arg–Arg, Arg–Lys at P1–P2 followed by a short chain amino acid (Gly, Ala, Ser) at the P1' position. The protease mediates cleavages at the junctions between NS2A-2B, NS2B-3, NS3-4A and NS4B-5,[72,73,151–153] as well as several cleavages that occur internal to both structural and nonstructural proteins.

An as-yet-unidentified membrane-bound host cell protease located in the endoplasmic reticulum mediates cleavage at the NS1-2A junction.[66] This cleavage occurs following a conserved octapeptide motif (Met/Leu-Val-X-Ser-X-Val-X-Ala) located at the carboxy terminus of NS1 and is also dependent upon the presence of downstream NS2A residues.[154,155]

Host cell signalases mediate cotranslational cleavages at C-prM, prM-E, E-NS1 and NS4A-4B junctions. The prM, E and NS1 proteins are translocated to the lumen of the endoplasmic reticulum, where signalase cleavages occur. They are subsequently transported through the Golgi apparatus, where carbohydrate side chains are added and further trimming occurs. Other proteins remain in the cell cytoplasm. Efficient signalase cleavage at the C-prM junction requires the presence of the viral NS2B/NS3 protease,[156–158] and deletions in the helicase domain of NS3 lead to inhibition of virus secretion (despite efficient replication of viral RNA), possibly as a result of inefficient cleavage at the dibasic cleavage site at the C terminus of the C protein.[114] Viral protease cleavage in the carboxy terminus of the C protein gives rise to C_{vir}, but is not strictly necessary for upregulation of C-prM signalase processing *in vitro*.[159] Rather, association of the protease complex with the polyprotein precursor seems to be sufficient for allowing efficient processing.[160] The apparent complexity of processing at the C-prM junction has led to the suggestion that this region might play a pivotal role in regulation of flavivirus replication.

Signalase cleavage at the NS4A/4B junction is also dependent on prior cleavage by the viral NS3/2B protease at a conserved site located upstream from the signalase cleavage site.[161]

In addition to the viral-protease-mediated cleavage of C protein and NS4A, several other internal cleavages have been reported for different flaviviruses, although these have not been universally identified. The NS3 protein of the dengue virus is cleaved at an internal site in the putative helicase domain, giving rise to a truncated form of NS3, designated NS3′.[162,163] Possible significance of this cleavage may lie in regulation of viral replication through disruption or alteration of NS3 protease/helicase functions, although this remains to be determined experimentally.

RNA replication

The most extensive analyses of flavivirus replication have been undertaken using KUNV, and have demonstrated that the viral replication complex consists of a double-stranded RNA template in association with most of the viral nonstructural proteins located within virus-induced membrane structures.[114,164–166] These membrane structures have been shown to contain markers suggesting an origin in either the *trans*-Golgi network or the intermediate compartment.[166] RNA synthesis occurs via a semiconservative, asymmetric process involving double-stranded replicative form (RF) RNA, and replicative intermediates (RI's) that contain regions of double-stranded and nascent single-stranded RNA. Similar findings have been reported for dengue and other flaviviruses.[94,167–170] Flavivirus positive-sense RNA seems to be synthesized preferentially, with an approximately tenfold excess compared to negative-strand RNA observed in dengue-virus-infected cells.[167] Positive-sense RNA templates can then be utilized for further negative strand synthesis or polyprotein translation, or may be packaged into newly forming virions.[171]

Based on analyses with KUNV, the following model for induction of flavivirus RNA replication has been proposed.[104,114] During polyprotein translation and prior to cleavage, NS3 (possibly in association with NS2A) binds to conserved sequences in the N terminal half of NS5. Following completion of translation, this partially assembled complex binds to an adjacent 3′ NCR via conserved secondary stem loop structures, and becomes membrane-associated via interactions between NS2A and the membrane-associated NS4A. NS4A also interacts with NS1, located in the lumen of the endoplasmic reticulum, via hydrophilic extensions into the ER. Late in the replication cycle this complex is capable of continuing RNA synthesis in the absence of further viral polyprotein synthesis, suggesting that the same protein components can be efficiently recycled in a stable membrane-associated complex.[164,172]

Analyses using in vitro replication systems for dengue viruses have shown that both the 5′ and 3′ NCR's are essential for genome replication, and that interactions between these two sequences (presumably mediated by the conserved cyclization motifs) lead to a conformational change in the RNA template which is essential for self-primed RNA synthesis at the 3′ end of the 3′ NCR by the viral replicase.[173] Subsequently, *in vivo* experiments using KUNV or YFV replicons confirmed the importance of complementary 5′ and 3′ interactions in replication competence.[174,175] Although predicted secondary structures in the 3′ NCR's of dengue and other flaviviruses are well conserved, substitution of WNV 3′ stem loop (SL) structures for those in a DENV-2 infectious clone significantly impaired replication of the derived virus, suggesting that specific DENV-2 sequence elements, particularly certain bulges contained within the long 3′ SL, were required for interactions with virus-derived proteins and/or other regions of the viral genome.[176,177] In particular, interactions with the viral NS3/NS5 replicase complex or with conserved cyclization motifs in the 5′ NCR are likely to be affected by substitution of nonhomologous sequences in the 3′ SL.

In addition to the viral nonstructural proteins listed above, host-cell-derived proteins are involved in flavivirus replication. At least three BHK cell proteins were reported to bind to SL structures in the 3′ NCR of WNV.[178] Subsequently, one of these proteins was identified as translation elongation factor 1α (EF-1α), a protein whose usual function is in binding of amino acid–tRNA complexes to the ribosome during translation.[13] More recently, the same group has also shown that TIA-1 and TIAR, two members of the RNA recognition motif family of RNA binding proteins, bind to the 3′ NCR of the negative-sense RNA intermediate of WNV.[179] Eight proteins from C6/36 mosquito cells, including EF-1α, La (a nuclear protein associated with protein transport from the nucleus to cytoplasm that may act as an RNA structural chaperone) and PTB (associated with RNA splicing) were shown to bind to the DENV-4 3′ NCR.[180] Possible roles for EF-1α in targeting viral RNA to intracellular membranes or in promoting protein–protein and/or protein–RNA interactions necessary for the assembly of the replication complex were proposed. Another study using mouse brain tissue preparations described three proteins (molecular weights 32, 35 and 50 kDa) that bound to the 3′ NCR of JEV RNA.[181] It was speculated that the 50 kDa protein may be EF-1α and the 35 kDa protein was identified as Mov34, a protein involved in the cell cycle cascade and the regulation of cellular transcription, translation and protein degradation.

Virion assembly and release

Assembly of flavivirus virions occurs at intracellular membranes of infected cells, with progeny virions observable in vesicles of the perinuclear ER within 8–12 hours of infection, suggesting that assembly occurs relatively rapidly. From studies with replication-competent and defective KUNV cDNA clones, it appears that RNA replication and virion assembly are closely linked, possibly to limit amplification and subsequent propagation of defective RNA virus genomes.[182]

The viral C protein associates with positive-sense RNA to form nucleocapsid structures. The membrane-bound orientation of C suggests that RNA encapsidation probably occurs in association with the membranes of the ER, rather than in a "free-floating" state in the cytoplasm.[20] Consistent with this, nucleocapsids have been observed on the cytosolic side of ER membranes.[183]

The viral prM and E proteins are translocated to the lumen of the ER, where prM/E heterodimers can be detected within 20 minutes of infection.[29] Multimeric prM/E complexes, possibly representing intermediate structures formed during virion assembly, can be detected in DENV-infected Vero cells within 2 hours of infection.[50] The cleavage and release of prM from the polyprotein has been reported to be relatively slow (half-maximal formation takes 15–20 minutes) and may be the rate-limiting step for prM/E heterodimer formation.[29] As has been discussed previously, cleavage at the C-prM junction is a complex process and may be an important regulatory point in flavivirus virion morphogenesis. The efficiency of secretion for DENV-virus-like particles is also significantly modulated by mutations within the prM/M and E stem helices, suggesting that interactions between these regions and the lipid envelope and/or the E ectodomain are critical for stabilizing the assembling particle.[184]

The prM and E proteins move through the ER and Golgi apparatus, where they are further modified by addition of N-linked glycans and glycan trimming. Inhibition of trimming in dengue-virus-infected cells has been shown to inhibit correct folding of the E protein and destabilize the formation of prM/E heterodimers, leading to large reductions in the yields of the progeny virus.[29]

It has been suggested that complete assembly of virions may occur through budding of the nucleocapsids into the lumen of the ER, thereby acquiring their prM/E/lipid envelope.[185] Virus particles are then transported via secretory pathways to the cell surface and released. As the virus is transported through exocytic vesicles, and immediately prior to release, the prM protein is cleaved by a furin-like protease to its mature M protein form, thus allowing the formation of E protein homodimers and "activating" the E protein for the pH-dependent conformational changes which occur during subsequent attachment and entry into cells.[8,33]

Conclusions

In recent years there havc been considerable advances in our knowledge of the mechanisms involved in the replication and infectivity of dengue and other flaviviruses. Molecular techniques have also provided an additional arm to the traditional, serologically based methods for identifying and classifying members of the family *Flaviviridae*. Further progress in our understanding of the molecular mechanisms involved in the function of dengue viruses should provide deeper insights into the basis of dengue disease and offer targets and strategies for the development of therapeutic agents, including recombinant vaccines and antiviral drugs.

References

1. Calisher CH, Karabatsos N, Dalrymple J *et al.* Antigenic relationships among flaviviruses as determined by cross-neutralization testd with polyclonal antisera. *J Virol* 1989;70:37–43.
2. Kuno G, Chang G-J, Tsuchiya KR, Karabatsos N, Cropp CB. Phylogeny of the genus *Flavivirus*. *J Virol* 1998;72:73–83.

3. Gaunt MW, Sall AA, Lamballerie XD *et al.* Phylogenetic relationships of flaviviruses correlate with their epidemiology, disease association and biogeography. *J Gen Virol* 2001;82:1867–1876.
4. Cook S, Holmes EC. A multigene analysis of the phylogenetic relationships among the flaviviruses (family: *Flaviviridae*) and the evolution of vector transmission. *Arch Virol* 2006;151:309–325.
5. Heinz FX, Collett MS, Purcell RH *et al.* Family *Flaviviridae*. In: van Regenmortel MH, Fauquet CM, Bishop DHL *et al.* (eds.) *Virus Taxonomy: 7th Report of the International Committee for the Taxonomy of Viruses.* 2000, pp. 859–878.
6. Kuhn RJ, Zhang W, Rossmann MG *et al.* Structure of dengue virus: implications for flavivirus organization, maturation and fusion. *Cell* 2002;108: 717–725.
7. Zhang W, Chipman PR, Corver J *et al.* Visualization of membrane protein domains by cryo-electron microscopy of dengue virus. *Nat Struct Biol* 2003;10:907–912.
8. Rey FA, Heinz FX, Mandl C, Kunz C, Harrison SC. The envelope glycoprotein from tick-borne encephalitis virus at 2 Å resolution. *Nature* 1995;375:291–298.
9. Zhang Y, Corver J, Chipman PR *et al.* Structures of immature flavivirus particles. *EMBO J* 2003;22:2604–2613.
10. Heinz FX, Roehrig JT. Flaviviruses. In: van Regenmortel MHV, Neurath AR (eds.) *Immunochemistry of Viruses. II: The Basis for Serodiagnosis and Vaccines*, 1990, pp. 289–305.
11. Henchal EA, Putnak JR. The dengue viruses. *Clin Microbiol Rev* 1990;3: 376–396.
12. Rice CM. *Flaviviridae*: the viruses and their replication. In: Fields BN, Knipe DM, Howley PM (eds.) *Fields Virology*, 3rd ed. Lippincott-Raven, 1996, pp. 931–959.
13. Blackwell JL, Brinton MA. Translation elongation factor-1 alpha interacts with the 3′ stem-loop region of West Nile virus genomic RNA. *J Virol* 1997;71:6433–6444.
14. Alvarez DE, Lodeiro MF, Luduena SJ, Pietrasanta LI, Gamarnik AV. Long-range RNA–RNA interactions circularize the dengue virus genome. *J Virol* 2005;79:6631–6643.

15. Alvarez DE, De Lella Ezcurra AL, Fucito S, Gamarnik AV. Role of RNA structures present at the 3′UTR of dengue virus on translation, RNA synthesis, and viral replication. *Virology* 2005;339:200–212.
16. Zhao B, Mackow E, Buckler-White A *et al.* Cloning full-length dengue type 4 viral DNA sequences: analysis of genes coding for structural proteins. *Virology* 1986;155:77–88.
17. Eckels KH, Dubois DR, Summers PL *et al.* Immunization of monkeys with baculovirus-dengue type-4 recombinants containing envelope and non-structural proteins: evidence of priming and partial protection. *Am J Trop Med Hyg* 1994;50:472–478.
18. Henchal EA, Henchal LS, Schlesinger JJ. Synergistic interactions of anti-NS1 monoclonal antibodies protect passively immunized mice from lethal challenge with dengue 2 virus. *J Gen Virol* 1988;69:2101–2107.
19. Khromykh AA, Westaway EG. RNA binding properties of core protein of the flavivirus Kunjin. *Arch Virol* 1996;141:685–699.
20. Markoff L, Falgout B, Chang A. A conserved internal hydrophobic domain mediates the stable membrane integration of the dengue virus capsid protein. *Virology* 1997;233:105–117.
21. Ma L, Jones CT, Groesch TD, Kuhn RJ, Post CB. Solution structure of dengue virus capsid protein reveals another fold. *Proc Nat Acad Sci USA* 2004;101:3414–3419.
22. Tadano M, Makino Y, Fukunaga T, Okuno Y, Fukai K. Detection of dengue 4 virus core protein in the nucleus. I. A monoclonal antibody to dengue 4 virus reacts with the antigen in the nucleus and cytoplasm. *J Gen Virol* 1989;70:1409–1415.
23. Bulich R, Aaskov JG. Nuclear localization of dengue 2 virus core protein detected with monoclonal antibodies. *J Gen Virol* 1992;73:2999–3003.
24. Wang S-H, Syu W-J, Huang K-J *et al.* Intracellular localization and determination of a nuclear localization signal of the core protein of dengue virus. *J Gen Virol* 2002;83:3093–3102.
25. Chang C-J, Luh H-W, Wang S-H *et al.* The heterogeneous nuclear ribonucleoprotein K (hnRNP K) interacts with dengue virus core protein. *DNA Cell Biol* 2001;20:569–577.
26. Russell PK, Brandt WE, Dalrymple JM. Chemical and antigenic structure of flaviviruses. In: Schlesinger RW (ed.) *The Togaviruses: Biology, Structure, Replication*. Academic, New York, 1980, pp. 503–529.

27. Wengler G, Wengler G. Cell-associated West Nile flavivirus is covered with E+prM protein heterodimers which are destroyed and reorganized by proteolytic cleavage during virus release. *J Virol* 1989;63:2521–2526.
28. Konishi E, Mason PW. Proper maturation of the Japanese encephalitis virus envelope glycoprotein requires cosynthesis with the premembrane protein. *J Virol* 1993;67:1672–1675.
29. Courageot M-P, Frenkiel M-P, dos Santos CD *et al.* α-Glucosidase inhibitors reduce dengue virus production by affecting the initial steps of virion morphogenesis in the endoplasmic reticulum. *J Virol* 2000;74: 564–572.
30. Guirakhoo F, Bolin RA, Roehrig JT. The Murray Valley encephalitis virus prM protein confers acid resistance to virus particles and alters the expression of epitopes within the R2 domain of E glycoprotein. *Virology* 1992;191:921–931.
31. Nowak T, Wengler G. Analysis of the disulphides present in the membrane protein of West Nile flavivirus. *Virology* 1987;156:127–137.
32. Rice CM, Strauss EG, Strauss JH. Structure of the flavivirus genome. In: Schlesinger S, Schlesinger MJ, (eds.) *The Togaviridae and Flaviviridae.* Plenum, New York 1986, pp. 279–326.
33. Stadler K, Allison SL, Schalich J, Heinz FX. Proteolytic activation of tick-borne encephalitis virus by furin. *J Virol* 1997;71:8475–8481.
34. Guirakhoo F, Heinz FX, Mandl CW, Holzmann H, Kunz C. Fusion activity of flaviviruses: comparison of mature and immature (prM-containing) tick-borne encephalitis virions. *J Gen Virol* 1991;72:1323–1329.
35. Elshuber S, Allison SL, Heinz FX, Mandl CW. Cleavage of protein prM is necessary for infection of BHK-21 cells by tick-borne encephalitis virus. *J Gen Virol* 2003;84:183–191.
36. Monath TP. Pathobiology of the Flaviviruses. In: Schlesinger S, Schlesinger MJ (eds.) *The Togaviridae and Flaviviridae*. New York, Plenum Press, 1986, pp. 375–440. *Virology* 1986;78:2287–2291.
37. Putnak R, Feighny R, Burrous J *et al.* Dengue-1 virus envelope glycoprotein gene expressed in recombinant baculovirus elicits virus-neutralizing antibody in mice and protects them from virus challenge. *Am J Trop Med Hyg* 1991;45:159–167.
38. Anderson R, King AD, Innis BL. Correlation of E protein binding with cell susceptibility to dengue 4 virus infection. *J Gen Virol* 1992;73:2155–2159.

39. Chen Y, Maguire T, Marks RM. Demonstration of binding of dengue virus envelope protein to target cells. *J Virol* 1996;70:8765–8772.
40. McMinn PC. The molecular basis of virulence of the encephalitogenic flaviviruses. *J Gen Virol* 1997;78:2711–2722.
41. Mandl CW, Guirakhoo F, Holzmann H, Heinz FX, Kunz C. Antigenic structure of the flavivirus envelope protein E at the molecular level, using tick-borne encephalitis virus as a model. *J Virol* 1989;63:564–571.
42. Guirakhoo F, Heinz FX, Kunz C. Epitope model of tick-borne encephalitis virus envelope glycoprotein E: analysis of structural properties, role of carbohydrate side chain, and conformational changes occurring at acidic pH. *Virology* 1989;169:90–99.
43. Kimura T, Gollins SW, Porterfield JS. The effect of pH on the early interaction of West Nile virus with P388D1 cells. *J Gen Virol* 1986;67: 2423–2433.
44. Roehrig JT, Johnson AJ, Hunt AR, Bolin RA, Chu MC. Antibodies to dengue 2 virus E-glycoprotein synthetic peptides identify antigenic conformation. *Virology* 1990;177:668–675.
45. Allison SL, Schalich J, Stiasny K, Mandl CW, Heinz FX. Mutational evidence for an internal fusion peptide in the flavivirus envelope protein E. *J Virol* 2001;75:4268–4275.
46. Heinz FX, Mandl CW, Holzmann H *et al.* The flavivirus envelope protein E: isolation of a soluble form from tick-borne encephalitis virus and its crystallization. *J Virol* 1991;65:5579–5583.
47. Modis Y, Ogata S, Clements D, Harrison SC. A ligand-binding pocket in the dengue virus envelope protein. *Proc Natl Acad Sci USA* 2003;100: 6986–6991.
48. Zhang Y, Zhang W, Ogata S *et al.* Conformational changes of the flavivirus E glycoprotein. *Structure* 2004;12:1607–1618.
49. Modis Y, Ogata S, Clements D, Harrison SC. Variable surface epitopes in the crystal structure of dengue virus type 3 envelope glycoprotein. *J Virol* 2005;79:1223–1231.
50. Wang S, He R, Anderson R. PrM- and cell-binding domains of the dengue virus E protein. *J Virol* 1999;73:2547–2551.
51. Ryman KD, Ledger TN, Weir RC, Schlesinger JJ, Barrett ADT. Yellow fever virus envelope protein has two discrete type-specific neutralizing epitopes. *J Gen Virol* 1997;78:1353–1356.

52. Allison SL, Schalich J, Stiasny K *et al.* Oligomeric rearrangement of tick-borne encephalitis virus envelope proteins induced by an acidic pH. *J Virol* 1995;69:695–700.
53. Modis Y, Ogata S, Clements D, Harrison SC. Structure of the dengue virus envelope protein after membrane fusion. *Nature* 2004;427:313–319.
54. Stiasny K, Allison SL, Marchler-Bauer A, Kunz C, Heinz FX. Structural requirements for low-pH-induced rearrangements in the envelope glycoprotein of tick-borne encephalitis virus. *J Virol* 1996;70:8142–8147.
55. Stiasny K, Allison SL, Mandl CW, Heinz FX. Role of metastability and acidic pH in membrane fusion by tick-borne encephalitis virus. *J Virol* 2001;75:7392–7398.
56. Winkler G, Maxwell SE, Ruemmler C, Stollar V. Newly synthesized dengue-2 virus non-structural protein NS1 is a soluble protein but becomes partially hydrophobic and membrane-associated after dimerization. *Virology* 1989;171:302–305.
57. Mason PW, McAda PC, Dalrymple JM, Fournier MJ, Mason TC. Expression of Japanese encephalitis virus antigens in *Escherichia coli*. *Virology* 1987;158:361–372.
58. Chambers TJD, McCourt DW, Rice CM. Production of yellow fever virus proteins in infected cells: identification of discrete polyprotein species and analysis of cleavage kinetics using region-specific polyclonal antisera. *Virology* 1990;177:159–174.
59. Blitvich BJ, Scanlon D, Shiell, BJ, MacKenzie JS, Hall RA. Identification and analysis of truncated and elongated species of the flavivirus NS1 protein. *Virus Res* 1999;60:67–79.
60. MacKenzie JS, Jones MK, Young PR. Immunolocalization of the dengue virus nonstructural glycoprotein NS1 suggests a role in viral RNA replication. *Virology* 1996;220:232–240.
61. Muylaert IR, Chambers TJ, Galler R, Rice CM. Mutagenesis of the N-linked glycosylation sites of the yellow fever virus NS1 protein: effects on virus replication and mouse neurovirulence. *Virology* 1996;222:159–168.
62. Muylaert IR, Galler R, Rice CM. Genetic analysis of the yellow fever virus NS1 protein: identification of a temperature-sensitive mutation which blocks RNA accumulation. *J Virol* 1997;71:291–298.
63. Lindenbach B, Rice CM. *Trans*-complementation of yellow fever virus NS1 reveals a role in early RNA replication. *J Virol* 1997;71:9608–9617.

64. Crabtree MB, Kinney RM, Miller BR. Deglycosylation of the NS1 protein of dengue 2 virus, strain 16681: construction and characterization of mutant viruses. *Arch Virol* 2005;150:771–786.
65. Fan W, Mason PW. Membrane association and secretion of the Japanese encephalitis virus nonstructural protein NS1 from cells expressing NS1 cDNA. *Virology* 1990;177:470–476.
66. Falgout B, Markoff L. Evidence that flavivirus NS1-NS2A cleavage is mediated by a membrane-bound host protease in the endoplasmic reticulum. *J Virol* 1995;69:7232–7243.
67. Hall RA, Khromykh AA, MacKenzie JS *et al.* Loss of dimerisation of the nonstructural protein NS1 of Kunjin virus delays viral replication and reduces virulence in mice, but still allows secretion of NS1. *Virology* 1999;264:66–75.
68. Pryor MJ, Wright PJ. The effects of site-directed mutagenesis on the dimerization and secretion of the NS1 protein specified by dengue virus. *Virology* 1993;194:769–780.
69. Flamand M, Megret F, Mathieu M *et al.* Dengue virus type 1 nonstructural glycoprotein NS1 is secreted from mammalian cells as a soluble hexamer in a glycosylation-dependent fashion. *J Virol* 1999;73: 6104–6110.
70. Pryor MJ, Wright PJ. Glycosylation mutants of dengue virus NS1 protein. *J Gen Virol* 1994;75:1183–1187.
71. Jacobs MG, Robinson PJ, Bletchly C, MacKenzie JS, Young PR. Dengue virus nonstructural protien 1 is expressed in a glycosyl-phosphatidylinositol-linked form that is capable of signal transduction. *FASEB J* 2000;14: 1603–1610.
72. Chambers TJ, Grakoui A, Rice CM. Processing of the yellow fever virus non-structural polyprotein: a catalytically active NS3 protease domain and NS2B are required for cleavages at dibasic sites. *J Virol* 1991;65: 6042–6050.
73. Falgout B, Pethel Y-M, Lai C-J. Both nonstructural proteins NS2B and NS3 are required for proteolytic processing of dengue virus nonstructural proteins. *J Virol* 1991;65:2467–2475.
74. Falgout B, Miller RH, Lai CJ. Deletion analysis of dengue virus type 4 nonstructural protein NS2B: identification of a domain required for NS2B–NS3 proteinase activity. *J Virol* 1993;67:2034–2042.

75. Li H, Clum S, You S, Ebner KE, Padmanabhan R. The serine protease and RNA-stimulated nucleoside triphosphatase and RNA helicase functional domains of dengue virus type 2 NS3 converge within a region of 20 amino acids. *J Virol* 1999;73:3108–3116.
76. Yusof R, Clum S, Wetzel M, Krishna Murthy HM, Padmanabhan R. Purified NS2B/NS3 serine protease of dengue virus type 2 exhibits cofactor NS2B dependence for cleavage of substrates with dibasic amino acids *in vitro*. *J Biol Chem* 2000;275:9963–9969.
77. Bazan JF, Fletterick RJ. Detection of a trypsin-like serine protease domain in flaviviruses and pestiviruses. *Virology* 1989;171:637–639.
78. Chambers TJ, Wier RC, Grakoui A *et al.* Evidence that the N-terminal domain of yellow fever virus NS3 is a serine protease responsible for site-specific cleavages in the viral polyprotein. *Proc Natl Acad Sci USA* 1990;87:8898–8902.
79. Wengler G, Czaya G, Farber PM, Hegemann JH. *In vitro* synthesis of West Nile virus proteins indicates that the amino-terminal segment of the NS3 protein contains the active centre of the protease which cleaves the viral polyprotein after multiple basic amino acids. *J Gen Virol* 1993;72:851–858.
80. Pugachev KV, Nomokonova NY, Dobrikova EY, Wolf YI. Site-directed mutagenesis of the tick-borne encephalitis virus NS3 gene reveals the putative serine protease domain of the NS3 protein. *FEBS Lett* 1993; 328:115–118.
81. Murthy HMK, Clum S, Padmanabhan R. Dengue virus NS3 serine protease: crystal structure and insights into interaction of the active site with substrates by molecular modeling and structural analysis of mutational effects. *J Biol Chem* 1999;274:5573–5580.
82. Valle RP, Falgout B. Mutagenesis of the NS3 protease of dengue virus type 2. *J Virol* 1998;72:624–632.
83. Niyomrattanakit P, Winoyanuwattikun P, Chanprapaph S *et al.* Identification of residues in the dengue virus type 2 NS2B cofactor that are critical for NS3 protease activation. *J Virol* 2004;78:13708–13716.
84. Erbel P, Schiering N, D'Arcy A *et al.* Structural basis for the activation of flaviviral NS3 proteases from dengue and West Nile virus. *Nat Struct Mol Biol* 2006;13:372–372.
85. Westaway EG, Mackenzie JM, Kenney MT, Jones MK, Khromykh AA. Ultrastructure of Kunjin virus-infected cells: colocalization of NS1 and

NS3 with double-stranded RNA, and of NS2B with NS3, in virus-induced membrane structures. *J Virol* 1997;71:6650–6661.

86. Wengler G, Wengler G. The carboxy-terminal part of the NS3 protein of the West Nile flavivirus can be isolated as a soluble protein after proteolytic cleavage and represents an RNA-stimulated NTPase. *Virology* 1991;184:707–715.
87. Wengler G, Wengler G. The NS3 nonstructural protein of flavivirus contains an RNA triphosphatase activity. *Virology* 1993;197:265–273.
88. Bartholomeusz AI, Wright PJ. Synthesis of dengue virus RNA *in vitro*: initiation and the involvement of proteins NS3 and NS5. *Arch Virol* 1993;128:111–121.
89. Matusan AE, Pryor MJ, Davidson AD, Wright PJ. Mutagenesis of the *Dengue Virus Type 2* NS3 protein within and outside helicase motifs: effects on enzyme activity and virus replication. *J Virol* 2001;75:9633–9643.
90. Bartelma G, Padmanabhan R. Expression, purification and characterization of the RNA 5′-triphosphatase activity of the dengue virus type 2 nonstructural protein 3. *Virology* 2002;299:122–132.
91. Xu T, Sampath A, Chao A *et al.* Structure of the dengue virus helicase/nucleoside triphosphatase catalytic domain at a resolution of 2.4 A. *J Virol* 2005;79:10278–10288.
92. Cui T, Sugrue RJ, Xu Q *et al.* Recombinant dengue virus type 1 NS3 protein exhibits specific viral RNA binding and NTPase activity regulated by the NS5 protein. *Virology* 1998;246:409–417.
93. Kapoor M, Zhang L, Ramachandra M *et al.* Association between NS3 and NS5 proteins of dengue virus type 2 in the putative RNA replicase is linked to differential phosphorylation of NS5. *J Biol Chem* 1995;270:19100–19106.
94. Chen CJ, Kuo MD, Chien LJ *et al.* RNA–protein interactions: involvement of NS3, NS5, and 3′ noncoding regions of Japanese encephalitis virus genomic RNA. *J Virol* 1997;71:3466–3473.
95. Utama A, Shimizu H, Morikawa S *et al.* Identification and characterization of the RNA helicase activity of Japanese encephalitis virus NS3 protein. *FEBS Lett* 2000;465:74–78.
96. Raviprakash K, Sinha M, Hayes CJ, Porter KR. Conversion of dengue virus replicative form RNA (RF) to replicative intermediate (RI) by nonstructural proteins NS-5 and NS-3. *Am J Trop Med Hyg* 1998;58:90–95.

97. Koonin EV. The phylogeny of RNA-dependent RNA polymerases of positive-strand RNA viruses. *J Gen Virol* 1991;72:2197–2206.
98. Koonin EV Computer-assisted identification of a putative methyltransferase domain in NS5 protein of flaviviruses and lambda 2 protein of reovirus. *J Gen Virol* 1993;74:733–740.
99. Egloff M-P, Benarroch D, Selisko B, Romette J-L, Canard B. An RNA cap (nucleoside-2'-O)-methyltransferase in the flavivirus RNA polymerase NS5: crystal structure and functional characterization. *EMBO J* 2002;21: 2757–2768.
100. Forwood JK, Brooks A, Briggs LJ *et al.* The 37-amino-acid interdomain of dengue virus NS5 protein contains a functional NLS and inhibitory CK2 site. *Biochem Biophys Res Commun* 1999; 257:731–737.
101. Brooks AJ, Johansson M, John AV *et al.* The interdomain region of dengue NS5 protein that binds to the viral helicase NS3 contains independently functional importin β1 and importin α/β-recognized nuclear localization signals. *J Biol Chem* 2002;277:36399–36407.
102. Tan B-H, Fu J, Sugrue RJ *et al.* Recombinant dengue type 1 virus NS5 protein expressed in *Escherichia coli* exhibits RNA-dependent RNA polymerase activity. *Virology* 1996;216:317–325.
103. Khromykh AA, Kenney MT, Westaway EG. *Trans*-complementation of flavivirus RNA polymerase gene NS5 by using Kunjin virus replicon-expressing BHK cells. *J Virol* 1998;72:7270–7279.
104. Khromykh AA, Sedlak PL, Westaway EG. *Trans*-complementation analysis of the flavivirus Kunjin NS5 gene reveals an essential role for translation of its N-terminal half in RNA replication. *J Virol* 1999;73:9247–9255.
105. Munoz-Jordan JL, Sanchez-Burgos GG, Laurent-Rolle M, Garcia-Sastre A. Inhibition of interferon signaling by dengue virus. *Proc Natl Acad Sci USA* 2003;100:14333–14338.
106. Clum S, Ebner KE, Padmanabhan R. Cotranslational membrane insertion of the serine proteinase precursor NS2B-NS3(Pro) of dengue virus type 2 is required for efficient *in vitro* processing and is mediated through the hydrophobic regions of NS2B. *J Biol Chem* 1997;272:30715–30723.
107. Brinkworth RI, Fairlie DP, Leung D, Young PR. Homology model of the dengue 2 virus NS3 protease: putative interactions with both substrate and NS2B cofactor. *J Gen Virol* 1999;80:1167–1177.

108. Yamshchikov VF, Compans RW. Processing of the intracellular form of the West Nile virus capsid protein by the viral NS2B–NS3 protease: an *in vitro* study. *J Virol* 1994;68:5765–5771.
109. MacKenzie JM, Khromykh AA, Jones MK, Westaway EG. Subcellular localization and some biochemical properties of the flavivirus Kunjin nonstructural proteins NS2A and NS4A. *Virology* 1998;245:203–215.
110. Kummerer BM, Rice CM. Mutations in the yellow fever virus nonstructural protein NS2A selectively block production of infectious particles. *J Virol* 2002;76:4773–4784.
111. Westaway EG, Khromykh AA, Kenney MT, MacKenzie JM, Jones MK. Proteins C and NS4B of the flavivirus Kunjin translocate independently into the nucleus. *Virology* 1997;234:31–41.
112. Miller S, Sparacio S, Bartenschlager R. Subcellular localization and membrane topology of the dengue virus type 2 non-structural protein 4B. *J Biol Chem* 2006;281:8854–8863.
113. Munoz-Jordan JL, Laurent-Rolle M, Ashour J *et al*. Inhibition of alpha/beta interferon signaling by the NS4B protein of flaviviruses. *J Virol* 2005;79:8004–8013.
114. Khromykh AA, Sedlak PL, Westaway EG. *Cis* and *trans*-acting elements in flavivirus RNA replication. *J Virol* 2000;74:3253–3263.
115. Chang Y-S, Liao C-L, Tsao C-H *et al*. Membrane permeabilization by small hydrophobic nonstructural proteins of Japanese encephalitis virus. *J Virol* 1999;73:6257–6264.
116. Gollins SW, Porterfield JS. Flavivirus infection enhancement in macrophages: an electron microscopic study of viral cellular entry. *J Gen Virol* 1985;66:1969–1982.
117. Marianneau P, Steffan A, Royer C *et al*. Differing infection patterns of dengue and yellow fever viruses in a human hepatoma cell line. *J Infect Dis* 1998;178:1270–1278.
118. Hung JJ, Hsieh MT, Young MJ *et al*. An external loop region of domain III of dengue virus type 2 envelope protein is involved in serotype-specific binding to mosquito but not mammalian cells. *J Virol* 2004;78:378–388.
119. Hase T, Summers PL, Eckels KH. Flavivirus entry into cultured mosquito cells and human peripheral blood monocytes. *Arch Virol* 1989;104:129–143.

120. Randolph VB, Stollar V. Low pH-induced cell fusion in flavivirus-infected *Aedes albopictus* cell cultures. *J Gen Virol* 1990;71:1845–1850.
121. Hung S-L, Lee P-L, Chen H-W *et al.* Analysis of the steps involved in dengue virus entry into host cells. *Virology* 1999;257:156–167.
122. Chen Y, Maguire T, Hileman RE *et al.* Dengue virus infectivity depends on envelope protein binding to target cell heparan sulfate. *Nat Med* 1997;3: 866–871.
123. Hilgard P, Stockert R. Heparan sulfate proteoglycans initiate dengue virus infection of hepatocytes. *Hepatology* 2000;32:1069–1077.
124. Lee E, Lobigs M. Substitutions at the putative receptor-binding site of an encephalitic flavivirus alter virulence and host cell tropism and reveal a role for glycosaminoglycans in entry. *J Virol* 2000;74:8867–8875.
125. Lin Y-L, Lei H-Y, Lin Y-S *et al.* Heparin inhibits dengue-2 virus infection of five human liver cell lines. *Antiviral Res* 2002;56:93–96.
126. Bielefeldt-Ohmann H. Analysis of antibody-independent binding of dengue viruses and dengue virus envelope protein to human myelomonocytic cells and B lymphocytes. *Virus Res* 1998;57:63–79.
127. Bielefeldt-Ohmann H, Meyer M, Fitzpatrick DR, MacKenzie JS. Dengue virus binding to human leukocyte cell lines: receptor usage differs between cell types and virus strains. *Virus Res* 2001;73:81–89.
128. Mandl CW, Kroschewski H, Allison SL *et al.* Adaptation of tick-borne encephalitis virus to BHK-21 cells results in the formation of multiple heparan sulfate binding sites in the envelope protein and attenuation *in vivo*. *J Virol* 2001;75:5627–5637.
129. Kroschewski H, Allison SL, Heinz FX, Mandl CW. Role of heparan sulfate for attachment and entry of tick-borne encephalitis virus. *Virology* 2003;308:92–100.
130. Daughaday CC, Brandt WE, McCown JM, Russell PK. Evidence for two mechanisms of dengue virus infection of adherent human monocytes: trypsin-sensitive virus receptors and trypsin-resistant virus immune complex receptors. *Infect Immun* 1981;32:469–473.
131. Ramos-Castaneda J, Imbert JL, Barron BL, Ramos C. A 65-kDa trypsin-sensible membrane cell protein as a possible receptor for dengue virus in cultured neuroblastoma cells. *J Neurovirol* 1997;3:435–440.

132. Martinez-Barragan JJ, del Angel RM. Identification of a putative coreceptor on Vero cells that participates in dengue 4 virus infection. *J Virol* 2001;75:7818–7827.
133. Salas-Benito JS, del Angel RM. Identification of two surface proteins from C6/36 cells that bind dengue type 4 virus. *J Virol* 1997;71:7246–7252.
134. Mendoza MY, Salas-Benito JS, Lanz-Mendoza H, Hernandez-Martinez S, del Angel RM. A putative receptor for dengue virus in mosquito tissues: localization of a 45-kDa glycoprotein. *Am J Trop Med Hyg* 2002;67:76–84.
135. Reyes-Del Valle J, Chavez-Salinas S, Medina F, del Angel RM. Heat shock protein 90 and heat shock protein 70 are components of dengue virus receptor complex in human cells. *J Virol* 2005;79:4557–4567.
136. Chen Y-C, Wang S-Y, King C-C. Bacterial lipopolysaccharide inhibits dengue virus infection of primary human monocytes/macrophages by blockade of virus entry via a CD14-dependent mechanism. *J Virol* 1999;73:2650–2657.
137. Tassaneetrithep B, Burgess TH, Granelli-Piperno A *et al.* DC-SIGN (CD209) mediates dengue virus infection of human dendritic cells. *J Exp Med* 2003;197:823–829.
138. Sakuntabhai A, Turbpaiboon C, Casademont I *et al.* A variant in the CD209 promoter is associated with severity of dengue disease. *Nat Genet* 2005;37:507–513.
139. Lozach PY, Burleigh L, Staropoli I *et al.* Dendritic cell-specific intercellular adhesion molecule 3-grabbing non-integrin (DC-SIGN)-mediated enhancement of dengue virus infection is independent of DC-SIGN internalization signals. *J Biol Chem* 2005;280:23698–23708.
140. Pokidysheva E, Zhang Y, Battisti AJ *et al.* Cryo-EM reconstruction of dengue virus in complex with the carbohydrate recognition domain of DC-SIGN. *Cell* 2006;124:485–493.
141. Halstead SB, O'Rourke EJ. Dengue viruses and mononuclear phagocytes. I. Infection enhancement by non-neutralizing antibody. *J Exp Med* 1977;146:201–217.
142. Peiris JSM, Gordon S, Unkeless JC, Porterfield JS. Monoclonal anti-Fc receptor IgG blocks antibody-dependent enhancement of viral replication in macrophages. *Nature* 1981;289:189–191.

143. Kliks S. Antibody-enhanced infection of monocytes as the pathogenetic mechanism for severe dengue illness. *AIDS Res Hum Retroviruses* 1990;6:993–998.
144. Littaua R, Kurane I, Ennis FA. Human IgG Fc receptor II mediates antibody-dependent enhancement of dengue virus infection. *J Immunol* 1990;144:3183–3186.
145. Halstead SB. Pathogenesis of dengue: challenges to molecular biology. *Science* 1988;239:476–481.
146. Kurane I, Ennis FA. Cytokines in dengue virus infections: roles of cytokines in the pathogenesis of dengue hemorrhagic fever. *Sem Virol* 1994;5:443–448.
147. Edgil D, Diamond MS, Holden KL, Paranjape SM, Harris E. Translation efficiency determines differences in cellular infection among dengue virus type 2 strains. *Virology* 2003;317:275–290.
148. Edgil D, Polacek C, Harris E. Dengue virus utilizes a novel strategy for translation initiation when cap-dependent translation is inhibited. *J Virol* 2006;80:2976–2986.
149. Chiu WW, Kinney RM, Dreher TW. Control of translation by the 5′- and 3′-terminal regions of the dengue virus genome. *J Virol* 2005;79:8303–8315.
150. Clyde K, Harris E. RNA secondary structure in the coding region of dengue virus type 2 directs translation start codon selection and is required for viral replication. *J Virol* 2006;80:2170–2182.
151. Preugschat F, Yao C-W, Strauss JH. *In vitro* processing of dengue virus type 2 nonstructural proteins NS2A, NS2B and NS3. *J Virol* 1990;64:4364–4374.
152. Lai CJ, Pethel M, Jan LR *et al.* Processing of dengue type 4 and other flavivirus nonstructural proteins. *Arch Virol* 1994;Suppl. 9:359–368.
153. Jan LR, Yang CS, Trent DW, Falgout B, Lai CJ. Processing of Japanese encephalitis virus non-structural protein NS2B–NS3 complex and heterologous proteases. *J Gen Virol* 1995;76:573–580.
154. Falgout B, Lai CJ. Proper processing of dengue virus nonstructural glycoprotein NS1 requires the N-terminal hydrophobic signal sequence and the downstream nonstructural protein NS2a. *J Virol* 1989;63:1852–1860.
155. Pethel M, Falgout B, Lai CJ. Mutational analysis of the octapeptide sequence motif at the NS1–NS2A cleavage junction of dengue type 4 virus. *J Virol* 1992;65:4749–4758.

156. Lobigs M. Flavivirus premembrane protein cleavage and spike heterodimer secretion require the function of the viral proteinase NS3. *Proc Natl Acad Sci USA* 1993;90:6218–6222.
157. Yamschikov VF, Compans RW. Regulation of the late events in flavivirus protein processing and maturation. *Virology* 1993;192:38–51.
158. Amberg SM, Nestorowicz A, McCourt DW, Rice CM NS2B-3 proteinase-mediated processing in the yellow fever virus structural region: *in vitro* and *in vivo* studies. *J Virol* 1994;68:3794–3802.
159. Yamschikov VF, Compans RW. Formation of the flavivirus envelope: role of the viral NS2B–NS3 protease. *J Virol* 1995;69:1995–2003.
160. Yamschikov VF, Trent DW, Compans RW. Upregulation of signalase processing and induction of prM-E secretion by the flavivirus NS2B–NS3 protease: roles of protease components. *J Virol* 1997;71: 4364–4371.
161. Lin C, Amberg SM, Chambers TJ, Rice CM. Cleavage at a novel site in the NS4A region by the yellow fever virus NS2B-3 proteinase is a prerequisite for processing at the downstream 4A/4B signalase site. *J Virol* 1993;67:2327–2335.
162. Arias CF, Preugschat F, Strauss JH. Dengue 2 virus NS2B and NS3 form a stable complex that can cleave NS3 within the helicase domain. *Virology* 1993;193:888–899.
163. Teo KF, Wright PJ. Internal proteolysis of the NS3 protein specified by dengue virus 2. *J Genl Virol* 1997;78:337–341.
164. Chu PWG, Westaway EG. Characterization of Kunjin virus RNA-dependent RNA polymerase: re-initiation of synthesis *in vitro*. *Virology* 1987;151: 330–337.
165. Chu PWG, Westaway EG. Molecular and ultrastructural analysis of heavy membrane fraction associated with the replication of Kunjin virus RNA. *Arch Virol* 1992;125:177–191.
166. MacKenzie JM, Joens MK, Westaway EG. Markers for *trans*-Golgi membranes and the intermediate compartment localize to induced membranes with distinct replication functions in flavivirus-infected cells. *J Virol* 1999;73:9555–9567.
167. Cleaves GR, Ryan TE, Schlesinger RW. Identification and characterization of type 2 dengue virus replicative intermediate and replicative form RNAs. *Virology* 1981;111:73–83.

168. Grun JB, Brinton MA. Dissociation of NS5 from cell fractions containing West Nile virus-specific polymerase activity. *J Virol* 1987;61:3641–3644.
169. Takegami T, Hotta S. *In vitro* synthesis of Japanese encephalitis virus (JEV) RNA: membrane and nuclear fractions of JEV-infected cells possess high levels of virus-specific RNA polymerase activity. *Virus Res* 1989;13: 337–350.
170. Grief C, Galler R, Cortes LMC, Barth OM. Intracellular localisation of dengue-2 RNA in mosquito cell culture using electron microscopic *in situ* hybridisation. *Arch Virol* 1997;142:2347–2357.
171. Brinton MA. Replication of flaviviruses. In: Schlesinger S, Schlesinger MJ (eds.) *The Togaviridae and Flaviviridae*. New York, Plenum, 1987, pp. 327–374.
172. Westaway EG, Khromykh AA, MacKenzie JM. Nascent flavivirus RNA colocalized *in situ* with double-stranded RNA in stable replication complexes. *Virology* 1999;258:108–117.
173. You S, Padmanabhan R. A novel *in vitro* replication system for dengue virus: initiation of RNA synthesis at the 3′-end of exogenous viral RNA templates requires 5′- and 3′-terminal complementary sequence motifs of the viral RNA. *J Biol Chem* 1999;274:33714–33722.
174. Khromykh AA, Meka H, Guyatt KJ, Westaway EG. Essential role of cyclization sequences in flavivirus RNA replication. *J Virol* 2001;75: 6719–6728.
175. Corver J, Lenches E, Smith K *et al*. Fine mapping of a *cis*-acting sequence element in the yellow fever virus RNA that is required for RNA replication and cyclization. *J Virol* 2003;77:2265–2270.
176. Zeng L, Falgout B, Markoff L. Identification of specific nucleotide sequences within the conserved 3′-SL in the dengue type 2 virus genome required for replication. *J Virol* 1998;72:7510–7522.
177. Yu L, Markoff L. The topology of bulges in the long stem of the flavivirus 3′ stem-loop is a major determinant of RNA replication competence. *J Virol* 2005;79:2309–2324.
178. Blackwell JL, Brinton MA. BHK cell proteins that bind to the 3′ stem loop structure of the West Nile virus genome RNA. *J Virol* 1995;69:5650–5658.
179. Li W, Li Y, Kedersha N *et al*. Cell proteins TIA-1 and TIAR interact with the 3′ stem-loop of the West Nile virus complementary minus-strand RNA and facilitate virus replication. *J Virol* 2002;76:11989–12000.

180. De Nova-Ocampo M, Villegas-Sepulveda N, del Angel RM. Translation elongation factor-1α, La and PTB interact with the 3′ untranslated region of dengue 4 virus RNA. *Virology* 2002;295:337–347.
181. Ta M, Vrati S. Mov34 protein from mouse brain interacts with the 3′ non-coding region of Japanese encephalitis virus. *J Virol* 2000;74:5108–5115.
182. Khromykh AA, Varnavski AN, Sedlak PL, Westaway EG. Coupling between replication and packaging of flavivirus RNA: evidence derived from the use of DNA-based full-length cDNA clones of Kunjin virus. *J Virol* 2001;75:4633–4640.
183. Ishak R, Tovey DG, Howard CR. Morphogenesis of yellow fever virus 17D in infected cell cultures. *J Gen Virol* 1988;69:325–335.
184. Purdy DE, Chang GJ. Secretion of noninfectious dengue virus-like particles and identification of amino acids in the stem region involved in intracellular retention of envelope protein. *Virology* 2005;332:239–250.
185. Chang G-J. Molecular biology of dengue viruses. In: Gubler DJ, Kuno GK (eds.) *Dengue and Dengue Hemorrhagic Fever*. CAB International, Cambridge, 1997, pp. 175–198.
186. Heinz FX, Allison SL. Structures and mechanisms in flavivirus fusion. *Adv Virus Res* 2000;55:231–269.

3

Epidemiology

Scott B. Halstead

Introduction

A comprehensive treatment of the epidemiology of dengue viruses will be presented by considering the bionomics of their principal mosquito vectors, *Aedes aegypti* and *Aedes albopictus*, and the biology of dengue virus interactions with this species. Next to be considered will be how human behavior impacts on the survival of this mosquito, and megaphenomena that control vector-borne infectious diseases, such as climate and weather. Finally, viral and host factors will be presented, such as infection force including viremia enhancement, herd immunity, crowding and transportation of hosts and vector mosquitoes, and the evolution and selection of dengue viruses. Early events in human dengue infections, especially immune status, are sufficiently important to require separate treatment in Chap. 7.

Ecology of Vector Mosquitoes

All the known vectors of the four dengue viruses belong to the genus *Aedes*. Attributes of dengue vectors in this genus, such as vector competence and vectoral capacity, have been studied in the laboratory and show wide variation among those subpopulations available for laboratory studies.[1,2] These data and epidemiological considerations support the possibility that major variations in these attributes must also exist in the huge and widely dispersed natural populations of *Aedes aegypti* and *Aedes albopictus*. However, whether any of these differences affect transmission of dengue viruses in nature remains unknown.

The subgenus *Stegomyia* contains the most important vectors of dengue viruses: *Ae aegypti*, *Ae albopictus* and *Ae polynesiensis*.

Aedes aegypti

History of dispersal

This species evolved in Africa, where a sylvatic, ancestral form, *Ae formosus*, is enzootic in East and Central Africa.[3] *Ae formosus* is nonanthropophilic. Lighter forms have become domesticated and adapted to human habitats and are abundant in villages and cities in West Africa, where, known as *Ae aegypti,* it has been the vector of urban epidemics of yellow fever. Due to its ability to breed in water stored for drinking or washing purposes and to its dessication-resistant eggs, *Ae aegypti* was able to survive on the long voyages of the era of sailing ships. Undoubtedly trade between West Africa and the rest of the world resulted in the introduction of *Ae aegypti* into the Asian and American tropics. It is likely that this began after the Portuguese voyages of discovery that led to the establishment of trading routes around the Cape of Good Hope from Europe to India and the Far East. This was followed by European colonial expansion into America during the 16th century and at the beginning of the slave trade. The near-simultaneous importation of yellow fever virus and *Ae aegypti* into the American tropics led to crippling yellow fever epidemics — dramatic examples of an early emergence event. Dengue virus(es) could well have been introduced into the American hemisphere during the same period as yellow fever. There is good evidence that each of the four dengue viruses evolved as parasites in subhuman primates, but evolutionary analysis of a small collection of zoonotic strains has not yet established their geographic origin.[4] Dengue 2 virus is enzootic in subhuman primate populations in Africa, and all four viruses in SE Asia.[5] Dengue viruses are also enzootic in the Indian subcontinent, but poorly studied. SE Asia is attractive as the site for the evolution of four antigenically distinct viruses from a common ancestor.[6] During the most recent Ice Age, the Indonesian archipelago was connected as a single land mass. With the end of the Ice Age, sea levels rose and created many islands, populated today with distinct species of subhuman primates. It seems reasonable to assume that once *Ae aegypti* was established in Asian cities each of the four sylvatic dengue viruses separately adapted to an urban cycle. It is interesting that the DENV-2 recovered from African primates is more closely related to the American genotype DENV-2 than to Asian DENV-2 viruses. It is possible that only the American genotype of DENV-2 moved from Africa to the New World. And the other three dengue viruses moved from Asia to both the American hemisphere and Africa during the 20th century.

Habitat

Eggs are laid individually by the female mosquito on the walls of both artificial and natural water containers. They resist desiccation for weeks to months and hatch when submerged in water. Larvae and pupae prefer clean water in many different types of artificial containers: water storage containers (tanks, jars, cisterns, pots), ornamental containers (flower holders, ant traps, shrine objects), discarded items (rubber tires, plastic containers, bottles). The species also occasionally uses natural larval habitats, such as bromeliads and tree holes. Containers located outside dwellings can be filled with rain water and are productive during the rainy season. Indoor or sheltered containers may produce pupae throughout the year. When extensive use is made of outdoor larval habitats, the prevalence of larvae and adults is often subject to marked seasonal variation. The duration of the larval stages is 7–9 days at 25°C and that of the pupal stage is 2–3 days at the same temperature.

Bionomics

The ecology of adult *Ae aegypti* in a domestic urban environment is characterized by strong anthropophilia and diurnal feeding, usually with two peaks — one at midmorning and the other in the late afternoon. It seems likely that most females can feed twice or even three times during a single gonotrophic cycle.[3,7,8] The preferred resting sites of adults are sheltered dark spaces inside houses. The average lifespan for females is about 8–15 days and that for males about 3–6 days. Spontaneous dispersal of adults is usually limited, averaging about 30–50 m a day for females, which means that a female rarely visits more than two or three houses during her lifetime. In some locales, such as Puerto Rico, female flight distance may be related to the availability of oviposition sites and so might be much longer. Passive dispersal of eggs and larvae is common, including by trains, boats and aircraft. Because of the weak spontaneous dispersal of the species and its easy passive dispersal, the International Sanitary Regulations require that the area within 400 m of international ports and airports be kept free of *Ae aegypti*.

Two main factors regulate *Ae aegypti* populations: climate and the availability of breeding sites. Population changes may or may not correlate with weather. Daily, seasonal and interannual variability in temperature, atmospheric moisture and rainfall all influence mosquitoes in a variety of ways. The following examples are illustrative of the role of rainfall on *Ae aegypti* populations.

In the late 1950's, the World Health Organization, at the request of the government of Thailand, set up an *Aedes* Research Unit to study the ecology and control of *Ae aegypti*. The hypothesis being tested was that seasonal changes in the density of the vector were correlated with annual outbreaks of dengue. A series of year-long studies was conducted in 1966–1968, in the residential compound of a Buddhist temple. Temple housing was similar to residences outside the compound. Also typical of other Bangkok residences were the types of water-filled containers, primarily large 100–200-liter water storage jars, flowerpot plates and ant traps. *Ae aegypti* was the only mosquito breeding in the great majority of containers.[7,9,10] Throughout the study period there were around 100 jars, 50 flowerpot plates and 50 ant traps in the temple, with around 53% of them occupied by *Ae aegypti*. The number of water-filled containers and the proportion with mosquitoes were remarkably constant. With the exception of a portion of the ant traps, all containers were manually filled and not influenced by rainfall. It was observed that in the mosquito population under study there were no fluctuations in adult production and densities in response to rainfall. Instead, there was a seasonal increase in adult survival due to temperature and atmospheric moisture. It was concluded that the association of dengue epidemics with rainfall could be explained by increases in adult survival and feeding activity of the vector mosquito (see below).

In contrast to the Bangkok temple study, longitudinal studies in Puerto Rico showed a positive correlation between rainfall and vector abundance, with the correlation being strongest in the dry, south coastal portions of the island.[11] Adult abundance varied not with temperature but with the abundance and productivity of water holding containers. In contrast with household breeding in Bangkok, most breeding in Puerto Rico occurs outdoors and in rain-filled containers, primarily animal watering dishes and discarded tires. Moore *et al.* describe the relationship between rainfall, vector abundance and transmission as follows: "At least in southern Puerto Rico *Ae aegypti* densities rise quickly with the onset of rains in July and August. This relationship further leads to a rather close correspondence between seasonal rainfall and dengue fever incidence, the peak of which occurs about 6 to 8 weeks after the peak in rainfall…and that rainfall patterns seem to be a reasonably effective predictor of time of peak dengue transmission." Container productivity is limited not by temperature or oviposition but by larval survival, ultimately driven by the amount of food present or formed photosynthetically within the container.[11]

Similar data linking an increase in mosquito populations to the onset of the rainy season were obtained in studies performed at the SEATO Medical Research Laboratory in Bangkok, Thailand. During 1962, five areas in Bangkok were monitored for mosquito populations with human-baited traps. Collectors were present in each of these areas from 0400 to 1300 and from 1500 to 2300 five days a week for the entire year. As they attempted to feed, 3674 *Ae aegypti* were captured by collectors (Scanlon JE, personal communication; Annual Report, SEATO Medical Research Laboratory, 1963). These mosquitoes yielded 21 strains of the dengue virus and 7 strains of the chikungunya virus, a togavirus transmitted by *Ae aegypti* which was endemic in Bangkok at that time. The numbers of *Ae aegypti* females collected in human-baited traps are shown in Fig. 1 in relation to hospitalized DHF/DSS cases, mean temperature and rainfall.

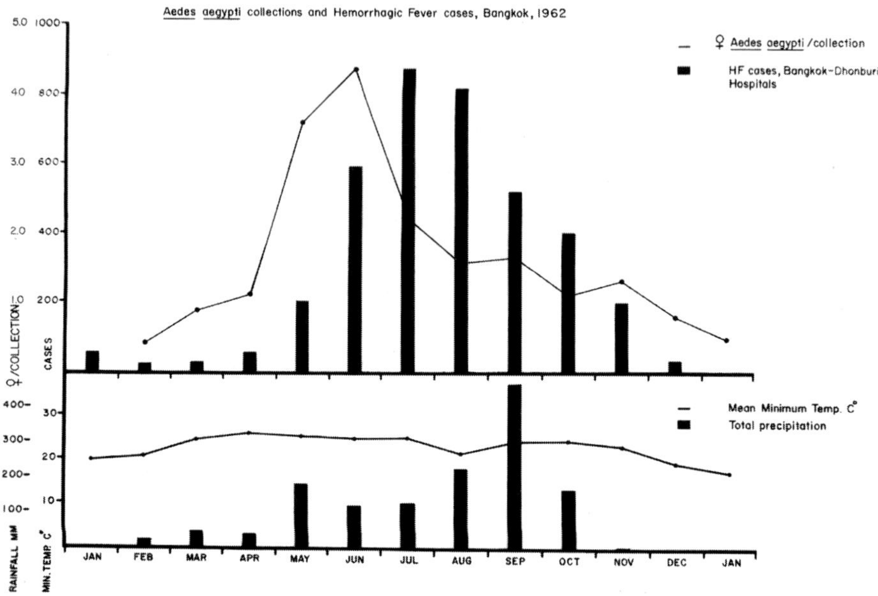

Fig. 1. Monthly admissions of dengue hemorrhagic fever cases to Bangkok, Thailand hospitals, captures of female *Aedes aegypti* in human-baited traps, rainfall and temperature, 1962.[134]

Daily survival rate, longevity and gonotrophic cycle

Longevity of *Aedes aegypti* depends on many factors, such as temperature, humidity and nutrition. Data from the limited number of field studies on female *Ae aegypti* showed longevity ranging from 8.5 days[10] to 42 days.[3] Longevity was 9 days under natural conditions in East Africa.[12] In computer simulations, survival rates ranging from 0.895 to 0.91 with an average of 0.908 were used.[13] Under laboratory conditions, however, at 27 +/– 1°C, females that are given both a 10% sucrose solution and blood meals survive on average for 55.6 days, with a maximum of 100 days.[14] Females that are fed a 5% sucrose solution and kept at ambient temperature in the tropics transmit dengue 75 days after feeding on a viremic human.[15] *Aedes albopictus* females are known to survive for up to 122 days in the laboratory,[16] but there are no reliable data from the field, where daily mortality rates may vary from 7.9 to 15%, depending on the month.

A small change in the daily survival rate has a considerable impact on transmission. For example, in a mathematical model of chikungunya virus transmission by *Ae aegypti*, a change in the daily survival rate from 0.89 to 0.94 altered the course of the disease from a relatively brief, self-limiting epidemic to a stable endemic state.[17] Most, if not all, investigators working on mathematical models of arthropod borne infections have accepted the assumption of Macdonald that the probability of mosquito survival is constant throughout a mosquito's life. This idea was supported by a more recent publication.[18] According to Smith, however, that is unlikely and further field studies were recommended.[19]

Temperature affects not only longevity but also the length of the gonotrophic cycle, contributing factors correlated with the seasonality of dengue in Thailand.[7]

Temperature

Under moist, tropical field conditions where the major mortality sources are accidents such as encountering a spider's web, the probability of surviving a single day is constant and independent of temperature. Experiments to measure this parameter in the field are notoriously noisy but a consensus value is somewhere between 0.87 and 0.91 attrition per day in most locations in the dengue endemic countries of SE Asia.[13] The integral of the survival time (Sa) provides the average lifespan of the female; for Sa = 0.89 the average lifespan is about 8.6 days. Keeping in mind that the population is declining exponentially with age it is easily seen that, while the majority of females die at an early age, the tail of this distribution contains the rare but older individuals with the potential to transmit dengue viruses. When a female takes an infectious bite on her first day of life and if the length of time required for infection to disseminate is one day, a substantial portion of the average lifespan will have passed. Most females will not be capable of transmission prior to their deaths. Once disseminated to salivary glands, the probability of transmitting the virus varies with biting frequency, which is related to the length of the gonotrophic development period. Figure 2 presents an estimate of the average number of potentially infectious replete feeds per newly emerged female as a function of temperature and daily survival probability. This

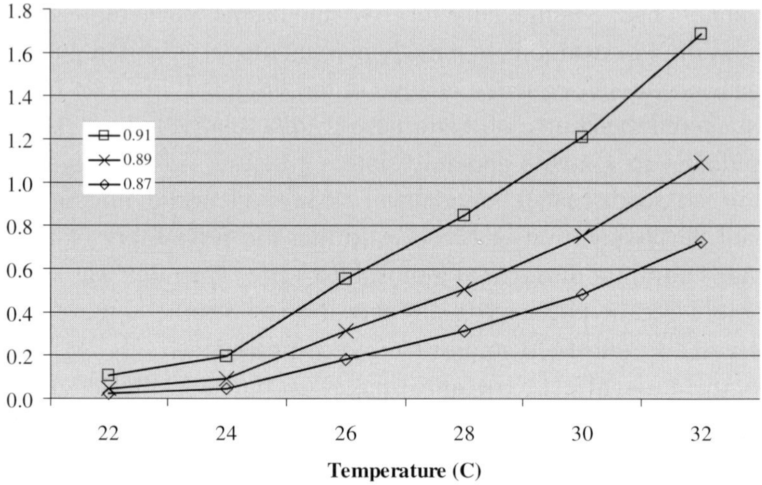

Fig. 2. Average number of potentially infectious replete feeds per newly emerged female as a function of temperature and daily survival probability, assuming that all mosquitoes take an infectious blood at one day of age. The actual number of potentially infectious bites per replete feed is unknown and may be as high two or three or more interrupted feeding attempts with resumption on the same or a different host.[45]

figure is based upon the assumption that all mosquitoes take an infectious blood meal at one day of age. *Ae aegypti* females are observed to take multiple potentially infectious bites per replete feed, potentially as high as two or three interrupted feeding attempts with resumption on the same or a different host. From an epidemiological perspective the increased number of interrupted feeding attempts per replete feed, which is the consequence of 2°-or-3°-warmer temperatures, is equivalent to a doubling of the density of *Ae aegypti*. Under circumstances where a majority of breeding sites are indoor, warm temperature and high moisture contribute to increasing adult survival together with the effect of warmer temperatures on shortening the extrinsic incubation period (EIP) contributing to the occurrence of dengue outbreaks during the hot, rainy season of SE Asia. Watts *et al.* observed the EIP to be 12 days in mosquitoes fed on viremic rhesus monkeys and then held at 30°C.[20] When mosquitoes were held at 32°C the EIP was shortened to 7 days. The EIP also varied directly with the infection load. Mosquitoes infected on monkeys with low viremia titers and then held at 30°C were unable to transmit the virus until 25 days after their blood meal.

In summary, the widely observed association between mean daily temperature and seasonal rainfall and dengue epidemics as observed in Fig. 1 can be explained in several ways. Increased water from rainfall can increase outdoor mosquito breeding sites. Also, the increased moisture accompanying a rainy season can lengthen the lifespan of adult mosquitoes. High ambient temperature has an even more powerful effect on vector mosquito populations, by increasing the lifespan of adult mosquitoes, increasing the biting activity as well as shortening the gonadotrophic period. Finally, the EIP is shortened as average temperatures rise, resulting in a higher proportion of the dengue-virus-infected adult mosquitoes that survive long enough to transmit.

Feeding behavior

That female mosquitoes engage in multiple feedings before completing a gonotrophic cycle has been recognized by many investigators.[21,22] The classic experiments by Siler demonstrated that female mosquitoes, once infected, remain infective throughout the adult stage, even after repeated bites on humans. Thus, in a susceptible population, multiple feedings undoubtedly contributes to the explosive spread of dengue viruses by a very small number of mosquitoes. Importantly, the virus acquired by mosquitoes in a second meal from a viremic person may be neutralized by the first meal if it contained neutralizing antibodies and the second meal was taken within 6 hours of the first.[23] In endemic settings, where a majority of individuals within a household circulate dengue antibodies, multiple feedings may not be very important in spreading viruses within a household.

Flight range of vector mosquitoes

It has been established by mark–release–recapture methods that the flight range of *Ae aegypti* is from 2.5 km per day in an open environment,[24] to a few hundred meters,[25] to less than 25 m in an urban environment.[26] Most of these measurements run counter to epidemiological evidence showing sharp localization of infected *Ae aegypti*. For example, during a small outbreak of dengue in an isolated boarding school in Malaysia, none of the approximately 600 students in a building only 50 m from the outbreak was affected.[27] Similarly, during an outbreak in Puerto Rico, a cluster of homes that had similar human and vector densities only 30 m away from other homes with dengue cases remained free of infection for more than 2 months.[28]

More relevant data have been obtained by measuring mean daily flight range after the first blood meal or oviposition rather than the distance traveled after release. In an East African village, the majority of marked *Ae aegypti* remained in the house where they were captured. Most females visited only one or two houses. Only 0.7% visited five houses or more.[3] The movement of *Ae aegypti* in urban environments has not been well studied.

Dengue virus — mosquito interactions

Analyses of individual human and mosquito hosts infected with the dengue virus (DENV) have found levels of nonsynonymous and synonymous genetic variation comparable to those seen in other, highly variable RNA viruses.[29–32] On the other hand, large-scale phylogenetic studies have revealed that arboviruses have significantly lower rates of nucleotide substitution than other RNA viruses, particularly at nonsynonymous sites,[33] and that they are subject to significantly less positive selection pressure (and hence lower levels of amino acid diversity) on their surface structural genes.[34] Moreover, while a variety of experimental studies confirm that selection pressures on arboviruses differ when the viruses are grown in insect and mammalian cells, there is conflicting evidence as to whether alternating replication among hosts produces negative fitness tradeoffs such that traits favored in one host are selected against in another.[35,36] It has been estimated that over 90% of nonsynonymous mutations observed in dengue viruses recovered from human blood are deleterious — that is, the viruses do not survive in mosquitoes to be reintroduced into the human population. Over time and distance, analyses of dengue virus genomes demonstrate a very slow rate of the introduction of stable mutations. Important to an understanding of the evolution of arthropod-borne RNA viruses is the observation that due to intrinsic constraints associated with dual replication in mammalian and invertebrate hosts, these viruses evolve more slowly than RNA viruses transmitted by other routes.[37]

Mosquito infection dynamics

How many infected mosquitoes are required to sustain transmission? Unfortunately, results of studies are difficult to compare because study design and mosquito collection techniques were highly variable. For example, in one study in Thailand, on average, 20 females were found per room,[22] while in another study in Puerto Rico, 5–10 females were found per house.[38] In the latter study, the number of female *Ae aegypti* per person (more than 1.5) was found to be a significant risk factor for transmission.[39] Large numbers of female mosquitoes per residence are not always associated with epidemic transmission. In a localized school outbreak in Malaysia, only three *Ae aegypti* females were collected in one hostel, where 20 students were infected.[27] Similarly, in a study in India, in a district where more than 43% of the residents suffered from dengue, only three female mosquitoes were found in 74 houses.[40] During an outbreak in Taiwan, the female density was only 0.07 per house.[41] While indoor mosquito density results may be biased by completeness of mosquito collections, the data strongly suggest that an epidemic can be sustained even where the mean adult female density per house is much less than 1.

A critical factor concerning the mosquito density required for transmission is the proportion of infected females. For many studies a calculation of percent infected mosquitoes has been impossible because the virus was recovered from mosquito pools. In Singapore, dengue virus isolation rates from *Ae aegypti* and *Ae albopictus* in four sentinel areas were, on average, 0.51 and 0.59 per 1000 mosquitoes, respectively,[42] while the corresponding figures in Thailand were 61 and 5 per 1000, respectively.[43] Mosquito transmission for arboviral diseases in which contact between infectious vector and human occurs outdoors is expressed by the minimum infection ratio. For dengue, it is preferable to calculate transmissibility by enumerating female mosquitoes captured in a residential unit — the number of infected female mosquitoes per house or premises. In a study in three villages in India, virus-infected females / total number of females captured / number of houses during an epidemic were 1/15/20, 1/18/25 and 1/18/32, respectively.[44] The number of infected females found per residence during dengue transmission was small. In a computer-simulated dengue epidemic in a 40 ha community with an initial population of 10,000, the projected estimate of infected female mosquitoes was determined to be roughly 1 per 1000 persons or about 2% of total female mosquitoes.[45]

Extrinsic incubation period

For transmission to continue, infected female mosquitoes must live from the period after emergence prior to feeding and for the duration of the extrinsic incubation period (EIP). It has been reported that the EIP ranges between 10 and 14 days.[15,46] The EIP is affected by ambient temperature, humidity, viremia level in the human host and the virus strain. Generally, higher temperature results in a shorter EIP, while lower temperature increases the EIP, which in turn decreases transmission.[20,47] The lowest daily temperature, rather than the average temperature, is thought to be a more important determinant of dengue transmission seasonality in Bangkok.[22]

Vector competence

Laboratory experiments have demonstrated a wide variation in oral infection rates among geographic strains of *Ae aegypti* and *Ae albopictus*.[1,48] Early laboratory experiments established that *Ae aegypti* is a rather inefficient vector, since oral feeding requires an artificial blood meal containing as much as 10^6 or more of infectious units per milliliter. This was interpreted by some to suggest that the vector required high levels of viremia to become infected. However, in most experiments, tests were conducted by using artificial blood meals. As reported for other flaviviruses, West Nile and St. Louis encephalitis viruses, the threshold of infection by pledget feeding is often more than 3 dex higher than that by feeding on viremic mammals with a similar virus titer.[49,50] In an evaluation of a DEN-2 candidate vaccine, a small number of mosquitoes acquired infection by biting a volunteer with a viremic titer of less than 1 PFU per k0.2 ml.[51] On the other hand, an attempt to isolate the virus by mosquito feeding (*Ae aegypti* and *Ae Tabu*) on viremic patients in Tonga yielded few viruses.[52] These contrasting observations may be partly explained by the fact that many factors regulating oral feeding that lead to infection of mosquitoes under natural conditions are still poorly understood.

Aedes albopictus

History

Aedes albopictus belongs to the *scutellaris* group of the subgenus *Stegomyia*. It is an Asian species that has spread with the traffic in used tires from a broad area in Asia that included the Indian Ocean islands, the Indian subcontinent, SE Asia, Taiwan, China, Korea and Japan, and the Hawaiian Islands to the continental United States, most of Central and South America, parts of Africa and Europe. Originally a forest mosquito feeding on a variety of vertebrates and breeding in tree holes, *Ae albopictus* also has adapted to the human environment without acquiring the same degree of domestication as *Ae aegypti*. Like the latter species, populations of *Ae albopictus* show a degree of genetic variation in enzyme profiles in susceptibility to infection with dengue viruses.[1] Populations exhibit biologic differences with respect to adaptation to cold. *Ae albopictus* can persist at locations as far north as Beijing or Chicago.

Bionomics

The eggs of *Ae albopictus* can resist desiccation for several months. Besides natural containers, such as tree holes, plant axils, cut bamboo stumps and opened coconuts, which constituted the original larval habitats of the species, larvae now use outdoor artificial containers such as rubber tires, water storage barrels, and glass and plastic bottles and jars. This diversity of larval habitats explains the abundance of this species in rural areas as well as in urban localities.

Ae albopictus populations show a close correlation with rainfall, as during the rainy season larger numbers of potential larval habitats are productive than may be the case for *Ae aegypti*. Adults are zoophilic as well as anthropophilic and feed outdoors during the day. Their longevity is comparable with that of *Ae aegypti*. The active maximum dispersal range is 400–600 m.

Mosquito–virus interactions

A notable outbreak of DENV-1 occurred in 2001–2002 in the state of Hawaii.[53] Cases were reported on Oahu, Kauai and Maui. The affected areas had high rainfall with dense tropical vegetation and high *Ae albopictus* densities and landing rates. A total of 122 cases of locally transmitted dengue fever were observed during the period from May 2001 to February 2002. There were an additional 43 cases of dengue in visitors from other Pacific islands, the Philippines or peninsular SE Asia. Viruses were independently imported into all three islands, probably from a large outbreak on Tahiti. A group of more than 30 Maui residents visited Tahiti in April–May 2001 and returned with dengue. Other visitors brought dengue to Oahu and Kauai over a period of several months. Most cases occurred among residents of rural communities. Of these, 80 occurred in the eastern Maui area (with less than 0.3% of the state's population), where the outbreak was first detected and the virus transmission rate was 40%. All virus isolations during the epidemic were DEN-1 and genetic analysis identified the virus strain as the Tahiti Pacific genotype. This outbreak was largely confined to an area with large mosquito populations and a human population without prior exposure to dengue. As opposed to *aegypti*-borne outbreaks, which are sharp and explosive, this epidemic smoldered for nearly one year. The most likely explanation of the differences in outbreak characteristics lies in the low efficacy of *Ae albopictus* as an epidemic vector of dengue viruses. This mosquito may feed on various vertebrates, thus decreasing the probability of transmission between human beings. Furthermore, it rarely takes blood from more than one host during a single gonadotrophic cycle, which also decreases the probability of transmission between humans. Plaque reduction neutralization test results on sera from residents suggested that the community on Maui may have experienced an autochthonous but silent DENV-2 outbreak within the last nine years.

Other species of the subgenus *Stegomyia*

Aedes polynesiensis

The *scutellaris* group includes about 40 species that have similar bionomics and are often difficult to differentiate. All species occur in the Oriental and Australian regions and have distinct geographic distributions which sometimes overlap. Aside from *Ae albopictus*, whose geographic expansion has been referred to above, some species have vast distributions whereas others are very localized. Thus, *Ae polynesiensis* is present in many archipelagos (the Society Islands, the Tuamotu Islands, the Austral Island, the Cook Islands, Samoa and Fiji), whereas *Ae hebrideus* is known only from Vanuatu and several islands north of New Guinea, and *Ae cooki* is found only in Niue.

All of the species have dessication-resistant eggs. Their larvae are found in tree holes, leaf axils, opened coconuts and, sometimes, crab holes (e.g. *Ae polynesiensis* salinity-tolerating larvae in the holes of *Cardisoma carnifex*). Artificial containers are sometimes used by certain species (e.g. *Ae polynesiensis* and *Ae hebrideus*). In general the biology of *Aedes* of the *scutellaris* group resembles that of *Ae albopictus*. They are characteristically found in rural areas or small villages and their abundance depends on rainfall. Most feed readily on human beings, mainly outdoors.

This group contains several species that have been incriminated as vectors of dengue by epidemiological observations, followed, in some instances, by experimental transmission or infection. Usually, a species has come under suspicion because it was the only plausible vector present during a dengue epidemic when *Ae aegypti* was either absent or so localized that it could not have been responsible for all transmissions observed. Thus *Ae polynesiensis* (in French Polynesia, the Cook Islands and Futuna) and *Ae scutellaris* (in New Guinea) have been shown to be vectors by both epidemiological and transmission studies to man or monkeys. Others are suspected vectors based on epidemiological studies and laboratory infection

(*Ae cooki* on Niue) or epidemiological studies alone (*Ae hebrideus* on Espiritu Santo in Vanuatu, or *Ae rotumae* on Rotuma Island).

Aedes (Stegomyia) of the African continent

African species of the subgenus *Stegomyia*, other than *Ae aegypti*, that are implicated in the transmission of both dengue and yellow fever include *Ae leteocephalus*, *Ae africanus* and *Ae opok*. The latter are tree hole mosquitoes with dessication-resistant eggs. The adults are crepuscular and inhabit forests or forest galleries. While the adults are usually found in the forest canopy, they sometimes descend to ground level and thus can come into contact with man. They feed primarily on primates, but they can be anthropophilic as well. Their population densities are closely tied to rainfall. Numerous isolates of the dengue 2 virus from these species in West Africa show that they could be involved in an enzootic forest cycle in which they transmit the virus during the rainy season from monkey to monkey, and possibly from monkey to humans.[54–56] The evidence that some *Ae* (*Stegomyia*) species transmit the dengue virus vertically suggests that they could also serve as reservoirs of the virus during dry seasons. Their role in the epidemiology of dengue is not known in detail.

Aedes* of the subgenus *Diceromyia

The dengue 2 virus has been isolated from two tropical African species of this genus, *Ae furcifer* and *Ae taylori*. For many years, however, it was not possible to distinguish the females of this species and, therefore, it is not known to which the older data pertain. Both species inhabit forests or forest galleries. They have dessication-resistant eggs and breed in tree holes. The adults live primarily in the forest canopy, are diurnal, and feed on both animals and humans. Their abundance depends on rainfall. At least for *Ae furcifer*, some parous females often leave the forest to reach villages and bite man, even indoors. Like the *Stegomyia* species, these mosquitoes could play a role in sylvatic cycles and transmit the dengue virus from monkey to humans. The isolation of the dengue 2 virus from males in both Senegal and the Ivory Coast implies that they also transmit the dengue virus vertically and thus could serve as reservoirs of the virus during the dry seasons.

Aedes of the subgenus *Finlaya*

In tropical Asia this subgenus includes many species of the *niveus* group which are very difficult to differentiate. Their bionomics is poorly known, but they are mosquitoes of the forest canopy that breed in tree holes, and feed primarily on monkeys and occasionally on humans. The isolation of the dengue type 4 virus from mosquitoes of this group in Malaysia suggests that they may play a role in a forest cycle of dengue and possibly in the transmission of the virus from monkey to humans.[57]

Aedes of other subgenera

Aedes (*Gymnometopa*) *mediovittatus* is a forest mosquito that has adapted to the rural peridomestic environment in the Caribbean region (Puerto Rico, Cuba and other islands). It can be considered a potential vector of dengue in view of its anthropophilia and high degree of susceptibility to infection in the laboratory.[58,59] Another possible vector is *Ae triseriatus*, of the subgenus *Protomacleaya*, which has been shown to be capable of transmitting the dengue virus in the laboratory.[60]

Infectious Disease Epidemiology of *Aedes aegypti*–Borne Dengue

Intrinsic incubation period

The classic experiments on human volunteers by Siler and Simmons established that the intrinsic incubation period averaged between 4.5 and 7 days, with a small number of cases exceeding 10 days, and viremia may ensue 6–18 hours before the onset of fever. The symptomatic viremia is 4–5 days, but may be as long as 12 days.[15,61] The viremia that occurs during the period before the onset of symptoms and in the vast majority of individuals who experience viremia without important symptoms surely constitutes the majority of infectious blood meals, as it can be expected that such individuals will carry on all normal daytime activities, in contrast to persons who are sick and bedridden.

The threshold of viremia in humans required to infect mosquitoes has not been measured accurately. In natural infections, virus titers in humans rise as high as $10^{8.3}$ Mosquito Infectious Doses$_{50}$ (MID$_{50}$) per ml. In monkeys, dengue viremia was as high as 10^6 MID$_{50}$, yet 10–15% of *Ae albopictus* mosquitoes allowed to feed on these monkeys were infected. As for the impact of strain variation, a considerable degree of variation in the EIP has been observed. For example, with mosquitoes feeding on humans infected with an unadapted strain of dengue 1, the EIP was 14 days, while with those infected with the strains at low mouse passage levels, it was 22 days.[46] Long EIP's have been observed with dengue virus strains with attenuation characteristics. Often, although infection is established, mosquitoes fail to transmit the virus.[51,62]

Mosquito density required for dengue transmission

The minimum density of *Ae aegypti* which permits transmission of dengue viruses has long been sought and is a topic subjected to fierce debate. For example, a house index of 5 was selected as the objective for urban yellow fever control in Quitos, Ecuador. In Singapore, where the vector density had been held to a house index of less than 1 for many years, dengue infections occurred endemically during the long period of vector control. This transmission is not indigenous — undoubtedly the result of the constant introduction of viremic hosts into Singapore from nearby dengue-endemic countries.

Household transmission

Once an infective mosquito enters a house or a member of a household becomes infected, the probability of multiple infections in the household increases and may eventually result in clusters of dengue infections. In one study in Puerto Rico, among 9 households with one person each, none had any dengue case, 61.4% of the 57 households with 2–4 members had at least one case and accounted for 61 cases, and in 69 households with 5–9 members 76.8% had at least one dengue case.[28] In Honduras, 5 members of a family became ill from dengue; 4 of them had an onset within 4 days. In one extreme case, 29 of 30 members of a household in China were infected.[63] Multiple patients per household have also been documented in DHF. In Thailand, among 271 families in which at least one member contracted DHF, 35 (12.9%) there were one or more additional DHF patients per household.[64]

Basic reproduction number

The basic reproduction number is denoted by R_0. It is defined as the number of secondary infection in humans that results from a single infected human. The higher the basic reproduction number, the more explosive the transmission. Diseases with high numbers are characterized by high herd immunity thresholds. A herd immunity threshold is the maximum level of immunity beyond which transmission is eliminated.[65] Dengue infection is via the mosquito and the basic reproduction number is controlled by variables discussed above. The basic reproduction number is related to the initial fraction of the population that is susceptible and to the final fraction infected due to an epidemic (p). When the whole population is initially susceptible, $R_0 = -\ln(1-p)/p$. As the estimate of R_0 determines if the dengue virus can sustain its chain of transmission, it is also called the threshold parameter. For a simple, directly transmitted pathogen in a homogeneously mixing population, the minimum proportion of the population required to be effectively vaccinated for disease elimination is given by $p_c = 1 - 1/R_0$.[66] R_0 was derived from a mathematical discussion on the periodicity of yellow fever outbreaks[67] and a consideration of entomological indices and the duration of viremia.[38] Theoretical values are computed by using one of many formulas. Some formulas are designed to be used early in epidemics, and others for endemic settings. The application of these formulas is based upon assumptions: (a) that an epidemic is introduced by an infected individual into a virgin soil where all at-risk persons are susceptible, (b) that humans mix homogeneously, (c) that the population size is so large that accretion of immunes is negligible, and (d) that R_0 measures only the initial phase of an epidemic. Recent estimate of R_0 have been made using one of three methods: (1) a final size equation estimated from a serological measurement of an epidemic,[68] (2) the initial intrinsic growth rate using data from the epidemic curve at the very first stage,[69] and (3) the age distribution of antibodies.[70]

Basic reproduction numbers estimated for dengue range between 1.33 and 11.6. Early measurements were relatively low.[38,68,69] Ferguson *et al.* used an age-stratified measurement of dengue-neutralizing antibodies, deriving dengue-type-specific R_0's of 4.3–5.8.[70] Favier *et al.* analyzed epidemic data from several cities in Brazil, developing a range of values, 3.8–5.1,[71] while Massad used other epidemic data from Brazilian communities, obtaining values ranging from 2.7 to 11.6.[72]

Many of those measurements suffer from limitations in epidemiological and entomological data available or the methods used to calculate R_0. Studies that estimate the total number of infected persons during an epidemic using the hemagglutinatoin inhibition (HI) or the IgG ELISA tests are unable to measure type-specific dengue antibodies and may confuse infections with other flaviviruses with past dengue infection. For example, an HI antibody study in Mexico conducted in 1986 putatively measured the transmission of dengue 1, a virus that had been introduced into the Western hemisphere in 1977, but may also have measured previous dengue 4 and dengue 2 infections, these viruses having been introduced into the region in 1981.[68] Additionally, this study may have undermeasured dengue antibody prevalence because the HI test used is known to be insufficiently sensitive to detect the low antibody titers often generated in response to inapparent infections. Calculations of R_0 made by counting reported cases are likely to be biased by underreporting and by an unknown fraction of silent infections. Even dengue 1, a virus which produces a high rate of clinically apparent disease in adults, results in mild or inapparent infections in children. Thus, in Brazil, as much as 40% of the population may not participate in data used to calculate reproduction numbers.[69,71,72] More recent calculations, particularly the thorough study by Ferguson based on the prevalence of dengue-neutralizing antibodies in a dengue-endemic setting, yielded higher and more realistic values of R_0.[70]

Particularly in the American hemisphere, *Ae aegypti* infestations are not uniformly present throughout residential areas. Well understood by entomologists, the problem of spatial heterogeneity was identified by modelers during a virgin soil epidemic on Easter Island. Heterogeneity occurred because not all houses contained *Ae aegypti*, and as a result their occupants were at low risk of being infected by the dengue 1 virus. When the authors applied corrections for heterogeneity they improved simulated epidemic curves for the Easter Island outbreak as well as for outbreaks reported from Belem and Brasilia, Brazil.[73]

Herd immunity

Values of R_0 can be used to estimate herd immunity. The higher the value of R_0, the larger the fraction of the population that must be immune to stop transmission. As illustrated by Ferguson *et al.* (1999), approximate levels for herd immunity (or the desired level of immunization of the population using vaccines) p may be derived given the relationship $p > 1-1/R_0$.Where many dengue viruses are endemic, the vaccine will be a cocktail of antigens, and the magnitude of p_c will be set by the type with the highest R_0 value. For example, for a value of R_0 of 5.6, the magnitude of p_c is 0.82. In other words, to protect against continuing circulation of dengue viruses, roughly 85% of the population would have to be immunized.

The value of the reproduction number and herd immunity are also affected by heterogeneity of the residential distribution of vector mosquitoes. For example, in Cuba in 1977, dengue 1 was introduced into a population whose only previous dengue experience was with type 2 before World War II. This led to an islandwide epidemic which resulted in little or no vector control. The epidemic continued until it stopped spontaneously in 1979. An islandwide serological survey detected HI antibodies in 44.5% of the population.[74] For the prevailing epidemiological conditions, the herd immunity was 45%. But, assuming a basic reproduction number for dengue 1 of 5 or greater, it can be estimated that 85% or more of the population would have to be immunized to prevent dengue infections which occurred predominantly in the at risk subpopulation with *Ae aegypti* in the household. The difference observed is likely to be the result of heterogeneity. That is, among households with *Ae aegypti*, herd immunity was 85%, but nearly one half of households were not at risk of dengue infection and, therefore, the epidemic-infected persons in less than half of households.

Influence of interannual climate variation

The El Niño Southern Oscillation (ENSO) is an atmosphere–ocean coupled system which produces quasi-periodic short-term climate and sea surface temperature changes over the Pacific region with impacts on weather worldwide, including many countries in the Americas, Africa and Asia. This system oscillates between two extremes known as El Niño and La Niña, which are associated with approximately opposite disturbances to climate worldwide. A chief phenomenon of an El Niño phase is an eastward extension of warm surface waters situated off northwestern Australia toward the west coast of equatorial South America. During the cool phase, La Niña, equatorial westerlies result in an upwelling of cold abyssal water which is transported to the west, creating a tongue of abnormally cool surface waters extending toward Indonesia. Because convection rainfall in this region is limited to sea surface temperatures greater than about 26–27°C, the spatial distribution of rainfall is associated with equatorial sea surface temperature anomalies associated with the ENSO state. "Southern Oscillation" refers to the oscillation of atmospheric pressures between the eastern and the western Pacific. One of the summary or indicator statistics of the ENSO state is the Southern Oscillation Index (SOI), the normalized difference in pressure between Darwin and Tahiti. El Niño and La Niña are associated with negative and positive values of the SOI, respectively. It is not surprising that much of the innerannual variability in climate in the central Pacific is attributable to the state and intensity of ENSO.

Because many infectious disease systems are influenced by weather, the ENSO state and associated anomalies in rainfall, atmospheric moisture, and temperature have become topics of considerable interest.[75] Largely through the use of numerical simulation models of the ocean and atmosphere, forecasts of anticipated the ENSO state are increasingly skillful, such that it is reasonable to expect that useful early warning systems (EWS's) will be developed.[76] Initial efforts involved nothing more that simple attempts to demonstrate correlations between the ENSO state and outbreaks.[77] More sophisticated development of ENSO-based EWS's will require addressing two related problems: while skill in forecasting an ENSO event is currently adequate, predicting the strength of the oscillation is problematic and leads to a lack of skill in predicting regional weather anomalies. The second area in need of much attention is elucidation of the mechanisms whereby weather anomalies lead to anomalies in the disease system.[78] Several recent studies have shown temporal correlations between malaria epidemics and various indices of the ENSO state.[79–81]

Given that dengue incidence is a function of the interaction of the many factors outlined above, it is not surprising that dengue activity has been correlated with the ENSO state or one of its statistics, SOI, in regions (most clearly in the South Pacific region) where ENSO or the SOI is correlated with temperature and/or rainfall anomalies.[77] Unfortunately, this study does not identify the environmental risk factors unequivocally. At a recent World Health Organization dengue workshop in the SE Asian region, directors of national antidengue programs in Thailand, Vietnam and Indonesia expressed the operational need for an EWS, for that would provide sufficient lead time (1–3 months) to permit mobilization of control operations. In response, an attempt has been made to develop practical EWS's for Yogyakarta on the island of Java in Indonesia and for Bangkok, Thailand, based on logistic regression (Focks D, personal communication, 2003). The predictor variables are sea surface temperature (SST) anomalies (a 5-month running mean of spatially averaged SST anomalies over the tropical Pacific: 4°S–4°N, 150°W–90°W, as measured by the Japanese Meteorological Association) and past monthly cases of dengue in each city. Previous incidence of anomalously high or low cases is an indication of the interaction between the types of virus currently circulating and the nature of the immune status of the human population. SST anomalies are highly correlated with subsequent surface air temperature anomalies and may be correlated with atmospheric moisture as well. The predicted variable is the probability of an epidemic year forecast 1–3 months before the peak transmission season. The skill level for 3-month predictions for Bangkok was inadequate for an operational system (6 errors in 35 years); the 2- and 1-month forecasts had error rates of 3 and 2 per 35 years, respectively. The Java EWS, however, was sufficiently skillful to be put into use in Yogyakarta, Indonesia; 1-, 2-, and 3-month forecasts were without error for the 14-year period of the record. Note that this system does not use the ENSO state directly, but rather one of its indicators, i.e. SST anomalies.

A recent National Research Council report on the subject of EWS's is cautiously optimistic but concludes that substantially more research is needed to understand the relationships between climate, human behavior and infectious diseases. The report states that one goal of such research should be to support a transition from the current practice of "surveillance and response" to a more proactive "prediction and prevention" strategy.[78]

Transport of vector mosquitoes

Historically, *Aedes aegypti* was been transported intercontinentally by ships and more recently by airplanes or vehicles. *Ae aegypti*, and in some instances viremic individuals, were transported in sailing ships from Africa through the Middle East to tropical Asia and to the American tropics. The practice of storing water in casks provided ample breeding sites. Modern ships are less hospitable to mosquito breeding. But specialized cargo, such as used tires, has proven highly efficient in reintroducing *Ae aegypti* from the United States in 1965 to El Salvador, after the species had been previously eradicated. It is not clear where and when dengue viruses adapted from the sylvatic cycle to human beings. A popular scenario is that the four dengue viruses entered the urban cycle in Asian towns once *Ae aegypti* was well established there. According to this scenario, the dengue viruses spread from Asia to Africa, most likely via sustained shipboard outbreaks. The chikungunya virus, an *Ae aegypti*–borne agent, moves in the opposite direction. It is well established that at intervals over a very long period, the chikungunya virus has emerged from its sylvatic focus in South and East Africa and is transported to the Indian Ocean islands and onward to India and the Far East.[82,83] Similarly, *Ae albopictus* was imported into the United States in the mid-1980's from Asia via eggs deposited in used tires.[84]

Although it is clear that adult mosquitoes can be transported in cars, trucks and airplanes, in the current era both *Ae aegypti* and *Ae albopictus* are now very widely distributed around the world, making the transport of viremic humans the more important mechanism now responsible for the introduction of dengue viruses across national borders.

Dispersal of the virus

Traveling waves in endemic countries

Since the early vector control campaigns against urban yellow fever in the Americas, it was recognized that viruses were maintained in and spread from urban centers. It was proposed that when the *Ae aegypti* house index was above 5% the yellow fever virus could be maintained endemically in tropical cities with populations of 50,000 or above. In fact, when effective mosquito control programs were applied in urban areas of South and Central America, yellow fever disappeared, only to reveal an unsuspected sylvatic cycle that has continued to produce human cases to this day. Studies by Cummings *et al.* demonstrated that dengue viruses reside in and spread out of large urban centers in the highly endemic countries of SE Asia.[85] When hospital statistics from each of the 72 provinces of Thailand for the period 1983–1997 were analyzed, Bangkok, a metropolitan area of over 9 million people, was found to be the country's endemic center. Longitudinal studies on dengue cases hospitalized in Bangkok Children's Hospital from 1973 through 1999 demonstrated a roughly three-year periodicity between large outbreaks plus a pattern of successive predominance of different dengue viruses: DENV-1 in 1990–92, DENV-2 in 1973–86 and 1988–89, DENV-3 in 1987 and 1995–99, and DENV-4 in 1993–94. Following these periodic increases and decreases in hospitalization rates and successive predominating dengue viruses, it was observed that waves of severe cases progressed from the capital in all directions, moving at a speed of 148 km per month.

Importation and spread of dengue viruses

Much of the history of dengue viruses in the mid-20th century derives from the reintroduction of *Ae aegypti* and the subsequent importation of dengue viruses. The most important importation route has been from SE Asia to the American hemisphere. In the first half of the 20th century, the only dengue virus known to circulate in the hemisphere was American genotype DENV-2. A dengue 3 virus of Asian origin was imported into Puerto Rico, producing an epidemic in 1963.[28] After traversing the Caribbean islands and possibly producing outbreaks in Venezuela and Colombia, this virus apparently became extinct. In 1977, a dengue 1 of SE Asian origin was imported into the Americas. In 1981, the region's first epidemic of dengue hemorrhagic fever was reported, caused by an Asian strain of DENV-2 distinct from the American subtype circulating previously.[86,87] In addition, in 1981, DENV-4 subtype II of Asian origin was recovered in the Americas and caused epidemics of dengue fever throughout the region. Finally, in 1994, genotype II DENV-3 was first recognized in Nicaragua and subsequently circulated widely throughout the Caribbean and throughout South America.[88]

Despite the importance of these epidemics, little is known about the rates or determinants of viral spread among island and mainland populations, or their directions of movement. A Bayesian method with a coalescent approach was used to investigate transmission histories, and a parsimony method to assess patterns of strain migration of DENV-2 and DENV-4 after their introduction in 1981. Using isolates from 1981 and 2004, Carrington and others studied the dispersal of these two viruses.[89] For both types of viruses there was an initial invasion phase characterized by an exponential increase in the number of DENV lineages, after which levels of genetic diversity remained constant despite reported fluctuations in DENV-2 and DENV-4 activity. Strikingly, viral lineage numbers increased 16 times more rapidly in DENV-4 than in DENV-2, indicating a more rapid growth rate or a higher rate of geographic dispersal of DENV-4. The most obvious explanation for the more rapid spread of DENV-4 compared to DENV-2 is that the American population was fully susceptible to DENV-4 but partially immune from long exposure to American genotype DENV-2. It is also possible that immunity to either DENV-1 or -3 raised antibodies capable of enhancing DENV-4 viremias and thus accelerating epidemic spread; or that antibodies to either DENV-1 or -3 interacted with DENV-2 to retard spread.

Each of the four dengue viruses in the American region that are of Asian origin evolved from single strains after introductions in the late 1970's and early 1980's. The transmission dynamics of the remaining two serotypes of the dengue virus in the Americas might also reflect the immunological landscape. Epidemiological evidence suggests that when DENV-1 first appeared in the Americas its pattern of spread was similar to that of DENV-4. This serotype was first reported in Jamaica in 1977,[90] and within only one year it spread throughout the region, with at least 30 countries reporting activity.[91] Preliminary data on DENV-3 isolates from the Americas suggest that this virus, which was absent from the region for 17 years before being reintroduced in 1994, has expanded at a faster rate than DENV-2 but not as rapidly as DENV-4.[89] Hence, some protective immunity against DENV-3 may remain, although this conclusion is tentative since the time frame and overall host population size differ.

Repeated introductions

In Pacific islands, a more complex pattern of viral strain introduction has been found. Outbreaks of dengue due to DENV-1 occurred almost simultaneously in 2001 in Myanmar and at multiple sites almost 10,000 km away in the Pacific.[92] Phylogenetic analyses of the E protein genes of DENV-1 strains recovered from Asia and the Pacific revealed three major viral genotypes (I, II and III), with distinct clades within each. The majority of strains from the Pacific and Myanmar, and a number of other Asian strains, fell into genotype I. Genotype II comprised a smaller set of Asian and Pacific strains, while genotype III contained viruses from diverse geographical localities. Analyses suggested that the outbreak of DENV-1 during the period 2000–2003, in the Pacific, was due to multiple introductions of DENV-1 from several different Asian localities (Palau, Western Samoa — the Philippines; New Caledonia, Fiji – Myanmar/Thailand; New Caledonia, Tahiti — Malaysia; New Caledonia – Cambodia). For example, it was learned that approximately 2500 individuals from Oceania visit Thailand each year, providing a ready source of human hosts to transport dengue viruses back to their country of origin. It is remarkable that only DENV-1 viruses resulted in overt disease in Oceania despite the evidence that all four dengue viruses were circulating in SE Asian countries and thus should have resulted in the importation of viruses other than DENV-1. It is likely that other dengue serotypes were introduced to the Pacific that did not result in overt disease or did not result in sustained cycles of transmission, possibly because protective immunity might have prevented sustained transmission.[93]

Endemic dengue: generation of genetic diversity

At single locales in endemic countries, dengue viruses have shown significant and rapid changes in the genetic structure, typified by lineage replacement on phylogenetic trees.[32,92,94,95] The absence of a proof-reading capacity in RNA-dependent RNA polymerases gives rise to approximately one nucleotide change in a dengue virus genome during each cycle of replication.[96] Although the majority of mutations that arise within each host are likely to be deleterious, such variation may give rise to diverse viruses able to occupy new ecological niches or to adapt to selective pressures.[6] Analyses of the ratio of nonsynonymous to synonymous nucleotide changes per site (dN/dS) within and between DENV populations suggest that they are subject to strong purifying selection,[95,97] with relatively little evidence for adaptive evolution.[6,32,37] A lack of longitudinal studies of the evolution of DENV at single locations has hampered the understanding of how frequently extinctions of DENV genotypes might have occurred or what evolutionary processes are responsible. The extinction of a lineage of DENV-1 has been documented recently in Myanmar.[98] Phylogenetic analyses of the sequences of DENV-1 genomes confirmed that three distinct clades (from two different genotypes) circulated in Myanmar since the early 1970's, when dengue hemorrhagic fever was first recognized in that country,[99] and that one of these clades (clade A, genotype III) became extinct after 1998, to be replaced by genotype I viruses from two other clades (B and C). The replacement genotype I clades originated in Asia. In three other instances, sudden changes in the genotype of dengue viruses have been observed; the changes also appeared to be due to stochastic events resulting in population bottlenecks.[32,94,98]

Single introductions: generation of genetic diversity

Genetic changes in DENV-4 viruses which circulated at irregular intervals causing disease in Puerto Rico illustrate how dengue viruses evolve over short periods.[95] The genetic structures of 82 viruses from 1982, 1986–7,

1994, 1996 and 1998 were sequenced and analyzed. No significant changes were observed in structural (E) genes, meaning that host immunity played no role in the evolutionary shifts observed. This is in contrast to the hemagglutinin (HA) gene of influenza A, which appears to be under strong immunologic selection.[100] In the dengue virus, constraints on the E gene may be imposed by the need to preserve functions that condition survival in a two-host life cycle.[101,102] Rates of nucleotide substitution are lower in arboviruses than those seen in many other RNA viruses.[36,103] Against this conservative background, genomes steadily accumulated changes, often nonsynonomous changes in the nonstructural gene NS2A. As an example of this process, a dominant Puerto Rican lineage twice descended from earlier rare genotypes — a pattern suggesting that much of the lineage turnover is driven by selection on the viral genotype.

The changes described above for the DENV-1 and -4 viruses provide evidence for rapid selective and adaptive evolution of dengue viruses from a single introduction. This adaptive evolution is driven by the stochastic nature of the dengue virus life cycle, which favors survival of common variants. Genetic bottlenecks occur at every mosquito feeding event, during seasonal reductions in vector populations, and variations in the abundance of susceptible human hosts.[104]

An interesting example of the evolution of genetic changes was studied on DENV-2 isolated from the 1997 DHF/DSS outbreak in Santiago de Cuba. During this outbreak, case fatality rates had increased month by month, a phenomenon which also had been documented during the 1981 DENV-2 Cuba-wide epidemic.[105] It had been hypothesized that an observed month-by-month increase in disease severity might be the result of serial infection of DENV-1 immunes by Asian genotype DENV-2 viruses, each successive generation having been selected to "escape" cross-reactive neutralization from dengue-1 antibodies, and thus more susceptible to the enhancing effects of these same antibodies in the next infected host.[105] Available 1997 isolates were sequenced, looking for changes in envelope protein that might signify emergence of neutralization escape mutants. These were not found.[106] However, significant changes were observed in NS5 indicative of a clear pattern of virus evolution during the epidemic.[107] The origin and meaning of these changes require further study.

Virus titer and variation in viremic periods

The size of the virus inoculum, the product of the viral titer and the quantity of the blood meal influence the probability of the vector subsequently developing a disseminated infection with virus in the salivary glands.[1,2] It has been suggested and there is some evidence to support the hypothesis that the titer of the virus in the blood meal alone could influence the probability of subsequent infection. Moreover, the duration of the dissemination period, the extrinsic incubation period (EIP), can vary with the titer. Watts *et al.* reported that the EIP for dengue in *Ae aegypti* at 30°C was 12 and 25 days for mosquitoes infected with *high* and *low* doses, respectively.[20]

Enhancing antibodies

Cummings *et al.* examined the epidemiological impact of enhancing antibodies on the prevalence and persistence of viral serotypes.[108] Using a dynamical system model of *n* cocirculating dengue serotypes, it was observed that ADE provides a competitive advantage to those serotypes that undergo enhancement compared to those that do not, and that this advantage increases with increasing numbers of cocirculating serotypes. Paradoxically, there are limits to the selective advantage provided by increasing levels of ADE, as greater levels of enhancement induce large amplitude oscillations in incidence of all dengue virus infections, threatening the persistence of both the enhanced and nonenhanced serotypes. Their results suggest that enhancement is most advantageous in settings where multiple serotypes circulate, and where a large host population is available to support pathogen persistence during the deep troughs of ADE-induced large amplitude oscillations of virus replication.

Preinfection factors and early infection events

Various preinfection factors which may influence the outcome of dengue infections are discussed in Chap. 7. They include protective antibodies, age–sex–race nutritional status, other conditions, herd immunity and enhanced viremia.

Infection parity

The most fundamental epidemiological observation in dengue is that severe dengue disease, i.e. dengue hemorrhagic fever / dengue shock syndrome (DHF/DSS), occurs regularly in locales where two or more dengue viruses are simultaneously or sequentially epidemic. In these locales, DHF/DSS occurs in two immunological settings: (1) in individuals 1-year-old or older who are infected with two or more different dengue viruses at intervals of at least 1 to more than 20 years, and (2) in individuals under 1 year of age who circulate passively acquired dengue antibodies and who are infected with a dengue virus for the first time.

DHF/DSS most frequently accompanies a second dengue infection. Early evidence that supported this supposition was embedded in the shape of the curve describing the age distribution of hospitalized DHF/DSS cases in the highly dengue-endemic city of Bangkok. In 1970, Fischer and Halstead explored in a mathematical model the age distributions resulting from first, second, third and fourth dengue infections.[109] The "best fit" was a two-infection model, which when applied to actual data for the years 1958–64 reconstructed the 1964 age distribution of cases in Bangkok Children's Hospital almost exactly. At the time this study was undertaken, the intrinsic susceptibility to dengue vascular permeability syndromes was unknown. This is now known from the 1981 Cuban dengue outbreak, an occasion when all individuals from 2 to 50 years of age experienced infections in the same sequence and at the same rates — dengue 1 in 1977–9 and dengue 2 in 1981 (Fig. 3). In the 1970 paper, a limitation of five years had to be imposed to fit sequential infection data to the age distribution curve. Using data for the intrinsic age susceptibility to DHF, an even better curve is fitted to hospital data.[110] Many prospective cohort and seroepidemiological studies have identified second dengue infections as the likely cause of DHF. They are summarized in Chap. 7.

Finally, second infections have been associated with large, classical outbreaks of DHF/DSS. This is best illustrated in Cuba, where two different dengue viruses were introduced into a largely susceptible population. In 1977–79, dengue 1 circulated throughout the country in a population that had not been exposed to any dengue infections since just prior to

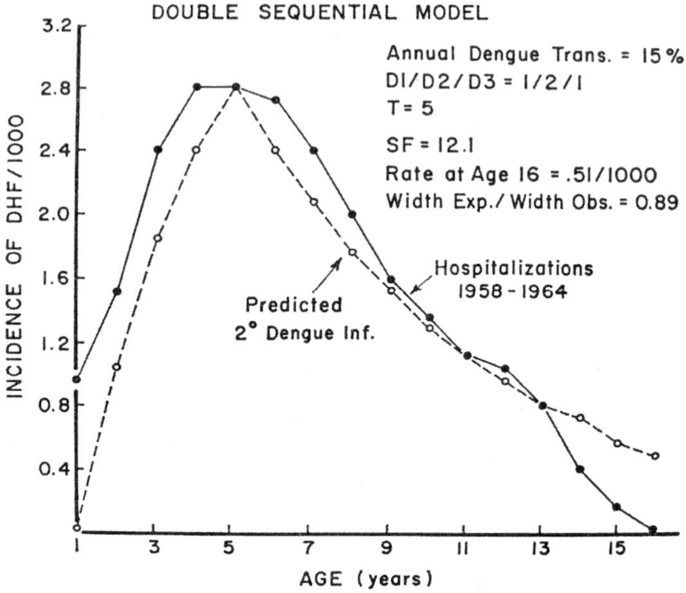

Fig. 3. Age-specific secondary dengue infection rate compared with observed age distribution of hospitalized dengue hemorrhagic fever cases, Bangkok Children's Hospital, 1962.[109] In the original model, two infections must occur within five years. A similar age distribution is observed when corrected for observed innate susceptibility of young children to developing vascular permeability during secondary dengue infections.[110]

World War II.[86] Neutralizing antibody studies showed the earlier dengue infection to be type 2, in all likelihood American genotype dengue 2. As many as 50% of individuals who were 40 years old and older were immune to dengue 2, and half of these were infected by dengue 1. None of these sequential dengue 2–dengue 1 infections produced severe disease. In 1981, a dengue 2 virus of SE Asian origin was introduced into Cuba, which produced DHF/DSS in those who were immune to dengue 1. Those immune to dengue 2 from pre–WWII infection were protected against DHF/DSS. In 1997, 20 years after the introduction of dengue 1, an Asian genotype dengue 2 again circulated in the city of Santiago de Cuba and its environs.[111] Dengue 1 immunes infected with dengue 2 developed overt disease, the most severe being DHF/DSS. Although the viruses causing the 1981 and 1997 epidemics were Asian genotypes, when attack rates and case fatality rates were corrected for age the 1997 epidemic was more severe than that in 1981.[87,106,112]

Viremic tertiary dengue infections have been reported in rhesus monkeys.[113,114] In a longitudinal study of dengue admissions in Bangkok, symptomatic tertiary dengue infections have been observed.[115] The fraction that tertiary dengue infections contribute to overall DHF/DSS morbidity is very small. Are fourth dengue infections possible? Studies on rhesus monkeys have shown that animals immune to three different dengue viruses are susceptible to infection with the fourth virus.[114] This implies that sequential infections with each of the four dengue viruses are theoretically possible. What fraction of human beings previously infected with three dengue viruses is susceptible to a fourth infection is unknown. Whether clinical disease accompanies a fourth dengue infection is not known either.

Cocirculation of multiple serotypes

The current pandemic of dengue and DHF/DSS originated in the Pacific and SE Asia in the 1940's and has subsequently spread to the Americas and Africa. Today most urban centers of SE Asia and many in Central and South America are hyperendemic for dengue, frequently with all four serotypes circulating simultaneously.[116] Given the significance of sequential infections in developing serious illness factors, regulating or influencing the spatial and temporal distributions of dengue serotypes also may be important in regulating or influencing the age-specific rates of infection and illness.[117]

An example of spatial and temporal variation in serotype abundance

Figure 4 presents an example of annual dengue serotype variability observed at a single institution over a long period of observation. Bangkok Children's Hospital (now the Queen Sirikit National Institute of Child Health) has a long-standing collaboration with the Virology Department, Armed Forces Research Institute of Medical Sciences, where comprehensive virological studies have been performed on all patients

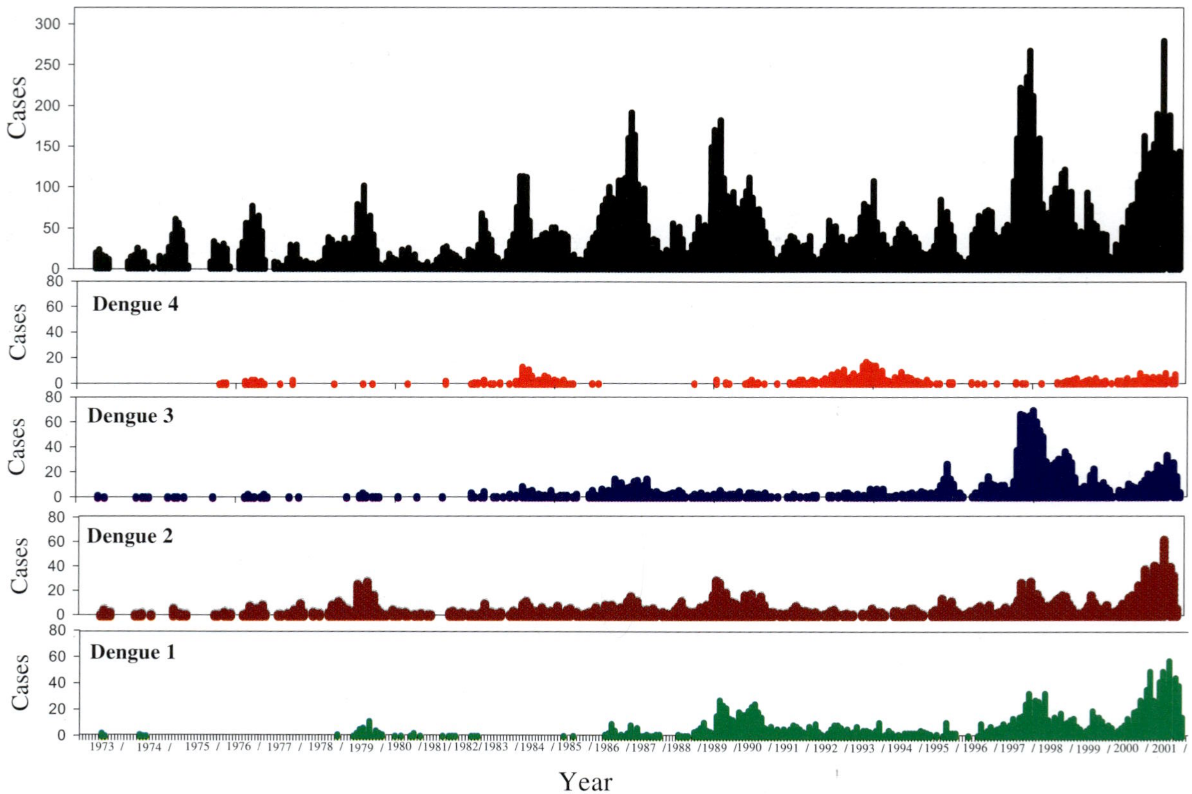

Fig. 4. Monthly dengue hemorrhagic fever admissions to the Queen Sirikit National Institute for Child Health, Bangkok, Thailand, 1977–2001. The lower panels show monthly dengue virus isolations.[118]

hospitalized for suspicion of dengue. The results of these observations have been published and discussed in a number of reports.[115,118–123] Reasons for the disappearance of dengue viruses or clades include herd immunity and the founder effect.

Founder or stochastic effect

In some regions, the mix of serotypes found each year simply reflects the mix in other endemic areas. In some areas dengue viruses may be lost during a halt in transmission due to cool season effects. In North Vietnam, for instance, each year dengue viruses are reintroduced from more southerly locations.[124] A similar situation is reported for small, relatively isolated Pacific islands that are too small to remain endemic; when the virus does arrive, the introduced strain may have been active earlier on larger islands or imported from mainland dengue-endemic countries.

Another factor in the distribution of serotypes is herd immunity. During the course of an epidemic not influenced by vector control or cooling temperatures, R_0 falls to ultimately less than 1 as a function of the rising proportion of immunes — increasingly, potentially infectious bites fall on refractory individuals and the epidemic dies out. Herd immunity, i.e. the proportion of individuals immune to a particular serotype of the virus, is therefore real, but not well studied. In acute, virgin soil epidemics, such as the DEN-1 outbreak in Cuba of 1977–1979, where some 44.5% of the urban population experienced infection in a single year, herd immunity was reflected by a similar prevalence of antibody in each age class.[125] This prevalence of dengue antibody does not mean that dengue 1 transmission stopped because 44.5% of the population was immune. Herd immunity applies only to persons at risk of dengue infection; for the most part these are individuals residing on premises with vector mosquitoes. However, in dengue-endemic areas where the premises index approaches 100%, there is ongoing circulation of multiple serotypes. This produces increasing dengue antibody prevalence with age. As a result, the nature of illness and the age-specific distribution of serious illness are a function of current and previous dengue activity. Here past activity (or lack of it) can influence the innate R_0 of the same serotype through the agency of herd immunity.

Sylvatic Cycle of Dengue Transmission

In the 1920s, studying the dengue 4 virus, Siler, Hall and Hitchens reported that most monkeys from lowland areas were immune to experimental dengue infection but that monkeys from the mountains of Luzon and others captured in Japan were susceptible, documenting an enzootic dengue virus cycle at lower altitudes in the Philippines.[15,126] Many workers reported the presence of dengue-neutralizing antibodies in wild-caught monkeys captured in SE Asian countries and sent to Europe or the United States for experimental studies.[127,128] In the Malaysian jungle, a maintenance cycle involving all four dengue viruses in several species of monkeys, probably transmitted by mosquitoes in the *Aedes niveus* group, has been documented.[129,130] From 1965 to 1975, Rudnick's group studied wild-caught and sentinel monkeys at ground level and in the canopy of mangrove swamp forests, freshwater peat swamp forests and primary dipterocarp forests. In these habitats *Macaca fascicularis* (crab eating macaque), *Presbytis melalophos* (banded leaf monkey), *Presbytis cristata* (silver leaf monkey) and *Macaca nemistrina* (pig-tailed macaque) circulated dengue-neutralizing antibodies. Caged, sentinel animals seroconverted to each of the four dengue viruses, but only in the canopy and not when held at ground level, and dengue viruses 1 and 2 were recovered from infected monkeys. The most abundant species biting monkeys in the canopy were members of a large and poorly defined group of mosquitoes, *Aedes* (*Finlaya*) *niveus*. Of more than 120 mosquito species captured in the Malaysian jungle, only *Ae niveus*, recovered from a monkey-baited canopy mosquito trap, yielded a strain of dengue 4. A similar enzootic cycle in West Africa, involving several species of *Aedes* mosquitoes, but restricted thus far to dengue 2, has been documented.[54,131] Dengue epizootics occur in the sub-Sudanese savannah across West Africa from Senegal to Ivory Coast. In Senegal, epizootics have been detected as a result of an ongoing program to monitor sylvatic yellow fever by attempting to isolate viruses from forest mosquitoes. In Senegal, epizootics involving the dissemination of dengue 2 viruses were recorded in 1974, 1980–82 and 1989–90.[132] In 1980, a large epizootic occurred in the Ivory Coast and Upper Volta, with 96 dengue 2 viruses recovered from *Aedes* (*Stegomyia*) *africanus*, *Ae luteocephalus*, *Ae opok*, *Ae taylori* and

Ae furcifer. The vertebrate cycle has not been well studied. Subhuman primates thought to be involved include *Erythrocebus patas* (patas monkeys), *Cercopithicus aethiops* (African green monkey) and *Papio anubis* (olive baboon).[54,55] In 1999–2000 an even larger epizootic occurred in an area of savannah and forest near Kedougou, Senegal.[133] Few infections of humans have been identified during these epizootics despite the fact that infected mosquitoes have been captured in or near villages.[131]

References

1. Gubler DJ. Variation among geographic strains of *Aedes albopictus* in susceptibility to infection with dengue viruses. *Am J Trop Med Hyg* 1976;25:319–325.
2. Rosen L, Roseboom LE, Gubler DJ, Lien JC, Chaniotis BN. Comparative susceptibility of mosquito species and strains to oral and parenteral infection with dengue and Japanese encephalitis viruses. *Am J Trop Med Hyg* 1985;34(3):603–615.
3. Trpis M, Haussermann W. Dispersal and other population parameters on *Aedes aegypti* in an African village and their possible significance in epidemiology of vector-borne diseases. *Am J Trop Med Hyg* 1986;35: 1263–1279.
4. Wang E, Ni H, Xu R, Barrett AD, Watowich SJ, Gubler DJ *et al.* Evolutionary relationships of endemic/epidemic and sylvatic dengue viruses. *J Virol* 2000;74(7):3227–3234.
5. Rudnick A, Lim TW. Dengue fever studies in Malaysia. *Inst Med Res Malaysia Bull* 1986;23:69–74.
6. Holmes EC, Twiddy SS. The origin, emergence and evolutionary genetics of dengue virus. *Infect Genet Evol* 2003;3:19–28.
7. Pant CP, Yasuno M. Field studies on the gonotrophic cycle of *Aedes aegypti* in Bangkok, Thailand. *J Med Entomol* 1973;10:219–223.
8. Hervy JP. Experience de marquage-lacher-recapture portant sur Aedes aegypti Linne, en zone de savane soudanienne ouest-africaine. 1. Le cycle gonotrophique. *Cah ORSTOM Ser Entomoil Med Parasitol* 1977;22:135–143.
9. Tonn RJ, Sheppard PM, Macdonald WW, Bang YH. Replicate surveys of larval habitats of *Aedes aegypti* in relation to dengue haemorrhagic fever in Bangkok, Thailand. *Bull World Health Organ* 1969;40:819–829.

10. Sheppard PM, MacDonald WW, Tonn RJ, Grab B. The dynamics of an adult population of *Aedes aegypti* in relation to dengue hemorrhagic fever in Bangkok. *J Anim Ecol* 1969;38:661–702.
11. Moore CG, Cline BL, Tiben ER. *Aedes aegypti* in Puerto Rico environmental determinants of larval abundance and relation to dengue virus transmission. *Am J Trop Med Hyg* 1978;27:1225–1231.
12. Trpis M, Haussermann W, Craig GB. Estimates of population size, dispersal, and longevity of domestic *Aedes aegypti* (*Diptera*: *Culicidae*) by mark–release–recapture in the villge of Shauri Moyo in eastern Kenya. *J Med Entomol* 1995;32:27–33.
13. Focks DA, Haile DG, Daniels E, Mount GA. Dynamic life table model for *Aedes aegypti* (*Diptera*: *Culicidae*): simulation results and validation. *J Med Entomol* 1993;30:1018–1028.
14. Briegel H, Kaiser C. Life-span of mosquitoes (*Diptera*: *Culicidae*) under laboratory conditions. *Gerontologia* 1973;19:240–249.
15. Siler JF, Hall MW, Hitchens AP. Dengue: its history, epidemiology, mechanism of transmission, etiology, clinical manifestations, immunity, and prevention. *Philippine J Sci* 1926;29:1–304.
16. Gubler DJ. Competitive displacement of *Aedes* (*Stegomyia*) *polynesiensis* (Marks) by *Aedes* (*Stegomyia*) *albopictus* (Skuse) in laboratory populations. *J Med Entomol* 1970;7:229–235.
17. de Moor PP, Steffens FE. A computer-simulated model of an arthropod-borne virus transmission cycle, with special reference to chikungunya virus. *R Soc Trop Med Hyg* 1970;64:927–934.
18. Clements AN, Paterson GD. The analysis of mortality and survival rates in wild populations of mosquitoes. *J Appl Ecol* 1981;18:373–399.
19. Smith CEG. The significance of mosquito longevity and blood-feeding behaviour in the dynamics of arbovirus infections. *Med Biol* 1975;53: 288–294.
20. Watts DM, Burke DS, Harrison BA, Whitmire RE, Nisalak A. Effect of temperature on the vector efficiency of *Aedes aegypti* for dengue 2 virus. *Am J Trop Med Hyg* 1987;36(1):143–152.
21. Gould DJ, Mount GA, Scanlon JE, Ford HR, Sullivan MF. Ecological control of dengue vectors on an island in the gulf of Thailand. *J Med Entomol* 1970;4:499–505.

22. Yasuno M, Tonn RJ. A study of biting habits of *Aedes aegypti* in Bangkok, Thailand. *Bull World Health Organ* 1970;43:319–325.
23. Chen WJ, Chen ER. Fate of ingested immunoglobulin in the mosquito *Aedes aegypti* (*Diptera*: *Culicidae*). *Kaohsiung J Med Sci* 1990;6:606–613.
24. Wolfinsohn M, Galun R. A method for determining the flight range of *Aedes aegypti* (Linn.). *Bull Res Council Israel* 1953;2:433–436.
25. Reiter P, Amador MA, Anderson RA, Clark GG. Short report: dispersal of *Aedes aegypti* in an urban area after blood feeding as demonstrated by rubidium-marked eggs. *Am J Trop Med Hyg* 1995;52(2):177–179.
26. Morlan HB, Hayes RO. Urban dispersal and activity of *Aedes aegypti. Mosq News* 1958;18:137–144.
27. Smith CEG. A localized outbreak of dengue fever in Kuala Lumpur: epidemiological and clinical aspects. *Med J Malaysia* 1956;10:289–303.
28. Neff JM, Morris L, Gonzalez-Alcover R, Coleman PH, Lyss, SB, Negron H. Dengue fever in a Puerto Rican community. *Am J Epidemiol* 1967;86:162–184.
29. Craig S, Thu HM, Lowry K, Wanng X-F, Holmes EC, Aaskov J. Diverse dengue type 2 virus populations contain recombinant and both parenteral viruses in a single mosquito host. *J Virol* 2003;77:4463–4467.
30. Wang W-K, Sung T-L, Lee C-N, Lin T-Y, King CC. Sequence diversity of the capsid gene and the nonstructural gene of NS2B of dengue-3 virus *in vivo*. *Virology* 2002;303:181–191.
31. Wang WK, Lin SR, Lee CM, King CC, Chang SC. Dengue type 3 virus in plasma is a population of closely related genomes: quasispecies. *J Virol* 2002;76(9):4662–4665.
32. Wittke V, Robb TE, Thu HM, Nisalak A, Nimmannitya S, Kalayanarooj S *et al*. Extinction and rapid emergence of strains of dengue 3 virus during an interepidemic period. *Virology* 2002;301:148–156.
33. Jenkins GM, Holmes EC. The extent of codon usage bias in human RNA viruses and its evolutionary origin. *Virus Res* 2003;92:1–7.
34. Woelk CH, Holmes EC. Reduced positive selection in vector-borne RNA viruses. *Mol Biol Evol* 2002;19:2333–2336.
35. Novella IS, Hershey CL, Escrmis C, Domingo E, Holland DJ. Lack of evolutionary stasis during alternating replication of an arbovirus in insect and mammalian cells. *J Mol Biol* 1999;287:459–465.

36. Weaver SC, Rico-Hesse R, Scott TW. Genetic and fitness changes accompanying adaptation of an arbovirus to vertebrate and invertebrate cells. *J Virol* 1992;73:99–117.
37. Holmes EC. Patterns of intra- and interhost nonsynonymous variation reveal strong purifying selection in dengue virus. *J Virol* 2003;77: 11296–11298.
38. Newton EAC, Reiter P. A model of the transmission of dengue fever with an evaluation of the impact of ultra-low volume (ULV) insecticide applications on dengue epidemics. *Am J Trop Med Hyg* 1992;47(6):709–720.
39. Rodriguez-Figueroa L, Rigau-Perez JG, Suarez EL, Reiter P. Risk factors for dengue infection during an outbreak in Yanes, Puerto Rico in 1991. *Am J Trop Med Hyg* 1995;52(6):496–502.
40. Ghosh SN, Pavri KM, Singh KRP, Sheikh BH, D'Lima LV, Mahadev PVM *et al.* Investigations on the outbreak of dengue fever in Ajmer city, Rajasthan state in 1969. Part 1. Epidemiological, clinical and virological study of the epidemic. *Indian J Med Res* 1974;62:511–522.
41. Chen WJ, Hwang JS, Guo YJ. Ecology and control of dengue vector mosquitoes in Taiwan. Kaohsiung *J Med Sci* 1994;10:578–587.
42. Chan YC, Ho BC, Chan KL. *Aedes aegypti* (L.) and *Aedes albopictus* (Skuse) in Singapore city. 5. Observations in relation to dengue haemorrhagic fever. *Bull World Health Organ* 1971;44:651–658.
43. Smith TJ, Winter PE, Nisalak A, Udomsakdi S. Dengue control on an island in the Gulf of Thailand. II. Virological studies. *Am J Trop Med Hyg* 1971;20(5):715–719.
44. Ilkal MA, Dhanda V, Hassan MM, Mavale M, Mahadev PV, Shetty PS *et al.* Entomological investigations during outbreaks of dengue fever in certain villages in Maharashtra state. *Indian J Med Res* 1991;93:174–178.
45. Focks DA, Daniels E, Haile DG, Keesling JE. A simulation model of the epidemiology of urban dengue fever: literature analysis, model development, preliminary validation, and samples of simulation results. *Am J Trop Med Hyg* 1995;53(5):489–506.
46. Sabin AB. Research on dengue during World War II. *Am J Trop Med Hyg* 1952;1:30–50.
47. Vithanomsat S, Watts DM, Nisalak A, Tharavanij S. The relationship of temperature to the replication and virulence of dengue viruses. *J Med Assoc Thai* 1983;66(9):530–541.

48. Gubler DJ, Nalim S, Tan R, Saipan H, Saroso JS. Variation in susceptibility to oral infection with dengue viruses among geographic strains of *Aedes aegypti*. *Am J Trop Med Hyg* 1979;28(6):1045–1052.
49. Jupp PG. The susceptibility of four South African species of *Culex* to West Nile and Sindbis viruses by two different infecting methods. *Mosq News* 1976;36:166–173.
50. Meyer RP, Hardy JL, Presser SB. Comparative vector competence of *Culex tarsalis* and *Culex quinquefasciatus* from the Coachella, Imperial, and San Joaquin Valleys of California for St. Louis encephalitis virus. *Am J Trop Med Hyg* 1983;32:305–311.
51. Bancroft WH, Scott RM, Brandt WE, McCown JM, Eckels KH, Hayes DE *et al*. Dengue-2 vaccine: infection of *Aedes aegypti* mosquitoes by feeding on viremic recipients. *Am J Trop Med Hyg* 1982;31(6): 1229–1231.
52. Gubler DJ, Reed D, Rosen L, Hitchcock JC, Jr. Epidemiological, clinical, and virologic observations on dengue in the kingdom of Tonga. *Am J Trop Med Hyg* 1978;27(3):581–589.
53. Effler PV, Pang L, Kitsutani P, Vorndam V, Nakata M, Ayers T *et al*. Dengue fever, Hawaii, 2001–2002. *Emerg Infect Dis* 2005;11:742–749.
54. Cordellier K, Dochite B, Roche JC, Monteny N, Diaco B, Akoliba P. Circulation selvatique du virus Dengue 2 en 1980, dans les savanes sub-soudeaniennes de Cote d'Ivoire. *Cah ORSTOM Serie Entomol Med Parasitol* 1983;21:165–179.
55. Roche JC, Cordellier R, Hervy JP, Digoutte JP, Monteny N. Isolment de 96 souches de virus dengue 2 a partir de mostiques captures en Cote-D'Ivoire et Haute-Volta. *Ann Virol* (*Inst Pasteur*) 1983;134:233–244.
56. Hervy JP, Legros F, Roche JC, Monteny N, Diaco B. Circulation du virus Dengue 2 dans plusiers milieux boises des savanes soudaniennes de la region de Bobo-Dioulasso (Burkina Faso). *Cah ORSTOM Ser Entomol Med Parasitol* 1984;22:135–143.
57. Rudnick A, Lim TW. Dengue fever studies in Malaysia. *Inst Med Res Malaysia Bull* 1978;23:127–147.
58. Gubler DJ, Novak RJ, Vergne E, Colon NA, Velez M, Fowler J. *Aedes* (*Gymnometopa*) *mediovittatus* (*Diptera*: *Culicidae*), a potential maintenance vector of dengue viruses in Puerto Rico. *J Med Entomol* 1985;22:469–475.

59. Freier JE, Rosen L. Vertical transmission of dengue viruses by *Aedes mediovittatus*. *Am J Trop Med Hyg* 1988;39(2):218–222.
60. Freier JE, Grimstad PR. Transmission of dengue virus by orally infected *Aedes triseriatus*. *Am J Trop Med Hyg* 1983;32:1429–1434.
61. Simmons JS, St. John JH, Reynolds FHK. Experimental studies of dengue. *Philippine J Sci* 1931;44(1–2):1–252.
62. Miller BR, Beaty BJ, Aitken THG, Eckels KH, Russell PK. Dengue-2 vaccine oral infection, transmission and lack of evidence for reversion in the mosquito *Aedes aegypti*. *Am J Trop Med Hyg* 1982;31(6): 1232–1237.
63. Zhao ZG. Epidemiological study on dengue fever in Hainan Island. *Chung-Hua Liu Hsing Ping Hsueh Tsa Chih* (*Chin J Epidemiol*). 1986;7:29–32.
64. Halstead SB, Scanlon J, Umpaivit P, Udomsakdi S. Dengue and chikungunya virus infection in man in Thailand, 1962–1964. IV. Epidemiologic studies in the Bangkok metropolitan area. *Am J Trop Med Hyg* 1969;18(6):997–1021.
65. Fine PEM. Herd immunity: history, theory, practice. *Epidemiol Rev*. 1993;15:265–302.
66. Anderson RM, May RM. *Infectious Diseases of Humans: Dynamics and Control*, 2nd ed. Oxford University Press, 1991.
67. Dietz K. The estimation of the basic reproductive number for infectious diseases. *Stat Methods Med Res* 1993;2:23–41.
68. Koopman JS, Prevots DR, Vaca Marin MA, Gomez Dantes H, Zarate Aquino ML, Longini IM, Jr. *et al*. Determinants and predictors of dengue infection in Mexico. *Am J Epidemiol* 1991;133:1168–1178.
69. Marques CA, Forattini OP, Massad E. The basic reproduction number for dengue fever in Sao Paulo state, Brazil — 1990–1991 epidemic. *Trans R Soc Trop Med Hyg* 1994;88:58–59.
70. Ferguson NM, Donnelly CA, Anderson RM. Transmission dynamics and epidemiology of dengue: insights from age-stratified sero-prevalence surveys. *Philos Trans R Soc Lond B Biol Sci* 1999;354(1384):757–768.
71. Favier C, Degallier N, Rosa-Freitas MG, Boulanger JP, Costa Lima JR, Luitgards-Moura JF *et al*. Early determination of the reproductive number for vector-borne diseaes: the case of dengue in Brazil. *Trop Med Int Health* 2006;11:332–340.

72. Massad E, Burattini MN, Coutinho FA, Lopez LF. Dengue and the risk of urban yellow fever reintroduction in Sao Paulo state, Brazil. *Rev Saude Publica* 2003;37(4):477–484.
73. Favier C, Schmit D, Muller-Graf CDM, Cazelles B, Degallier N, Mondet B *et al.* Influence of spatial heterogeneity on an emerging infectious disease: the case of dengue epidemics. *Proc Biol Sci* 2005;272: 1171–1177.
74. Mas P. Dengue fever in Cuba in 1977: some laboratory aspects. In: Go W (ed.) *Dengue in the Caribbean, 1977; 1978–1979*. PAHO Montego Bay, Jamaica, 1978, pp. 1–186.
75. Bouma MJ, Kovats RS, Goubet SA, Cox JS, Haines A. Global assessment of El Nino's disaster burden. *Lancet* 1997;350:1435–1438.
76. Hales S, Weinstein P, Souares Y, Woodward A. El Nino and the dynamics of vectorborne disease transmission. *Environ Health Perspect* 1999;107(2):99–102.
77. Hales S, Weinstein P, Woodward A. Dengue fever epidemics in the South Pacific: driven by El Nino Southern Oscillation? [letter]. *Lancet* 1996;348:1664–1665.
78. Burke DS, Carmichael A, Focks DA. *Under the Weather: Exploring the Linkages between Climate, Ecosystems, Infectious Disease and Human Health.* National Research Council, Washington, D.C., 2001.
79. Bouma MJ, van der Kaay HJ. The El Nino Southern Oscillation and the historic malaria epidemics on the Indian subcontinent and Sri Lanka: an early warning system. *Trop Med Int Health* 1996;1:86–96.
80. Bouma MJ, Dye C. Cycles of malaria associated with El Nino in Venezuela. *J Am Med Assoc* 1997;278:1772–1774.
81. Akhtar R, McMichael AJ. Rainfall and malaria outbreaks in western Rajasthan. *Lancet* 1996;348:1457–1458.
82. Christie J. Remarks on "kidinga Pepo", a peculiar form of exantematous disease. *Br Med J* 1872;1:577–579.
83. Carey DE. Chikungunya and dengue: a case of mistaken identity? *J Hist Med Allied Sci* 1971;26(3):243–62.
84. Reiter P, Sprenger D. The used tire trade: a mechanism for the worldwide dispersal of container breeding mosquitoes. *J Am Mosq Control Assoc* 1987;3:494–501.

85. Cummings DA, Irizarry RA, Huang NE, Endy TP, Nisalak A, Ungchusak K *et al.* Travelling waves in the occurrence of dengue haemorrhagic fever in Thailand. *Nature* 2004;427:344–347.
86. Kouri G, Guzman MG, Bravo J. Hemorrhagic dengue in Cuba: history of an epidemic. *Bull Pan Am Health Organ* 1986;20:24–30.
87. Guzman MG, Deubel V, Pelegrino JL, Rosario D, Marrero M, Sariol C *et al.* Partial nucleotide and amino acid sequences of the envelope and the envelope/nonstructural protein-1 gene junction of four dengue-2 virus strains isolated during the 1981 Cuban epidemic. *Am J Trop Med Hyg* 1995;52:241–246.
88. Gubler D. The emergence of epidemic dengue fever and dengue hemorrhagic fever in the Americas: a case of failed public health policy. *Rev Panam Salud Publica* 2005;17(4):221–224.
89. Carrington CV, Foster JE, Pybus OG, Bennett SN, Holmes EC. The invasion and maintenance of dengue virus type 2 and type 4 in the Americas. *J Virol* 2005;79:14680–14687.
90. PAHO. *Dengue in the Caribbean, 1977; 1978–1979*. PAHO Montego Bay, Jamaica 1978, pp. 1–186.
91. Campione-Piccardo J, Ruben M, Vaughan H, Morris-Glasgow V. Dengue viruses in the Caribbean: twenty years of dengue virus isolated from the Caribbean Epidemiology Centre. *West Indian Med J* 2003;52: 191–198.
92. A-Nuegoonpipat A, Berlioz-Arthaud A, Chow VT, Endy T, Lowry K, le QM *et al.* Sustained transmission of dengue virus type 1 in the Pacific due to repeated introductions of different Asian strains. *Virology* 2004;329:505–512.
93. Kiedrzynski T, Souares Y, Stewart T. Dengue in the Pacific: an updated story. *Pac Health Dialog* 1998;5:129–136.
94. Sittisombut N, Sistayanarain A, Cardosa MJ, Salminen M, Damrongdachakul S, Kalayanarooj S *et al.* Possible occurrence of a genetic bottleneck in dengue serotype 2 viruses between the 1980 and 1987 epidemic seasons in Bangkok, Thailand. *Am J Trop Med Hyg* 1997;57(1): 100–108.
95. Bennett SN, Holmes EC, Chirivella M, Rodriguez DM, Beltran M, Vorndam V *et al.* Selection-driven evolution of emergent dengue virus. *Mol Biol Evol* 2003;20:1650–1658.

96. Steinhauer DA, Domingo E, Holland JJ. Lack of evidence for proofreading mechanisms associated with an RNA polymerase. *Gene* 1992;122:281–288.
97. Twiddy SS, Farrar JF, Chau NV, Wills B, Gould EA, Gritsun T *et al.* Phylogenetic relationships and differential selection pressures among genotypes of dengue-2 virus. *Virology* 2002;298:63–72.
98. Thu HM, Lowry K, Myint TT, Shwe TN, Han AM, Khin KK *et al.* Myanmar dengue outbreak associated with displacement of serotypes 2, 3 and 4 by dengue 1. *Emerg Infect Dis* 2004;10:593–597.
99. Ming CK, Thein S, Thaung U, Tin U, Myint KS, Swe T *et al.* Clinical and laboratory studies on haemorrhagic fever in Burma, 1970–72. *Bull World Health Organ* 1974;51:227–235.
100. Bush RM, Bender CA, Subbarao K, Cox NJ, Fitch WM. Predicting the evolution of human influenza A. *Science* 1999;286:1921–1925.
101. Beaty BJ, Trent DW, Roehrig JT. Virus variation and evolution. In: Monath TP (ed.) *The Arboviruses: Epidemiology and Ecology*. CRC, Boca Raton, Florida, 1988, pp. 59–85.
102. Strauss JH, Strauss EG. Evolution of RNA viruses. *Annu Rev Microbiol* 1988;42:657–683.
103. Jenkins GM, Rambaut A, Pybus OG, Holmes EC. Rates of molecular evolution in RNA viruses: a quantitative phylogenetic analysis. *J Mol Evol* 2002;54:152–161.
104. Gubler DJ. Dengue and dengue hemorrhagic fever in the Americas. *PR Health Sci J* 1987;6:107–111.
105. Guzman MG, Kouri G, Halstead SB. Do escape mutants explain rapid increases in dengue case-fatality rates within epidemics? *Lancet* 2000;355(9218):1902–1903.
106. Rodriguez-Roche R, Alvarez M, Gritsun T, Rosario D, Halstead SB, Kouri G *et al.* Dengue virus type 2 in Cuba, 1997: conservation of E gene sequence in isolates obtained at different times during the epidemic. *Arch Virol* 2005;150:415–425.
107. Rodriguez-Roche R, Alvarez M, Gritsun T, Halstead SB, Kouri G, Gould EA *et al.* Virus evolution during a severe dengue epidemic in Cuba, 1997. *Virology* 2005;334:154–157.
108. Cummings DAT, Schwartz IB, Billings L, Schaw LB, Burke DS. Dynamic effects of antibody-dependent enhancement on the fitness of viruses. *Proc Natl Acad Sci USA* 2005;102:15259–15264.

109. Fischer DB, Halstead SB. Observations related to pathogenesis of dengue hemorrhagic fever. V. Examination of age-specific sequential infection rates using a mathematical model. *Yale J Biol Med* 1970;42:329–349.
110. Guzman MG, Kouri G, Bravo J, Valdes L, Vazquez S, Halstead SB. Effect of age on outcome of secondary dengue 2 infections. *Int J Infect Dis* 2002;6:118–124.
111. Guzman MG, Kouri G, Valdes L, Bravo J, Alvarez M, Vazquez S *et al.* Epidemiologic studies on dengue in Santiago de Cuba, 1997. *Am J Epidemiol* 2000;152(9):793–799.
112. Guzman MG, Kouri G, Valdes L, Bravo J, Vazquez S, Halstead SB. Enhanced severity of secondary dengue 2 infections occurring at an interval of 20 compared with 4 years after dengue 1 infection. *PAHO J Epidemiol* 2002;81:223–227.
113. Halstead SB, Casals J, Shotwell H, Palumbo N. Studies on the immunization of monkeys against dengue. Protection derived from single and sequential virus infections. *Am J Trop Med Hyg* 1973;22(3):365–374.
114. Halstead SB, Palumbo NE. Studies on the immunization of monkeys against dengue: II. Protection following inoculation of combinations of viruses. *Am J Trop Med Hyg* 1973;22(3):375–381.
115. Gibbons RV, Kalanarooj S, Jarman RG, Nisalak A, Vaughn DW, Endy TP *et al.* analysis of repeat hospital admissions for dengue to estimate the frequency of third or fourth dengue infections resulting in admissions and dengue hemorrhagic fever, and serotype sequences. *Am J Trop Med Hyg* 2007;77(5):910–913.
116. Gubler DJ. Dengue and dengue hemorrhagic fever: its history and resurgence as a global public health problem. In: Gubler DJ, Kuno G (eds.) *Dengue and Dengue Hemorrhagic Fever*. CAB, New York 1997, pp. 1–22.
117. Halstead SB. Epidemiology of dengue and dengue hemorrhagic fever. In: Gubler DJ, Kuno, G (eds.) *Dengue and Dengue Hemorrhagic Fever*. CAB, Wallingford, 1997, pp. 23–44.
118. Nisalak A, Endy TP, Nimmannitya S, Kalayanarooj K, Thisyakorn U, Scott RM *et al.* Serotype-specific dengue virus circulation and dengue disease in Bangkok, Thailand from 1973 to 1999. *Am J Trop Med Hyg* 2003;68:191–202.
119. Adams B, Holmes EC, Zhang C, Mammen MP, Jr., Nimmannitya S, Kalayanarooj S *et al.* Cross-protective immunity can account for the

alternating epidemic pattern of dengue virus serotypes circulating in Bangkok. *Proc Natl Acad Sci USA* 2006;103(38):14234–14239.

120. Zhou Y, Mammen MP, Jr, Klungthong C, Chinnawirotpisan P, Vaughn DW, Nimmannitya S *et al.* Comparative analysis reveals no consistent association between the secondary structure of the 3′-untranslated region of dengue viruses and disease syndrome. *J Gen Virol* 2006;87(Pt 9): 2595–2603.
121. Zhang C, Mammen MMJ, Chinnawirotpisan P, Klungthong C, Rodpradit P, Monkongdee P *et al.* Clade replacements in dengue virus serotypes 1 and 3 are associated with changing serotype prevalence. *J Virol* 2005;79: 15123–15130.
122. Zhang C, Mammen MP, Jr, Chinnawirotpisan P, Klungthong C, Rodpradit P, Nisalak A *et al.* Structure and age of genetic diversity of dengue virus type 2 in Thailand. *J Gen Virol* 2006;87(Pt 4):873–883.
123. Klungthong C, Zhang C, Mammen MP, Jr, Ubol S, Holmes EC. The molecular epidemiology of dengue virus serotype 4 in Bangkok, Thailand. *Virology* 2004;329(1):168–179.
124. Nam VS, Yen NT, Holynska M, Reid JW, Kay BH. National progress in dengue vector control in Vietnam: survey for *Mesocyclops* (*Copepoda*), *Micronecta* (*Corixidae*), and fish as biological control agents. *Am J Trop Med Hyg* 2000;62(1):5–10.
125. Kouri GP, Guzman MG, Bravo JR, Triana C. Dengue haemorrhagic fever / dengue shock syndrome: lessons from the Cuban epidemic, 1981. *Bull World Health Organ* 1989;67:375–380.
126. Halstead SB. Etiologies of the experimental dengues of Siler and Simmons. *Am J Trop Med Hyg* 1974;23(5):974–982.
127. Hammon WM, Schrack WD, Sather GE. Serological survey for arthropod-borne viruses in the Philippines. *Am J Trop Med Hyg* 1958;7:323–328.
128. Smith CEG. The distribution of antibodies to Japanese encephalitis, dengue, and yellow fever viruses in five rural communities in Malaya. *Trans R Soc Trop Med Hyg* 1958;52:237–242.
129. Rudnick A. Studies of the ecology of dengue in Malaysia: a preliminary report. *J Med Entomol* 1965;2:203–208.
130. Rudnick A. Ecology of dengue virus. *Asian J Infect Dis* 1978;2:156–160.
131. Saluzzo JF, Cornet M, Adam C, Eyraud M, Digoutte JP. [Dengue 2 in eastern Senegal: serologic survey in simian and human populations; 1974–85]

Dengue 2 au Senegal oriental: enquete serologique dans les populations simiennes et humaines, 1974–1985. *Bull Soc Pathol Exot Filiales* 1986;79:313–322.

132. Diallo M, Thonnon J, Traore-Lamizana M, Fontenille D. Vectors of chikungunya virus in Senegal: current data and transmission cycles. *Am J Trop Med Hyg* 1999;60(2):281–286.
133. Diallo M, Ba Y, Sall AA, Diop OM, Ndione JA, Mondo M *et al.* Amplification of the sylvatic cycle of dengue virus type 2, Senegal, 1999–2000: entomologic findings and epidemiologic considerations. *Emerg Infect Dis* 2003;9:362–367.
134. Halstead SB. Dengue virus–mosquito interactions. *Ann Rev Entomol* 2008;53:273–291.

4

Resistance to Infection

David W. Vaughn, LTC John M. Scherer and Wellington Sun

Introduction

The current dengue pandemic is linked to the resurgence of the *Aedes aegypti* mosquito, population growth in tropical countries and the rapid movement of dengue viruses in mosquito vectors and infected people.[1,2] Vector control has been largely ineffective and the disease burden of dengue fever and dengue hemorrhagic fever (DHF) now rivals highly endemic diseases such as tuberculosis and malaria.[3,4] This chapter will summarize the current understanding of factors present in human hosts which prevent or mitigate dengue virus (DENV) infections or disease following bites by infected mosquitoes. Included is a brief review of the status of dengue vaccine development.

Innate Immunity

The innate immune system comprises a wide variety of nonspecific defensive measures to protect against pathogens. The most commonly described of these defenses include barriers, such as the skin, phagocytic cells, natural killer (NK) cells, acute phase proteins, chemokines and certain cytokines. In DENV infections, complement activation,[5–10] NK cell activation,[11] and production of tumor necrosis factor α[12,13] and interleukin 6[12] and interferon α[14] are all increased. Unfortunately, these observations in natural infection only demonstrate that an innate immune response to DENV occurs, but do not provide insight into the protective nature of these responses. While macrophages have been shown to become infected with DENV,[15] considerable research has been undertaken to understand how dengue viruses interact and influence a relatively recently identified class of antigen-presenting cells known as dendritic cells (DC's).

It is now recognized that dendritic cells are an important interface between the innate and adaptive immune responses. There is direct evidence that DC's in skin are infected by dengue viruses following mosquito bites. DC's are distributed as partially undifferentiated immune cells at strategic sites where pathogens may enter. There are at least four types of DC's: CD14+ blood monocyte-derived DC's (moDC's), CD34+ hematopoietic progenitor cell (HPC)–derived dermal or interstitial DC's (DDC-IDC's), CD34+ HPC-derived Langerhans cells (LC's) and plasmacytoid DC's (PDC's).[16] Both LC's and DDC-IDC's express typical myeloid-type DC markers, and LC's also express Langerin. Plasmacytoid DC's are instrumental in antiviral innate immunity and in shaping Th 1 adaptive immune responses. Peripheral blood DC's make alpha and beta IFN in response to microbial stimuli. LC's and immature DC's (iDC's) act as sentinels sensing the antigenic microenvironment and capturing antigens. On becoming infected with DENV, as demonstrated by infection of human cadaveric skin, iDC's become mature and migrate out of the dermis.[17,18] They travel to the paracortex of draining lymph nodes, where they present peptides in the context of major histocompatibility complex class I and class II molecules to T cells.[19] Interstitial DC's, but not LC's, have the ability to take up large amounts of antigens by the mannose receptors and to produce IL-10, which may contribute to naïve B cell activation and IgM production in the presence of CD40 ligand and IL-2. Infection of immature myeloid CD1a DC's results in activation and maturation of infected and neighboring cells. Activated DC's secrete TNF alpha and IFN alpha, and express maturation markers, such as HLA-DR, CD11b and CD83.[20,21]

On the surface of DC's is a DC-specific ICAM-3 grabbing nonintegrin (DC-SIGN), a type II membrane protein with a C-type lectin ectodomain. DC-SIGN binds to high mannose oligosaccharides. DENV grown in mosquitoes incorporate mannose at envelope protein positions

Asn-67 and Asn-153 and bind to DC's using DC-SIGN.[22,23] As an indication of the importance of this infection pathway, it has been observed that a 336A/G polymorphism in DC-SIGN is associated with a reduced risk of severe dengue disease in a case control study in Thailand.[24] DC's also contain toll-like receptor (TLR) molecules. Interactions between TLR's and microbial ligands trigger the activation of the nuclear transcription factor, inducing the expression of cytokine genes. TLR's are evolutionarily conserved proteins, of which 10 have been recognized in humans.[25] More than one TLR may be involved in recognizing DENV. Postinfection signaling involves the Toll/IL-1 receptor-like domain, which initiates a complex cascade of activation proteins, which in turn induce inflammatory cytokines.[16]

Not only has DENV been shown to infect and activate DC's, but it has been observed that the absolute number of circulating premyeloid DC's decreases early in the acute phase of DHF compared with patients with DF.[26] The exact reason for this decrease is not known and may be related to premyeloid DC's migrating to the lymph nodes, a decrease in their production in the bone marrow, apoptosis, or a combination of these factors.

One of the major secreted products induced by viral infection are type I interferons (IFN alpha and beta). IFN alpha is produced by many cell types to include DC's. IFN alpha is induced rapidly, usually within hours of viral infection, and induces an antiviral state by a wide range of mechanisms, from activating RNA nuclease genes to indirectly inducing T cell proliferation. *In vitro*, alpha, beta and gamma IFN can inhibit DENV infection in a large number of cell types,[27] and mice deficient in IFN receptors succumbed to infection with a mouse-adapted DENV.[28] In spite of this, DENV can achieve high titers in blood of humans even though there is an induction of high titers of alpha and gamma IFN's, suggesting that the virus has evolved mechanisms to counter the IFN response.[14,29] Nonstructural proteins, NS2A or NS4B, may be responsible for blocking the IFN receptor-janus/signal transducer and activator of the transcription signaling pathway in infected cells.[30] There is early evidence that DENV infection of Fc-receptor-bearing cells when complexed with dengue antibodies may suppress the interferon system.[31]

NK cells are capable of rapidly killing target cells (within hours). They constitute 5–15% of all circulating lymphocytes. They are rapidly recruited to infected foci by chemoattractants released from virus-infected cells and by activated macrophages or DC's that release alpha and beta IFN.[32] NK cells can kill virus-infected cells by using cytotoxic granules or by recognizing and inducing lysis of antibody-coated target cells using antibody-binding receptors, CD16 and CD56. NK activation is regulated through two families of receptors: the HLA class I specific receptors in the immunoglobulin Ig superfamily and an NK receptor specific for HLA-E.[33] Any ligand on an infected cell attributed directly or indirectly to pathogens may be sufficient to trigger cytotoxicity in NK cells.[34] In species that are relatively resistant to dengue infection, NK cells may help to contain early infection.[35] Midcourse of human dengue infections[11] and later, NK cells may contribute to the killing of virus-infected cells.[36] Infection by flaviviruses may lead to an upregulation of MHC class I on the surface of mouse cells.[16] Other viruses use a similar strategy to evade early NK cell recognition of non-MHC I-expressing cells.[37] NK cells are also involved in killing infected cells by antibody-dependent cellular cytotoxicity (ADCC, discussed below). Based on these observations, it is apparent that NK cells play a role in containing DENV infection throughout the course of infection.

Finally, new discoveries of DENV interaction with components of the complement system may lead to a better understanding of DENV pathogenicity. Studies in the 1970's documented the profound activation of complement in children with DHF.[38] Complement is activated both in children experiencing a second DENV infection and in infants who develop DHF during a primary DENV infection,[39] and complement components C3a and C5a are known to mediate vascular permeability. Interest has been rekindled and recent studies have demonstrated a possible role for dengue NS1 protein in activating complement by both the classical and alternative pathways.[40] NS1 circulates in disease-severity-related concentrations in acute phase blood of humans with dengue infection.[41]

Finally, apoptosis is a mechanism by which DENV infection causes cell death.[42] This might result in promoting the spread of DENV either by phagocytosis of infected cells or by lysing cells and releasing the virus early after replication. Dengue virus M and NS3 proteins have been shown to have proapoptotic activities.[43,44] In addition to triggering apoptosis, antibodies to DENV NS1 protein also bind to endothelial cells and platelets — a mechanism that some workers have postulated contributes to dengue disease.[45,46]

Antibody Responses in Natural Infection

Overview

Neutralizing antibodies to viruses serve as the most commonly used correlate of protection.[47] Dengue viruses are no exception. While there are instances where high-titer neutralizing antibody did not afford protection against reinfection with dengue viruses[48–50] or protection was observed in the absence of detectable neutralizing antibody,[51] passive transfer of neutralizing antibody has been sufficient to afford protection against DENV infection (discussed below). On the other hand, adoptive intravenous transfer of immunized mouse spleen cells to naïve mice did not protect against lethal challenge.[52] While neutralizing antibody may provide long-lasting protection against reinfection when challenged with homologous dengue viruses, T cells play an essential role in the formation of antibodies as immunoglobulin class-switching occurs only when T cell help is present.[53] Therefore, a logical goal of dengue vaccine development would seem to be to devise products that elicit humoral and cellular immune responses. This chapter will focus on the role of antibody in protection. Chapter 8 will focus on T cell responses.

Interpretation of antibody responses following DENV infection requires knowledge of previous flavivirus exposure. As dengue viruses are found in countries where nondengue flaviviruses circulate, prior infection or immunization against other flaviviruses, such as yellow fever virus or Japanese encephalitis virus, is commonly observed. If a flavivirus-immune individual is infected with DENV, an anamnestic response to antigenic determinants shared between the infecting viruses develops.[54–57] Many observations on anamnestic responses have been made in clinical DENV vaccine studies using yellow-fever-naïve and- immune individuals.[58,59] In one study, anamnestic responses to a DENV type 2 (DENV-2) vaccine were observed in yellow-fever-immune volunteers. The yellow-fever-immune group developed a geometric mean titer (GMT) of neutralizing antibodies five times greater than the GMT observed in DENV-2-immunized naïve volunteers. Also, the yellow-fever-naïve volunteers that seroconverted lost neutralizing antibodies to below a titer of 1:10 six months after vaccination, whereas yellow-fever-immune volunteers had detectable neutralizing antibody titers that persisted for at least three years.[58] The speed and level of IgG antibody raised may depend on the number of shared epitopes between the previous and current infecting viruses. If the previous infection was a more distantly related flavivirus, the ratio of IgM to IgG antibody may be nearly equal. When infecting viruses are closely related, such as sequential DENV infections, the IgG antibody response may be proportionally more robust.[60]

Antibody response and specificity in natural infection

Serologically, a primary DENV infection results in detectable levels of IgM antibodies by the third afebrile day following defervescence in clinically apparent cases. These IgM antibodies persist for 1–2 months following infection. As mentioned above, IgM antibodies predominate over IgG antibodies in primary DENV infections whereas IgG antibody levels are similar to or exceed IgM antibody levels early in the course of secondary DENV infections.[55] The IgM antibodies are predominately directed against the E protein; however, IgM antibody can be detected against nonstructural proteins as well.[61] While IgM antibodies produced during a DENV infection have the highest activity against the infecting DENV type, the IgM antibodies are also cross-reactive with other dengue viruses.[62] The predominant IgG antibodies that appear in the convalescent phase following primary natural DENV infection are against the premembrane (PrM) and envelope (E) proteins. While IgG antibodies against the capsid (C) and nonstructural (NS) proteins 1, 3 and 5 (NS1, NS3 and NS5) are sometimes detected after primary infection, they typically appear after a second infection of a different DENV.[63–67] Antibodies produced during DENV infection provide short-lived protection against infection with a heterologous DENV. In experimental inoculations conducted by Sabin,[68] cross-protection against disease following challenge with a second DENV lasted at least two months, though Sabin expressed the opinion that disease was somewhat modified when a second virus was administered up to nine months after infection with a first DENV. These studies were likely in the sequence of DENV-1 followed by DENV-2. A decline of cross-protection with time, as measured by the presence or absence of postchallenge viremia, was also observed in a monkey study.[69] Overall, neutralizing IgG antibodies produced to the infecting DENV last for decades. Such antibodies are believed to provide lifelong protection against reinfection from the same DENV.[70]

Other types of cross-reactive antibodies are also produced in DENV infections. Of particular interest are antibodies that may bind components of the coagulation cascade. In one study, 75% of 40 acutely ill Thai patients infected with DENV-1, -2, -3, or -4 produced anti-E antibodies that cross-reacted with a synthetic fibrinogen fragment by ELISA.[71] This cross-reaction was not detected in 15 other patients diagnosed with Japanese encephalitis. In a second study, human plasminogen cross-reactive antibodies were found in acute and convalescent serum of children with primary and secondary DENV infections. Plasminogen cross-reactive antibodies were twice as likely to be found in secondary versus primary DENV-infected children, but these cross-reactive antibodies did not correlate with DHF/DSS or thrombocytopenia.[72] While the role of the cross-reactive antibody to plasminogen is unclear, it is of interest that the E protein of DENV has been shown to activate the fibrinolytic process *in vitro*.[73] Another potentially harmful cross-reactive dengue antibody has been identified in mice. Several monoclonal antibodies directed against the NS1 protein cross-reacted with human fibrinogen, platelets and endothelial cells.[74] These same antibodies were also capable of producing hemorrhage in mice.[75] In summary, the role of these antibodies in human disease still remains to be established.

IgG subclasses

IgG subclass may also be important in contributing to protection versus disease following DENV infection. It is important to note that human, mouse and rat IgG subclasses differ in function. In humans, IgG1 and IgG3 are the most effective at activating complement and binding Fcγ receptors. In mice IgG2a and IgG2b are the most efficient in activating complement and binding the Fcγ receptors.[53] Viral infection in humans results in the production of IgG1 and IgG3 and in the mouse IgG2a.[53]

In individuals with DHF there was an increase in the amount of IgG1 antibody during the acute phase of illness compared to those with dengue fever.[76] Conversely, patients with dengue fever had higher levels of IgG2 that DHF patients.[76] No significant differences were observed in the levels of IgG3 and IgG4 between DHF and dengue fever. In mouse models, immunization with live DENV resulted predominately in IgG2a while a recombinant E protein resulted in the production of IgG1 antibodies.[77,78] In a PrM and E DNA vaccine construct, neutralization was found predominately in the IgG2a subclass, but the recombinant protein vaccine also induced neutralizing antibody of the IgG1 subclass.[79] Smuncy and colleagues found that recombinant-E-vaccinated mice raised neutralizing antibody only of the IgG2a subclass.[77] While different IgG isotypes have been observed in dengue fever versus DHF, the role these IgG subclasses play in protection and pathogenesis remains to be answered.

Antibodies and protection

Antibodies protect against viral infection through several mechanisms: opsonization, elimination of infected cells, and neutralization. These three mechanisms do not necessarily work independently, as neutralizing antibody can also opsonize pathogens for uptake by phagocytic cells.

Opsonization of pathogens targets them for uptake by Fcγ-bearing cells such as macrophages and neutrophils. Opsonization is an efficient way to eliminate pathogens that do not replicate in phagocytic cells, however, opsonization of viable dengue viruses (i.e. not neutralized) may be responsible for the immune enhancement phenomenon with subsequent heterologous DENV infection. This phenomenon is the basis for the antibody-dependent enhancement (ADE) theory (discussed in Chap. 9).

A second mechanism by which antibodies control viral infection is the elimination of infected cells. In one mechanism, antibodies bind cells which have viral antigens expressed on their surface, which then recruit components of the complement pathway to kill the infected cell. These cells can also be targeted by a process known as antibody-dependent cellular cytotoxicty (ADCC). ADCC is mediated by Fcγ-bearing effector cells, particularly NK cells. However, cytotoxic T cells and macrophages are also capable of ADCC.[80] Several *in vivo* studies have demonstrated that antibodies that result in the elimination of infected cells can provide partial protection against dengue viruses. In particular, studies that examined the effectiveness of DENV nonstructural protein vaccines are of key interest. Because nonstructural proteins are by definition not found in the mature virion, antibodies directed against these proteins cannot be neutralizing. In a passive transfer experiment in mice, anti-NS1 antibody provided partial protection against lethal DENV-2 intracerebral challenge. There was no correlation between protection and the ability of the NS1 antibody to fix complement.[81] This suggests that infection was partially controlled by an ADCC-dependent mechanism. *In vitro* data support this observation, as NK cells harvested from DENV naïve individuals could kill cells infected with dengue viruses by ADCC.[36,82] Several other studies have also demonstrated partial protection of mice by vaccination with recombinant or DNA NS1 or NS3 vaccines.[83–85] In two of the studies,[84,85] protection was also demonstrated in passive transfer of antibodies. These data suggest that NS1 or NS3 antibodies in the absence of memory T cells can provide some protection against the spread of DENV, presumably by eliminating infected cells.

The last mechanism is neutralization. As mentioned, neutralizing antibody is the critical determinant of protection against reinfection with most viruses. Antibodies that neutralize viruses do so by mechanisms not completely understood; however, it is believed that neutralizing antibodies block viral interaction with host cellular receptors, prevent uncoating, or both. Both of these mechanisms require antibodies to interact with structural components of the virus. In the case of dengue viruses these are the M and E proteins (the C protein would not be readily accessible to antibody as it is covered by the envelope). Neutralizing antibody directed against the E protein appears to be the pivotal antibody that mediates homologous protection against reinfection. However, in mice, PrM vaccines have also been shown to protect against lethal DENV challenge.[86] Recombinant and purified E protein in mice,[78,87–90] DNA E protein vaccines in mice,[91–95] a chimeric dengue 4 virus expressing the prM and E of dengue 1 and dengue 2 in monkeys,[96] and the 17D yellow fever vaccine virus with an inserted E protein in mice and monkeys[97,98] all provide significant protection against DENV challenge. As with the nonstructural protein experiments described above, passive transfer of antibody was also capable of protecting mice. However, unlike the nonstructural vaccine studies, E or PrM antibodies that were not neutralizing failed to protect mice against challenge in two studies[99,100] and only partially protected in another.[101] These data suggest that E or PrM antibodies must be neutralizing to afford protection against infection. In summary, the above data suggest that antibodies against nonstructural proteins may provide some protection against the spread of DENV *in vivo*, and while there are data that suggest the contrary, the key determinant of protection from reinfection with homologous serotypes of DENV appears to be neutralizing antibody to the E protein.

Vaccine Development

Introduction

The development of safe and effective dengue vaccines faces many challenges. Four vaccines must be developed without the benefit of a full understanding of the pathogenesis of severe dengue disease or an adequate animal disease model. Animal models have been limited to the demonstration of immunogenicity in mice and monkeys, and protection from viremia in monkeys (neither mice nor monkeys routinely show signs of disease when administered wild-type dengue viruses).[102–105] However, novel approaches to using mice as a disease model show promise.[28,106] For live virus approaches, reduced viremia in rhesus macaques (*Macaca mulatta*) has been associated with reduced reactogenicity in humans.[107] Recently, a human challenge model has been re-established (M. Mammen, personal communication) and affords a method to evaluate the protective efficacy of dengue vaccines prior to Phase 3 efficacy trials. While a licensed dengue vaccine is not yet available, the scope and intensity of dengue vaccine development has increased dramatically in recent years. Table 1 outlines the leading approaches to vaccine development active at the time of this writing. The order of approaches in the table is by virtue of the number of dengue gene products presented to the vaccine recipient. It is theorized that more gene products is better than fewer and that live approaches are superior to inactive approaches in eliciting broad, enduring protection. Technical barriers and safety issues are discussed for each approach.

Table 1. A partial list of approaches to dengue vaccine development.

Approach	Dengue genes per serotype	Examples	Status	References
Live attenuated viruses (traditional)	10	Mahidol University/Sanofi Pasteur — tetravalent	Phase 2 trials	116–125
		Walter Reed Army Institute of Research/GlaxoSmithKline — tetravalent	Phase 2 trials	105, 107, 126–129
Live attenuated viruses (molecular)	10	Food and Drug Administration mutation "F"	Preclinical	150, 152, 155, 156, 162, 228
		National Institutes of Health	Phase 2	145–149
Live chimeric viruses	2 structural genes plus backbone	National Institutes of Health using PrM and E genes inserted into an attenuated dengue 4 virus backbone	Phase 1	96, 142, 144, 158–162
		Centers for Disease Control and Prevention using *PrM* and *E* genes inserted into an attenuated dengue 2 virus backbone	Preclinical	163–167
		Acambis/Sanofi Pasteur using *PrM* and *E* genes inserted into a yellow fever vaccine virus backbone (tetravalent)	Phase 2	98, 137, 169–173

(Continued)

Table 1. (*Continued*).

Approach	Dengue genes per serotype	Examples	Status	References
DNA vaccines	2 or more	Naval Medical Research Center, Walter Reed Army Institute of Research, Kobe University School of Medicine, MaxyGen, Johns Hopkins University, CytoPulse and Powderject — several constructs	Phase 1	79, 91, 95, 175, 180–183, 187, 223
Inactivated whole virus	4	Walter Reed Army Institute of Research — purified inactivated virus (PIV) vaccine containing M, C, E and NS1 proteins	Phase 1	50, 202, 203
Recombinant subunit	<1	Hawaii Biotechnology — tetravalent 80% *E* gene and dengue 2 NS1 gene expressed by baculovirus in drosophilus cells	Preclinical	50, 222
		Kobe University — dengue 2 subviral extracellular particles produced in mammalian cells	Preclinical	221

Traditionally attenuated live virus vaccines

Most viral vaccines licensed in the United States are live-attenuated virus (LAV) vaccines (e.g. smallpox, yellow fever, measles, mumps, rubella, polio, adenovirus 4, adenovirus 7, varicella and zoster, influenza and rotovirus)[108] (http://www.fda.gov/cber/vaccine/licvacc.htm). In 1929, Blanc and Caminopetros were the first to describe an effort to produce a dengue vaccine.[109] Volunteers were administered doses of viremic blood treated with two concentrations of ox bile and then challenged with untreated virus ten days later. Although few of the challenged individuals developed dengue fever, there were no controls and the immune status of volunteers was unknown.

Both the Japanese and the Americans made significant contributions to dengue vaccinology during WWII.[68,110] Sabin used a dermal neutralization test to show that there were two different dengue viruses, the Hawaii and New Guinea strains B, C and D, later designated as DENV-1 and DENV-2 respectively.[68] The test was performed by mixing and then incubating convalescent serum with acute serum containing a known amount of the virus. The mixture was then injected intracutaneously into a human volunteer. The virus was considered neutralized if the subject did not develop local or systemic signs of infection. Sabin confirmed that type-specific antibody was protective and observed that heterologous cross-protection to classic dengue fever lasted for at least two months. He attempted to attenuate the Hawaii strain virus by serial passage in mouse brains. Tests conducted in human subjects showed that the virus became attenuated in humans after the seventh mouse brain passage.[68] Schlesinger and colleagues similarly adapted the New Guinea strain of DENV-2 to grow in mouse brain and tested resultant strains as a monovalent vaccine or in combination with the attenuated Hawaiian strain and yellow fever 17-D.[48] The Sabin DENV-1 virus attenuated by 18 serial passages in weanling mouse brain, underwent seven more passages in the brains of 3–5-day-old mice and then a further five limiting dilution passages. Fourteen human subjects receiving this strain at mouse passage 32 or 33, designated MD-1, were immunized without developing significant clinical signs or symptoms.[111] In 1963 Wisseman and colleagues[112] conducted a placebo-controlled field trial of this mouse-brain-adapted DENV-1 in

Puerto Rican adolescent males during a DENV-3 epidemic. Among the 561 vaccine recipients, the attack rate of dengue fever was 39% that of the placebo control group. Further work on mouse-brain-derived live virus vaccines was suspended due to safety concerns raised by the possibility of autoimmune reactions to mouse neural protein antigens or the inclusion of mouse adventitious agents.[113]

In 1971, a Dengue Task Force was formed by the Virus Commission, United States Armed Forces Epidemiology Board, to accelerate efforts to develop DENV-1, -2, -3 and -4 vaccines. Dengue virus vaccine candidates were derived using the serial cell culture passage strategy employed to develop other live-attenuated virus vaccines.[114] Over the next two decades, investigators from Mahidol University and the Walter Reed Army Institute of Research (WRAIR) independently evaluated the first monovalent and then tetravalent dengue vaccine candidates. Because no reliable *in vitro* marker of attenuation or adequate animal model that mimics human disease was available, progress has been slow. For example, a DENV-2 candidate, PR-159/S-1, was passed in primary green monkey kidney cells and selected for growth at 35°C but not 39°C, small plaque size, and decreased monkey neurovirulence and viremia. It elicited neutralizing antibody and protected monkeys from wild-type virus challenge. However, despite initial encouraging results in yellow-fever-immune individuals a large Phase 1 study in 98 US Army soldiers showed that 20% of seroconverters lost at least one day from duty due to side effects and only 60% of DENV-2 nonimmune individuals seroconverted.[115]

The Mahidol candidates, DENV-1 16007 PDK13, DENV-2 16681 PDK53, DENV-3 16562 PGMK 30/FRhL3 and DENV-4 1036 PDK48, were clinical isolates grown in either primary dog kidney (PDK) cells (DENV-1, -2 and -4) or African green monkey kidney (PGMK) cells and then in fetal rhesus lung (FRhL) cells (DENV-3). Similarly, the WRAIR candidates, DENV-1 45AZ5 PDK20/FRhL3, DENV-1 45AZ5 PDK27/FRhL3, DENV-2 16803 PDK50/FRhL3, DENV-3 CH53489 PDK20/FRhL3, DENV-4 341750 PDK6/ FRhL3 and DENV-4 341750 PDK20/ FRhL4, were all grown in PDK cells with terminal passages in FRhL cells.

The Mahidol University monovalent vaccines, licensed to Sanofi Pasteur were well tolerated in adult flavirus-naïve volunteers in Thailand[116–118] and the US.[119,120] Thai investigators found that fewer passages in PDK or PGMK cells increased vaccine immunogenicity. In a series of Phase 1 trials, passage levels were identified with minimal reactogenicity while inducing 100% seroconversion for all four monovalent components (some with a booster dose).[117] Bivalent and trivalent formulations using DENV-1, -2 and -4 vaccine candidates elicited uniform seroconversions in Thai subjects.[121] However, when all four serotypes were combined into a tetravalent vaccine, only DENV-3 viremia and neutralizing antibody were elicited.[120] Studies evaluating formulations using lower doses of DENV-3 with one or two boosters were subsequently pursued to overcome this apparent viral interference phenomenon. Indeed, a reformulated tetravalent vaccine (seven different formulations) resulted in tetravalent seroconversions in 71% of volunteers after two doses given at 0 and 6 months in a study in Thai adults, though moderate fever, headache and myalgia occurred in most subjects.[122] Phase 2 testing of select formulations was completed in children with tetravalent seroconversion seen in 100% of recipients following three doses of vaccine for one formulation.[123] However, further development of this vaccine has been deferred due to reactogenicity associated with the DENV-3 component.[124,125]

The WRAIR vaccine candidates were also selected based on results of small studies in human volunteers of the safety and immunogenicity of viruses from different PDK cell passage levels. Among the selected passage levels, the seroconversion rates were 100%, 92%, 46% and 58% for a single dose of DENV-1, -2, -3, and -4 respectively. The WRAIR DENV-2, -3 and -4 vaccine viruses were well tolerated by volunteers. The DENV-1 PDK20/FRhL3 monovalent candidate was associated with increased reactogenicity, with 40% developing fever and generalized rash. Vaccine-related reactions consisted of modified symptoms of dengue fever, including headache, myalgia and rash.[107,126,127]

Sixteen different dose formulations of the WRAIR tetravalent vaccine were tested in 64 adult volunteers.[128,129] They were derived by using undiluted vaccine (5–6 logs of the virus) or a 1:30 dilution for each virus serotype. Overall seroconversion rates after a single dose of tetravalent vaccine were 83%, 65%, 57% and 25% to DENV-1, -2, -3 and -4 respectively, similar to that seen with monovalent vaccines. Few additional seroconversions were seen following a booster dose one month after the first dose. Though group sample size was small, the trend was toward increased reactogenicity when a full dose of the DENV-1 component was combined with a 30-fold-lower dose of DENV-2 or DENV-4. Therefore, it appears that with these particular strains of the virus, at the doses evaluated, viral interference did not affect antibody response but may have modified reactogenicity. Recently, one such combination achieved an overall seroconversion rate in 16 adults of 69%, 100%, 81% and 94% to DENV-1, -2, -3, and -4 respectively after two doses.[130] This same tetravalent formulation was also tested in six flavivirus-naive Thai children and was found to be safe and induced tetravalent neutralizing antibody responses in all six children after two doses.[131] Further testing is in progress to determine the optimal dose and combinations of various PDK passage level vaccine viruses.

There are several important safety issues for live dengue vaccines. Principal among these concerns is the theoretical risk of enhanced disease following DENV vaccination. The rationale for a tetravalent vaccine is the perceived requirement to induce primary-type immune responses to all four dengue viruses simultaneously. The simultaneous production of neutralizing antibodies specific to each of the four dengue viruses is predicted to minimize the risk of disease enhancement following natural infection. However, antibody-dependent enhancement (see Chap. 9) appears to occur with neutralizing antibodies at subneutralizing concentrations, so a vaccine that induces protection for a period of time might later increase the risk for enhanced disease. This is particularly a concern for vaccines that induce low levels of neutralizing antibodies,[132] but might occur with any vaccine given enough time. Experiences with formalin-inactivated respiratory syncytial virus and measles vaccines suggest that low affinity of neutralizing antibodies, complement fixing antibodies, high CD4+ proliferation and absent CD8+ memory responses are associated with exaggerated disease with subsequent natural infection.[133,134] Skewing toward a Th-2 type response by vaccination may also play a role in disease enhancement.[135,136] Some hold that tetravalent seroconversion with ample CD8+ memory responses will preclude enhanced disease even many years following vaccination.

Enhanced vaccine reactogenicity is also possible when a live tetravalent DENV vaccine is administered to persons with pre-existing antiflavivirus antibody. Specifically, pre-existing monovalent anti-DENV antibody could neutralize one component of a live tetravalent vaccine and enhance the replication of one or more of the other attenuated viruses, resulting in severe dengue fever or dengue hemorrhagic fever. To date, enhanced disease has not been observed in yellow-fever-immune study subjects who have been given attenuated dengue viruses,[115,137] in volunteers receiving closely spaced sequential dengue viruses, or in volunteers who tested flavivirus-antibody-negative on a preimmunization screen yet experienced an anamnestic antibody response following vaccination.[59] In this last group, volunteers were screened using the plaque reduction neutralization assay for dengue viruses. If the pre-existing flavivirus antibody resulted from an infection with a flavirus other than dengue, such antibodies pose a lesser risk. Inadvertent administration of tetravalent vaccines to volunteers with documented pre-existing partial DENV immunity has yet to demonstrate an increased risk, though the numbers of volunteers tested have been small to date. Data from outpatient studies in Southeast Asia provide evidence that wild-type DENV-4 and DENV-2 viruses are naturally attenuated in susceptible children.[138,139] That wild-type viruses may be nonpathogenic in flavivirus-naive children poses difficult questions concerning the ability to assure the safety of vaccine candidates. This is a particularly acute problem in the context of secondary infections. Given the observation that secondary DENV infections are also largely subclinical,[140,141] large numbers of partially dengue-immune volunteers must be given candidate vaccines before this risk can be adequately assessed. As yet, increased reactogenicity attributed to candidate live-attenuated vaccines has not been seen in clinical trials conducted in Thailand and in the United States giving two doses six months apart (R Edelman, personal communication and Ref. 122). A better understanding of the immune mechanisms of vaccine protection and enhanced disease is needed to assess risks associated with Phase 3 vaccine efficacy trials.

Other safety concerns with LAV vaccines include cell-culture-derived adventitious agents, community spread of the vaccine virus by resident vector mosquitoes, vaccine virus neurovirulence, and the effects of vaccine administration to immunocompromised hosts.

Molecularly attenuated live virus vaccines

The cloning of a DENV-4 (WRAIR 814669, Dominica, 1981) by C. J. Lai and colleagues at the National Institutes of Health inaugurated a new era of dengue vaccine research. Dengue and other flavivirus genomes were readily altered genetically, resulting in attenuated variants.[142] As reviewed in Chap. 2, the flavivirus genome is a single-stranded, positive-sense RNA molecule of nearly 11 kilobases containing a single open reading frame. The RNA is translated into a polyprotein that is processed into at least 10 gene products: the 3 structural proteins capsid (C), premembrane (prM) and envelope (E), and 7 nonstructural (NS) proteins, NS1, NS2A, NS2B, NS3, NS4A, NS4B and NS5. The untranslated regions of the genome at the 5′ and 3′ ends are crucial for protein translation and minus strand transcription.[143]

The first DENV vaccine candidate derived using recombinant infectious cDNA technology produced decreased "vaccinemia" with a slightly decreased neutralizing antibody response in rhesus macaques.[144] This DENV-4 mutant (DENV-4 2AΔ30) with a 30-nucleotide deletion in the 3′ untranslated region (10478–10507), grown in Vero cells, has been evaluated in Phase 1 trials.[145,146] Twenty volunteers received the vaccine candidate (10^5 pfu in 0.5 ml subcutaneously). The vaccine was well tolerated by all recipients, with no injection site swelling or soreness and minimal subsequent symptoms. The most common effects of vaccination were neutropenia and a transient rash that was often unnoticed by the vaccine recipient. A mild increase in ALT levels has been seen in some volunteers. The vaccine has been immunogenic in doses from 10 to 10,000 pfu. Since attenuation of this virus is the result of a large deletion mutation, it is likely to remain genetically stable in vaccinees. While this monovalent vaccine has had promising results, some safety concerns exist. ALT elevations are of concern and should be evaluated in larger numbers of volunteers. Alternatively, if studies demonstrate liver tropism with any constructs, further genetic manipulation of candidate vaccines may be required. Additionally, because wild-type DENV-4 virus strains produce clinically inapparent infections, as discussed above,[138] safety of the deletion mutants must also be evaluated in partially dengue-immune volunteers. Full length infectious clones with the 30-nucleotide deletion and other changes are being produced[147–149] and a derivative of the DENV-4 2A Δ30 virus, the nonstructural gene segment of rDENV-4 Δ30, is being used as a backbone for the insertion of structural genes from the other three dengue viruses that can be combined to produce a tetravalent vaccine formulation (see below, regarding chimeric vaccine approaches).

In 1997, scientists at the US Food and Drug Administration (FDA) published a method to create a DENV-2 infectious clone (New Guinea C strain, NGC).[150] Next, they tested the role played by the highly conserved secondary structure of the terminal 3′ stem and loop structure in virus replication.[151] Portions of the terminal 90 nucleotides of the NGC cDNA were replaced with comparable sequences from the West Nile virus.[152] It had been shown previously that the this 3′ stem and loop structure bound to viral proteins NS3 and NS5, which are involved in virus replication, and to cellular proteins with a putative role in replication,[153,154] and that 3′ noncoding region deletions that extended into the structure were lethal for virus replication.[144] Further studies identified a mutant genome with a chimeric dengue/West Nile 3′ stem and loop structure that exhibited an altered host range, i.e. this virus, designated "mutant F" or "mutF," grew normally in mammalian LLC-MK2 cells and yet was severely restricted for growth in C6/36 insect cells. Introduction of the mutF mutations into an infectious DNA clone of the West Pacific DENV-1 virus genome[155] resulted in a DENV-1 virus with the same phenotype in cell culture as that exhibited by the DENV-2mutF virus (it grew well in mammalian cells but not in insect cells). The "DEN1mutF" was evaluated in rhesus macaque monkeys.[156] Vaccinemia in rhesus monkeys was greatly reduced compared to the DENV-1 West Pacific wild-type virus, suggesting that the virus will be less reactogenic in people,[107,157] while immunogenicity was similar to the wild-type virus (personal communication, Mammen Mammen). Monkeys challenged 12 or 17 months following a single dose of vaccine were protected from viremia.

Molecular clone-based strategies for a tetravalent dengue vaccine offer important advantages over traditional methods (empiric attenuation in cell culture). These include a reduced risk of adventitious agents that will also reduce product quality assurance costs, and a molecular explanation for attenuation. Interference observed when mixtures of four dengue viruses are inoculated in susceptible human volunteers must also be studied in mixtures of genetically modified vaccine viruses.

Chimeric live virus vaccines

The nonstructural genes of DENV-4 cDNA described above were used to construct chimeric viruses by substitution of the heterotypic genes coding for the respective viral structural proteins.[158–161] Chimeric DENV-1/DENV-4 (*C-PrM-E* genes placed in the DENV-4 backbone) or DENV-2/DENV-4 (*PrM-E* genes placed in the DENV-4 backbone) resulted in chimeric viruses with DENV-1 or DENV-2 antigenicity with reduced occurrence and duration of viremia compared to the wild-type dengue 4 virus. These chimeric viruses produced robust antibody responses and protection from challenge with wild-type viruses when given as monovalent vaccines or bivalent vaccines in rhesus monkeys.[96] In humans, using PrM-E inserts, the vaccines have been well tolerated to date, with high rates of seroconversion.[162]

The US Centers for Disease Control and Prevention has made progress toward developing a tetravalent dengue vaccine using the nonstructural portion of the cDNA derived from the DENV-2 component of the Mahidol University/Sanofi Pasteur LAV vaccine (DENV-2 16681 PDK-53).[163] This group reported that the attenuation markers for this vaccine virus are located in the nonstructural gene region. Nonetheless, it is planned to insert Pr-E genes from DENV-1, -3 and -4 from viruses demonstrated to be attenuated into this virus backbone.[164,165] The structural genes of the DENV-1 16007 virus appear to be more immunogenic in mice than those of the PDK-13 vaccine virus.[166] Studies on mice have demonstrated retention of phenotypic markers of attenuation and immunogenicity for each of the chimeric viruses and as tetravalent formulations.[167]

As described, chimeric dengue vaccine viruses can be derived by inserting serotype-specific dengue antigen genes into a single attenuated dengue genomic construct. A different approach was taken to insert dengue structural genes into the infectious cDNA backbone of the well-established yellow fever vaccine virus, strain 17D. It was developed by teams at the Washington University and the St. Louis University medical schools. These yellow fever chimeras are being developed commercially by Acambis, Inc.[168–170] Vero cells serve as the substrate for vaccine virus production. A dengue/yellow fever virus chimera has been developed for each of the four dengue virus serotypes using the *PrM-E* genes from DENV-1 PUO-359, DENV-2 PUO-218, DENV-3 PaH881 and DENV-4 1228 viruses.[171] When inoculated intracerebrally in weanling mice, neurovirulence was reduced compared with following the inoculation of the yellow fever 17D vaccine virus. In rhesus macaques viremia is similar in titer and duration to that of the yellow fever 17D virus and greatly reduced compared with wild-type dengue viruses.[172] Seroconversion in rhesus monkeys was 100% using a tetravalent formulation comprising three logs of the DENV-2 chimera and five logs of DENV-1, -3 and -4 chimeras.[98] Lower DENV-3 and DENV-4 neutralizing antibody titers were seen with five logs of the DENV-2 component, suggesting interference.[171] Rhesus monkeys were protected from viremia following challenge with any of four wild-type DENV serotypes[171] and mosquito studies indicate that spread from viremic vaccinees in unlikely.[173] A Phase 1 evaluation of the monovalent ChimeriVax-DENV-2 was recently completed, showing the vaccine to be safe, well tolerated and immunogenic regardless of pre-existing yellow-fever-immune status.[137] Initial clinical testing of a tetravalent formulation has been completed. The vaccine was well tolerated, although at the time of this writing immunogenicity data have not been released.

If rates of virus replication are controlled by nonstructural genes or noncoding regions of the genome, interference with the growth of one or more components of a tetravalent vaccine would not be expected with chimeric approaches which use a common genetic backbone for each component of a tetravalent vaccine. However, the experience with ChimeriVax-DEN, where one formulation provided a broader and more even neutralizing antibody response than another, suggests that interference may occur with chimeric vaccines due to a poorly understood mechanism. At any rate, should recombination occur among the components of tetravalent chimeric vaccines without the nonstructural genes of a DENV present, it is unlikely that the progeny virus will revert to a more virulent phenotype. It is not known whether chimeric vaccines will stimulate sufficient numbers of T cells to produce dengue serotype-specific immunologic memory that characterizes wild-type immunity.

DNA/RNA approach

Dengue DNA vaccines offer a possible method to raise protective immunity bypassing the problem of interference seen with multivalent live virus vaccines. DNA-based vaccine constructs can be produced to express a specific portion of genes that elicit protective immune responses with no requirement that gene products be assembled into complete virions. DNA vaccines are composed of a plasmid or plasmids containing dengue genes. These are reproduced to a high copy number in bacteria such as *E.* coli.[174] The plasmid contains a eukaryotic promoter and termination sequence to drive transcription in cells after being inoculated into a vaccine recipient. The transcribed RNA is translated to produce proteins that are processed and presented to the immune system in the context of the host's own MHC molecules. Adjuvant genes such as those that regulate intracellular trafficking or provide immunostimulatory sequences can be added to the plasmid. The target organism's immune system recognizes the antigen, and generates humoral and/or cell-mediated immune responses.

Preclinical evaluation of the first dengue DNA vaccine was published in 1997,[175] following the early successes in animals with DNA vaccines targeting HIV,[176] influenza,[177] and malaria.[178] The Naval Medical Research Center evaluated two eukaryotic plasmid expression vectors (pkCMVint-Polyli and pVR1012; Vical, Inc., San Diego, California) expressing the PrM protein and 92% of the E protein for DENV-2 virus (New Guinea C strain). Both constructs induced neutralizing antibody in all mice,[175] with a subsequent improvement seen with the addition of immunostimulatory CpG motifs (pUC 19, Gibco BRL, Gaithersburg, Maryland).[91] Konishi and colleagues successfully immunized mice with a similar DENV-2 vaccine construct using *C-PrM-E* genes in the pcDNA3 vector (Invitrogen, Corp., San Diego, California).[179] In subsequent experiments using genes from the West Pacific 74 strain of the DENV-1 virus and the pVR1012 plasmid, it was determined that the full length *E* gene with *PrM* served as a better immunogen[180] and was shown to reduced the frequency and duration of viremia in rhesus macaques following challenge with wild-type virus.[181] Gene shuffling and screening technologies have been used to produce chimeric DNA constructs that express antigens with shared epitopes from all four dengue viruses.[182] At present, the DNA approach has produced modest neutralizing antibody levels in nonhuman primates, with only a portion of the animals fully protected from viremia.[95,183] With the recognition that skin dendritic cells are high permissive for DENV replication,[17] ongoing efforts are focused on vaccine delivery systems to target these cells. Additional dengue gene sequences, immunostimulatory gene sequences, trafficking sequences and codelivery of plasmids encoding cytokine genes are being evaluated to increase vaccine potency.[184] Clinical trials have been initiated.

DNA vaccines used in combination with other approaches, such as the inactivated virus vaccines discussed below, may increase the complexity and effectiveness of the immune response.[184–186] Preliminary studies on mice at the Naval Medical Research Center and Kobe University School of Medicine suggest that this approach may be effective for dengue.[79,187] In mice, DNA vaccination alone elicited primarily an IgG2a antibody response while the recombinant protein used elicited an exclusive IgG1 antibody response. Vaccination with different products in sequence or simultaneously elicited a more balanced distribution of both antibody subclasses. However, for a vaccine targeted for one billion children in the tropics, an eight-component vaccine will present obstacles of cost and complexity.

In theory, DNA vaccines afford numerous advantages over conventional vaccines, including ease of production, stability and transport at room temperature, and they provide a possibility of vaccinating against multiple pathogens in a single vaccination. The immunology of DNA vaccines has recently been reviewed.[188] While the DNA approach offers unique advantages, no human vaccine against infectious diseases has been commercialized and the approach also carries unique risks.[189,190] These include the theoretical risk of nucleic acid integration into the host's chromosomal DNA to potentially inactivate tumor suppressor genes or activate oncogenes. This risk appears to be well below the spontaneous mutation frequency for mammalian cells.[191,192] However, if a mutation due to DNA integration is a part of a multiple hit phenomenon leading to carcinogenesis, it could take many years before this problem becomes evident. Another concern is that foreign DNA might induce anti-DNA antibodies, leading to autoimmune diseases such as systemic lupus erythematosis. However, studies on lupus-prone mice, normal mice, rabbits and people to date have not validated this concern[193,194] and, in fact, DNA vaccines are being proposed as an approach to the management of autoimmune diseases.[195–197]

Recently, infectious virus RNA derived directly from cDNA has been used to successfully immunize mice with nanogram amounts of the tick-borne encephalitis virus.[198,199] This approach mimics natural infection with the advantages of increased vaccine purity and stability over traditional cell-culture-based approaches. It also eliminates the risk of DNA integration and anti-DNA autoimmune disease.

Inactivated whole virus vaccines

Simmons, St. John and Reynolds developed the US military's first dengue vaccine in 1929.[200] These researchers conducted investigations in which 100% of 35 individuals experimentally infected with dengue were later found to be resistant to an attack accompanying the inoculation of the same virus 400 days later. This not only provided evidence of homotypic protective immunity following dengue infection but suggested the possibility that a prophylactic vaccine might be developed. Benefiting from the work of Siler, Hall and Hitchens, who had shown that DENV was transmitted by *A. aegypti*, measured the period of infectivity of dengue patients, and the extrinsic incubation period when the virus replicates in *A. aegypti*,[201] Simmons and colleagues prepared a "mosquito" vaccine by grinding 2010 infective *A. aegypti* mosquitoes in a sterile porcelain mortar with a salt solution and chemically pure phenol and formalin. The suspension was transferred to a sterile bottle after it had stood for 48 hours, with the addition of 10 cc of sterile physiologic salt solution to further dilute the preservatives. It stood another 8 days to destroy bacteria and 20 cc of salt solution was added. This solution was slowly centrifuged to eliminate remaining insoluble matter. Before use, the vaccine was examined for bacteria (10-day aerobic and anaerobic cultures) and bacterial toxins, and safety-tested in rabbits, guinea pigs and white mice, all of which remained normal. One-milliliter doses were given to each of the authors and then two volunteers. While these injections failed to prevent subsequent infection and disease, some of the volunteers developed only mild cases of dengue. There was no model to test the immunogenicity of the vaccine preparation. It is also possible that giving two doses of vaccine just four days apart and challenging a week after the second dose may not have provided sufficient time for a protective antibody response to develop.

A purified, inactivated DENV-2 vaccine (D2-PIV) compliant with good manufacturing practices has been manufactured by the Department of Biologics Research, WRAIR.[202,203] For preparation of D2-PIV, D2 strain S16803, Vero cell passage 3, was propagated in certified Vero cells, which were maintained in serum-free Eagle's Minimum Essential medium. The virus from the culture supernatant fluid was concentrated by ultrafiltration and purified on 15–60% sucrose gradients. The high-titer purified virus (approximately 9 $\log_{10}$ pfu/ml) was inactivated with 0.05% formalin at 22°C for 10 days. The immunogenicity and efficacy of D2-PIV formulated with four different adjuvants (alum and adjuvants SBAS4, SBAS5 and SBAS8, provided by GlaxoSmithKline Biologicals) was evaluated in rhesus monkeys in groups of three animals each.[50] Two doses of D2-PIV, D2 LAV or saline were given three months apart. The animals were challenged with the wild-type D2 S16803 parent virus three months after the second dose and their subsequent viremia and antibody responses were measured. All vaccines were safe and all but one vaccinated animal seroconverted after the first dose of vaccine. Moreover, all vaccinated animals had anamnestic antibody rises following the second dose of vaccine. After virus challenge, viremia was detected by cell culture in the saline group (3–5 days of viremia for all animals, mean 4 days), the D2-PIV/alum group (2 animals without viremia, 1 animal with 2 days of low-titer viremia) and the D2-PIV/SBAS4 (2 animals without viremia, 1 animal with 2 days of low-titer viremia). No viremia was detected in the animals that received D2-PIV/SBAS5 or D2-PIV/SBAS8. Additionally, a subset of monkeys were rechallenged with wild-type virus 1 year later once PRNT antibody titers had decreased; the immune response remained protective.[50]

Inactivated whole virus vaccines have two advantages when compared to LAV vaccines; inactivated vaccines cannot revert to a more pathogenic phenotype, and they are unlikely to interfere with each other in combination. Moreover, induction of cell-mediated and humoral immune responses has been demonstrated with inactivated flavivirus vaccines.[204]

On the other hand, inactivated or killed vaccines express only the part of the viral genome that encodes structural proteins. In the context of dengue immunity and immunopathology, raising antibodies that are not fully protective may lead to breakthrough infections or enhance infections with wild-type dengue viruses. Here, the lessons of the RSV and measles vaccines are sobering. Other potential disadvantages of these vaccines are the difficulty in manufacture of preparations of sufficient titer, their increased cost per dose and their usual requirement for multiple immunizations. Despite these apparent drawbacks, two successful inactivated flavivirus vaccines are safe, effective, and licensed for use to prevent Japanese encephalitis[205] and tick-borne encephalitis.[206] The critical requirement for a vaccine that prevents viral encephalitis is to block entry of the virus into the central nervous system. A wide range of inactivated vaccines have been shown to prevent flaviviral encephalitides, suggesting the possibility that even low levels of circulating antibody may be protective. If these concerns can be laid to rest, an inactivated dengue vaccine may be useful as a traveler's vaccine or as a part of a prime-boost strategy with live or replicating vaccines.

Recombinant subunit vaccines

T and B cell epitopes have been mapped to all DENV structural proteins and most nonstructural proteins.[207,208] The right combination of epitopes expressed in protein subunit vaccines could be the basis for an effective and safe vaccine at moderate cost.[209] Structural and nonstructural DENV proteins have been produced to adequate yields to serve as vaccines in a wide variety of expression systems to include *E.* coli,[210–212] baculovirus in *Spodoptera frugiperda* insect cells,[78,87,101,213–215] yeast,[216] vaccinia virus,[217–220] mammalian cells[221] and *Drosophila* cells.[50] The last approach, 80% *E* gene expression in *Drosophila* cells, was pioneered by Hawaii Biotechnology, Inc. One microgram of DENV-2 E protein with the SBAS5 adjuvant protected rhesus monkeys from viremia following challenge with the wild-type virus. A tetravalent formulation is undergoing evaluation in rhesus monkeys.[222] The DNA shuffling approach has also been used to express a single recombinant dengue envelope antigen capable of inducing neutralizing antibodies against all four dengue viruses.[223]

Synthetic peptides containing B and T cell epitopes are immunogenic in mice, and combinations of peptides could be effective as subunit vaccines.[224–227] However, peptides provide fewer epitopes than other approaches, lack conformational epitopes, and are therefore less desirable.

Recombinant subunit approaches offer advantages in anticipated minimal reactogenicity, freedom from adventitious agents, and low cost. However, incomplete posttranslational processing of proteins can lead to proteins that differ from native proteins, and antibody responses described to date favor IgG1 (in mice) for some subunit vaccines which may be less than optimal for long-term protection.[77] Production in mammalian cells may reduce some of these concerns.[221] Risk of enhanced disease upon exposure to wild-type viruses postvaccination would need to be assessed as for all other approaches to dengue vaccines. Vaccines that elicit cytotoxic T cell memory responses may lower this risk.

Conclusion

As demonstrated by Sabin and by the natural experience in which infants born to dengue-immune mothers are protected against dengue infections and disease for at least three months, DENV-neutralizing antibodies are the only immune substance shown to prevent dengue infections. Beyond this, the specifics of the role of antibody in affording protection or enhancing disease are unclear. There is growing evidence that natural viral infections result in life-long antibody responses and afford life-long protection. Translated into dengue, it is expected that infection, and by analogy vaccination, will provide serotype specific immunity, and it has provided a way forward for vaccine developers. Current knowledge suggests that the risk of enhanced disease will be less if, following vaccination, there is a broad immune response to the four dengue viruses, and there may be merit in targeting the establishment of robust cytotoxic T cell memory (see Chap. 5).

It has been more than 75 years since the first DENV vaccines were evaluated in volunteers.[109,200] Progress toward an effective licensed dengue vaccine has been slow, partially due to the concerns of disease enhancement and limited research funding for a disease that primarily affects the developing world. However, in the past few years there has been increased interest in dengue vaccines among governmental, non-governmental and commercial entities. LAV vaccines are moving into Phase 2 and Phase 3 testing. Vaccine development efforts are accelerating, due to an increased awareness of the dengue pandemic and the development of new molecular techniques.

References

1. Gubler DJ. The emergence of dengue/dengue hemorrhagic fever as a global public health problem. In: Saluzzo JF, Dodet B (eds.) *Factors in the Emergence of Arbovirus Diseases*. Elsevier, Paris, 1997, pp. 83–92.
2. Halstead SB. The XXth century dengue pandemic: need for surveillance and research. *World Health Stat Q* 1992;45:292–298.
3. Gubler DJ, Meltzer M. Impact of dengue/dengue hemorrhagic fever on the developing world. *Adv Virus Res* 1999;53:35–70.
4. Halstead SB. Is there an inapparent dengue explosion? [letter]. *Lancet* 1999;353(9158):1100–1101.
5. Churdboonchart V, Bhamarapravati N, Futrakul P. Crossed immunoelectrophoresis for the detection of split products of the third complement in dengue hemorrhagic fever I. Observations in patients' plasma. *Am J Trop Med Hyg* 1983;32(3):569–576.
6. Churdboonchart V, Bhamarapravati N, Yoksan S. Crossed immunoelectrophoresis for the detection of split products of the third complement component in dengue hemorrhagic fever. II. *In vitro* activation by dengue viral antigen. *Am J Trop Med Hyg* 1984;33(6):1239–1247.
7. Scott RM, Nimmannitya S, Bancroft WH, Mansuwan P. Shock syndrome in primary dengue infections. *Am J Trop Med Hyg* 1976;25(6):866–874.
8. Shaio MF, Chang FY, Hou SC. Complement pathway activity in serum from patients with classical dengue fever. *Trans R Soc Trop Med Hyg* 1992;86:672–675.
9. Kuberski T, Rosen L, Reed D, Mataika J. Clinical and laboratory observations on patients with primary and secondary dengue type 1 infections with hemorrhagic manifestations in Fiji. *Am J Trop Med Hyg* 1977;26(4): 775–783.
10. Malasit P. Complement and dengue haemorrhagic fever/shock syndrome. *Southeast Asian J Trop Med Public Health* 1987;18(3):316–320.

11. Green S, Pichyangkul S, Vaughn DW *et al.* Early CD69 expression on peripheral blood lymphocytes from children with dengue hemorrhagic fever. *J Infect Dis* 1999;180(5):1429–1435.
12. Hober D, Poli L, Roblin B *et al.* Serum levels of tumor necrosis factor-alpha (TNF-alpha), interleukin-6 (IL-6), and interleukin-1 beta (IL-1 beta) in dengue-infected patients. *Am J Trop Med Hyg* 1993;48:324–331.
13. Atrasheuskaya A, Petzelbauer P, Fredeking TM, Ignatyev G. Anti-TNF antibody treatment reduces mortality in experimental dengue virus infection. *FEMS Immunol Med Microbiol* 2003;35(1):33–42.
14. Kurane I, Innis BL, Nimmannitya S, Nisalak A, Meager A, Ennis FA. High levels of interferon-alpha in the sera of children with dengue virus infection. *Am J Trop Med Hyg* 1993;48:222–229.
15. Hotta H, Hotta S, Homma M. Effect of interferons on dengue virus multiplication in cultured monocytes/macrophages. *Biken J* 1984;27:189–193.
16. Navarro-Sanchez E, Despres P, Cedillo-Barron L. Innate immune responses to dengue virus. *Arch Med Res* 2005;36(5):425–435.
17. Wu SJ, Grouard-Vogel G, Sun W *et al.* Human skin Langerhans cells are targets of dengue virus infection. *Nat Med* 2000;6(7):816–820.
18. Palucka AK. Dengue virus and dendritic cells [news; comment]. *Nat Med* 2000;6(7):748–749.
19. Liu YJ. Dendritic cell subsets and lineages, and their functions in innate and adaptive immunity. *Cell* 2001;106(3):259–262.
20. Libraty DH, Pichyangkul S, Ajariyakhajorn C, Endy TP, Ennis FA. Human dendritic cells are activated by dengue virus infection: enhancement by gamma interferon and implications for disease pathogenesis. *J Virol* 2001;75(8):3501–3508.
21. Ho LJ, Wang JJ, Shaio MF *et al.* Infection of human dendritic cells by dengue virus causes cell maturation and cytokine production. *J Immunol* 2001;166(3):1499–1506.
22. Navarro-Sanchez E, Altmeyer R, Amara A *et al.* Dendritic-cell-specific ICAM3-grabbing non-integrin is essential for the productive infection of human dendritic cells by mosquito-cell-derived dengue viruses. *EMBO Rep* 2003;4(7):723–728.
23. Tassaneetrithep B, Burgess TH, Granelli-Piperno A et al. DC-SIGN (CD209) mediates dengue virus infection of human dendritic cells. *J Exp Med* 2003;197(7):823–829.

24. Sakuntabhai A, Turbpaiboon C, Casademont I *et al.* A variant in the CD209 promoter is associated with severity of dengue disease. *Nat Genet* 2005; 37(5):507–513.
25. Zhang S-Y, Jouanguy E, Sancho-Shimizu V, von Bernuth H, Yang K, Abel L, Picard C, Puel A, Casanova J-L. Human Toll-like receptor-dependent induction of interferons in protective immunity to viruses. *Immunol Rev* 2007;220(1):225–236.
26. Pichyangkul S, Endy TP, Kalayanarooj S *et al.* A blunted blood plasmacytoid dendritic cell response to an acute systemic viral infection is associated with increased disease severity. *J Immunol* 2003;171(10):5571–5578.
27. Diamond MS, Roberts TG, Edgil D, Lu B, Ernst J, Harris E. Modulation of dengue virus infection in human cells by alpha, beta, and gamma interferons. *J Virol* 2000;74(11):4957–4966.
28. Johnson AJ, Roehrig JT. New mouse model for dengue virus vaccine testing. *J Virol* 1999;73(1):783–786.
29. Vaughn DW, Green S, Kalayanarooj S *et al.* Dengue viremia titer, antibody response pattern, and virus serotype correlate with disease severity. *J Infect Dis* 2000;181(1):2–9.
30. Munoz-Jordan JL, Sanchez-Burgos GG, Laurent-Rolle M, Garcia-Sastre A. Inhibition of interferon signaling by dengue virus. *Proc Natl Acad Sci USA* 2003;100(24):14333–14338.
31. Chareonsirisuthigul T, Kalayanarooj S, Ubol S. Dengue virus (DENV) antibody-dependent enhancement of infection upregulates the production of anti-inflammatory cytokines, but suppresses anti-DENV free radical and pro-inflammatory cytokine production, in THP-1 cells. *J Gen Virol* 2007;88(Pt 2):365–375.
32. Robertson MJ. Role of chemokines in the biology of natural killer cells. *J Leukoc Biol* 2002;71(2):173–183.
33. Yokoyama WM, Kim S, French AR. The dynamic life of natural killer cells. *Annu Rev Immunol* 2004;22:405–429.
34. Brown MG, Dokun AO, Heusel JW *et al.* Vital involvement of a natural killer cell activation receptor in resistance to viral infection. *Science* 2001;292(5518):934–937.
35. Shresta S, Kyle JL, Robert Beatty P, Harris E. Early activation of natural killer and B cells in response to primary dengue virus infection in A/J mice. *Virology* 2004;319(2):262–273.

36. Kurane I, Hebblewaite D, Ennis FA. Characterization with monoclonal antibodies of human lymphocytes active in natural killing and antibody-dependent cell-mediated cytotoxicity of dengue virus-infected cells. *Immunology* 1986;58:429–436.
37. Arase H, Mocarski ES, Campbell AE, Hill AB, Lanier LL. Direct recognition of cytomegalovirus by activating and inhibitory NK cell receptors. *Science* 2002;296(5571):1323–1326.
38. Bokisch VA, Top FH, Jr, Russell PK, Dixon FJ, Muller-Eberhard HJ. The potential pathogenic role of complement in dengue hemorrhagic shock syndrome. *N Engl J Med* 1973;289:996–1000.
39. Pathogenetic mechanisms in dengue haemorrhagic fever: report of an international collaborative study. *Bull World Health Organ* 1973;48(1): 117–133.
40. Avirutnan P, Punyadee N, Noisakran S *et al.* Vascular leakage in severe dengue virus infections: a potential role for the nonstructural viral protein NS1 and complement. *J Infect Dis* 2006;193(8):1078–1088.
41. Libraty DH, Young PR, Pickering D *et al.* High circulating levels of the dengue virus nonstructural protein NS1 early in dengue illness correlate with the development of dengue hemorrhagic fever. *J Infect Dis* 2002; 186(8):1165–1168.
42. Courageot MP, Catteau A, Despres P. Mechanisms of dengue virus-induced cell death. In: Chambers TJ, Monath TP (eds.) *Advances in Virus Research* Elsevier, San Diego, California, 2003, pp. 157–186.
43. Catteau A, Courageot MP, Despres P. Flaviviruses and apoptosis regulation. *Prog Mol Subcell Biol* 2004;36:171–189.
44. Lin CF, Lei HY, Shiau AL *et al.* Endothelial cell apoptosis induced by antibodies against dengue virus nonstructural protein 1 via production of nitric oxide. *J Immunol* 2002;169(2):657–664.
45. Lin CF, Lei HY, Liu CC *et al.* Generation of IgM anti-platelet autoantibody in dengue patients. *J Med Virol* 2001;63(2):143–149.
46. Lin CF, Chiu SC, Hsiao YL *et al.* Expression of cytokine, chemokine, and adhesion molecules during endothelial cell activation induced by antibodies against dengue virus nonstructural protein 1. *J Immunol* 2005;174(1): 395–403.
47. Plotkin SA. Immunologic correlates of protection induced by vaccination. *Pediatr Infect Dis J* 2001;20(1):63–75.

48. Schlesinger RW, Gordon I, Frankel JW, Winter JW, Patterson PR, Dorrance WR. Clinical and serologic response of man to immunization with attenuated dengue and yellow fever viruses. *J Immunol* 1956;77(5): 352–364.
49. Whitehead RH, Chaicumpa V, Olson LC, Russell PK. Sequential dengue virus infections in the white-handed gibbon (*hylobates lar*). *Am J Trop Med Hyg* 1970;19(1):94–102.
50. Putnak R, Coller BA, Voss G *et al.* An evaluation of dengue type-2 inactivated, recombinant subunit, and live-attenuated vaccine candidates in the rhesus macaque model. *Vaccine* 2005;23(35):4442–4452.
51. Halstead SB, Palumbo NE. Studies on the immunization of monkeys against dengue. II. Protection following inoculation of combinations of viruses. *Am J Trop Med Hyg* 1973;22(3):375–381.
52. Chaturvedi UC, Tandon P, Mathur A, Kumar A. Host defence mechanisms against dengue virus infection of mice. *J Gen Virol* 1978;39:293–302.
53. Snapper CM, Finkelman FD. Immunoglobulin class switching. In: Paul WE (ed.) *Fundamental Immunology*, 3rd ed. Raven, New York, 1993, pp. 837–863.
54. Halstead SB, Yamarat C. Recent epidemics of hemorrhagic fever in Thailand: observations related to pathogenesis of a "new" dengue disease. *J Am Pub Health Assoc* 1965;55(9):1386–1395.
55. Innis BL, Nisalak A, Nimmannitya S *et al.* An enzyme-linked immunosorbent assay to characterize dengue infections where dengue and Japanese encephalitis co-circulate. *Am J Trop Med Hyg* 1989;40(4):418–427.
56. Halstead SB, Rojanasuphot S, Sangkawibha N. Original antigenic sin in dengue. *Am J Trop Med Hyg* 1983;32(1):154–156.
57. Okuno Y, Fukunaga T, Srisupaluck S, Kasemsarn P, Dharakul C, Sangkawibha N. Serological and virological studies on patients with dengue hemorrhagic fever (DHF) in Chanthaburi province, Thailand. I. Serological studies on paired sera from DHF patients by neutralization (N), hemagglutination inhibition (HI) and staining tests. *Biken J* 1980;23: 113–121.
58. Scott RM, Eckels KH, Bancroft WH *et al.* Dengue 2 vaccine: dose response in volunteers in relation to yellow fever immune status. *J Infect Dis* 1983; 148(6):1055–1060.
59. Kanesa-thasan N, Hernandez L, Lyons A *et al.* A phase I study of the WRAIR tetravalent live attenuated dengue vaccine in flavivirus-immune

adult volunteers. In 5th *Annual Conference on Vaccine Research*, 6–8 May 2002, Baltimore, Maryland.

60. Vaughn DW, Green S, Kalayanarooj S *et al.* Dengue in the early febrile phase: viremia and antibody responses. *J Infect Dis* 1997;176: 322–330.
61. Innis BL. Dengue and dengue hemorrhagic fever. In: Porterfield JS (ed.) *Kass Handbook of Infectious Diseases: Exotic Virus Infections*, 1st ed. Chapman & Hall, London, 1995, pp. 103–146.
62. Nawa M, Yamada KI, Takasaki T, Akatsuka T, Kurane I. Serotype-cross-reactive immunoglobulin M responses in dengue virus infections determined by enzyme-linked immunosorbent assay. *Clin Diagn Lab Immunol* 2000;7(5):774–777.
63. Se-Thoe SY, Ng MM, Ling AE. Retrospective study of Western blot profiles in immune sera of natural dengue virus infections. *J Med Virol* 1999;57(3):322–330.
64. Churdboonchart V, Bhamarapravati N, Peampramprecha S, Sirinavin S. Antibodies against dengue viral proteins in primary and secondary dengue hemorrhagic fever. *Am J Trop Med Hyg* 1991;44(5):481–493.
65. Kuno G, Vorndam AV, Gubler DJ, Gomez I. Study of anti-dengue NS1 antibody by Western blot. *J Med Virol* 1990;32:102–108.
66. Valdes K, Alvarez M, Pupo M, Vazquez S, Rodriguez R, Guzman MG. Human dengue antibodies against structural and nonstructural proteins. *Clin Diagn Lab Immunol* 2000;7(5):856–857.
67. AbuBakar S, Azmi A, Mohamed-Saad N, Shafee N, Chee HY. Antibody responses of dengue fever patients to dengue 2 (New Guinea C strain) viral proteins. *Malays J Pathol* 1997;19(1):41–51.
68. Sabin AB. Research on dengue during World War II. *Am J Trop Med Hyg* 1952;1:30–50.
69. Halstead SB, Casals J, Shotwell H, Palumbo N. Studies on the immunization of monkeys against dengue. I. Protection derived from single and sequential virus infections. *Am J Trop Med Hyg* 1973;22(3): 365–374.
70. Halstead SB. Etiologies of the experimental dengues of Siler and Simmons. *Am J Trop Med Hyg* 1974;23(5):974–982.
71. Markoff LJ, Innis BL, Houghten R, Henchal LS. Development of cross-reactive antibodies to plasminogen during the immune response to dengue virus infection. *J Infect Dis* 1991;164:294–301.

72. Chungue E, Poli L, Roche C, Gestas P, Glaziou P, Markoff LJ. Correlation between detection of plasminogen cross-reactive antibodies and hemorrhage in dengue virus infection. *J Infect Dis* 1994;170:1304–1307.
73. Monroy V, Ruiz BH. Participation of the dengue virus in the fibrinolytic process. *Virus Genes* 2000;21(3):197–208.
74. Lin CF, Lei HY, Shiau AL *et al*. Antibodies from dengue patient sera cross-react with endothelial cells and induce damage. *J Med Virol* 2003;69(1): 82–90.
75. Falconar AK. The dengue virus nonstructural-1 protein (NS1) generates antibodies to common epitopes on human blood clotting, integrin/adhesin proteins and binds to human endothelial cells: potential implications in haemorrhagic fever pathogenesis. *Arch Virol* 1997;142(5):897–916.
76. Thein S, Aaskov J, Myint TT, Shwe TN, Saw TT, Zaw A. Changes in levels of anti-dengue virus IgG subclasses in patients with disease of varying severity. *J Med Virol* 1993;40:102–106.
77. Smucny JJ, Kelly EP, MacArthy PO, King AD. Murine immunoglobulin G subclass responses following immunization with live dengue virus or a recombinant dengue envelope protein. *Am J Trop Med Hyg* 1995;53:432–437.
78. Staropoli I, Frenkiel MP, Megret F, Deubel V. Affinity-purified dengue-2 virus envelope glycoprotein induces neutralizing antibodies and protective immunity in mice. *Vaccine* 1997;15(17-18):1946–1954.
79. Simmons M, Murphy GS, Kochel T, Raviprakash K, Hayes CG. Characterization of antibody responses to combinations of a dengue-2 DNA and dengue-2 recombinant subunit vaccine. *Am J Trop Med Hyg* 2001;65(5):420–426.
80. Berke G. The functions and mechanisms of action of cytolytic lymphocytes. In: Paul WE (ed.) *Fundamental Immunology*, 3rd ed. Raven, New York, 1993, pp. 965–1014.
81. Henchal EA, Henchal LS, Schlesinger JJ. Synergistic interactions of anti-NS1 monoclonal antibodies protect passively immunized mice from lethal challenge with dengue 2 virus. *J Gen Virol* 1988;69:2101–2107.
82. Kurane I, Hebblewaite D, Brandt WE, Ennis FA. Lysis of dengue virus-infected cells by natural cell-mediated cytotoxicity and antibody-dependent cell-mediated cytotoxicity. *J Virol* 1984;52(1):223–230.

83. Qu XY, Chen WB, Maguire T, Austin F. Immunoreactivity and protective effects in mice of a recombinant dengue-2 Tonga virus-NS1 protein produced in a baculovirus expression system. *J Gen Virol* 1993;74:89–97.
84. Falgout B, Bray M, Schlesinger JJ, Lai CJ. Immunization of mice with recombinant vaccinia virus expressing authentic dengue virus nonstructural protein NS1 protects against lethal dengue virus encephalitis. *J Virol* 1990;64(9):4356–4363.
85. Tan CH, Yap EH, Singh M, Deubel V, Chan YC. Passive protection studies in mice with monoclonal antibodies directed against the non-structural protein NS3 of dengue 1 virus. *J Gen Virol* 1990;71:745–749.
86. Kaufman BM, Summers PL, Dubois DR *et al.* Monoclonal antibodies for dengue virus prM glycoprotein protect mice against lethal dengue infection. *Am J Trop Med Hyg* 1989;41:576–580.
87. Delenda C, Frenkiel MP, Deubel V. Protective efficacy in mice of a secreted form of recombinant dengue-2 virus envelope protein produced in baculovirus-infected insect cells. *Arch Virol* 1994;139:197–207.
88. Delenda C, Staropoli I, Frenkiel MP, Cabanie L, Deubel V. Analysis of C-terminally truncated dengue 2 and dengue 3 virus envelope glycoproteins: processing in insect cells and immunogenic properties in mice. *J Gen Virol* 1994;75:1569–1578.
89. Putnak R, Feighny R, Burrous J *et al.* Dengue-1 virus envelope glycoprotein gene expressed in recombinant baculovirus elicits virus-neutralizing antibody in mice and protects them from virus challenge. *Am J Trop Med Hyg* 1991;45(2):159–167.
90. Feighny R, Burrous J, McCown JM, Hoke CH, Jr, Putnak R. Purification of native dengue-2 viral proteins and the ability of purified proteins to protect mice. *Am J Trop Med Hyg* 1992;47(4):405–412.
91. Porter KR, Kochel TJ, Wu SJ, Raviprakash K, Phillips I, Hayes CG. Protective efficacy of a dengue 2 DNA vaccine in mice and the effect of CpG immuno-stimulatory motifs on antibody responses. *Arch Virol* 1998;143(5):997–1003.
92. Bray M, Zhao BT, Markoff L, Eckels KH, Chanock RM, Lai CJ. Mice immunized with recombinant vaccinia virus expressing dengue 4 virus structural proteins with or without nonstructural protein NS1 are protected against fatal dengue virus encephalitis. *J Virol* 1989;63(6):2853–2856.

93. Men RH, Bray M, Lai CJ. Carboxy-terminally truncated dengue virus envelope glycoproteins expressed on the cell surface and secreted extracellularly exhibit increased immunogenicity in mice. *J Virol* 1991;65(1): 1400–1407.
94. Zhang YM, Hayes EP, McCarty TC *et al.* Immunization of mice with dengue structural proteins and nonstructural protein NS1 expressed by baculovirus recombinant induces resistance to dengue virus encephalitis. *J Virol* 1988;62:3027–3031.
95. Putnak R, Fuller A, Vanderzanden L, Innis BL, Vaughn DW. Vaccination of rhesus macaques against dengue-2 virus with a plasmid DNA vaccine encoding the viral prM and E genes. *Am J Trop Med Hyg* 2003;68(4): 469–476.
96. Bray M, Men R, Lai CJ. Monkeys immunized with intertypic chimeric dengue viruses are protected against wild-type virus challenge. *J Virol* 1996;70:4162–4166.
97. Monath TP, McCarthy K, Bedford P *et al.* Clinical proof of principle for ChimeriVax: recombinant live, attenuated vaccines against flavivirus infections. *Vaccine* 2002;20(7-8):1004–1018.
98. Guirakhoo F, Pugachev K, Arroyo J *et al.* Viremia and immunogenicity in nonhuman primates of a tetravalent yellow fever – dengue chimeric vaccine: genetic reconstructions, dose adjustment, and antibody responses against wild-type dengue virus isolates. *Virology* 2002;298(1):146–159.
99. Simantini E, Banerjee K. Epitope mapping of dengue 1 virus E glycoprotein using monoclonal antibodies. *Arch Virol* 1995;140:1257–1273.
100. Aaskov JG, Williams L, Fletcher J, Hay R. Failure of a dengue 1 sub-unit vaccine to protect mice against a lethal dengue virus infection. *Am J Trop Med Hyg* 1988;39(5):511–518.
101. Feighny R, Burrous J, Putnak R. Dengue type-2 virus envelope protein made using recombinant baculovirus protects mice against virus challenge. *Am J Trop Med Hyg* 1994;50(3):322–328.
102. Halstead SB, Shotwell H, Casals J. Studies on the pathogenesis of dengue infection in monkeys. I. Clinical laboratory responses to primary infection. *J Infect Dis* 1973;128(1):7–14.
103. Kraiselburd E, Gubler DJ, Kessler MJ. Quantity of dengue virus required to infect rhesus monkeys. *Trans R Soc Trop Med Hyg* 1985;79: 248–251.

104. Lin YL, Liao CL, Chen LK *et al.* Study of dengue virus infection in SCID mice engrafted with human K562 cells. *J Virol* 1998;72(12):9729–9737.
105. Sun W, Nisalak A, Gettayacamin M *et al.* Protection of rhesus monkeys against dengue virus challenge after tetravalent live attenuated dengue virus vaccination. *J Infect Dis* 2006;193(12):1658–1665.
106. An J, Kimura-Kuroda J, Hirabayashi Y, Yasui K. Development of a novel mouse model for dengue virus infection. *Virology* 1999;263(1):70–77.
107. Edelman R, Tacket CO, Wasserman SS *et al.* A live attenuated dengue-1 vaccine candidate (45AZ5) passaged in primary dog kidney cell culture is attenuated and immunogenic for humans. *J Infect Dis* 1994;170: 1448–1455.
108. Plotkin S, Orenstein W, Offit P (eds.). *Vaccines*, 3rd ed. W. B. Saunders, Philadelphia, 1999.
109. Blanc G, Caminopetros. *J Bull Acad Med* 1929;102(26):47–50.
110. Kimura R, Hotta S. Studies on dengue: anti-dengue active immunization experiments in mice. *Jpn J Bacteriol* 1944;1:96–99.
111. Wisseman CL, Jr, Sweet BH, Rosenzweig EC, Eylar OR. Attenuated living type 1 dengue vaccines. *Am J Trop Med Hyg* 1963;12:620–623.
112. Bellanti JA, Bourke RTA, Rosenzweig EC, Eylar OR. Report of dengue vaccine trial in the Caribbean, 1963: a collaborative study. *Bull World Health Organ* 1966;35:93.
113. Wisseman CL, Jr. Prophylaxis of dengue, with special reference to live virus vaccine. *Proc Japan Soc Trop Med* 1966;7:51–56.
114. Halstead SB. Studies on the attenuation of dengue 4. *Asian J Infect Dis* 1978;2:112–117.
115. Bancroft WH, Scott RM, Eckels KH *et al.* Dengue virus type 2 vaccine: reactogenicity and immunogenicity in soldiers. *J Infect Dis* 1984;149(6):1005–1010.
116. Bhamarapravati N, Yoksan S, Chayaniyayothin T, Angsubphakorn S, Bunyaratvej A. Immunization with a live attenuated dengue-2-virus candidate vaccine (16681-PDK 53): clinical, immunological and biological responses in adult volunteers. *Bull World Health Organ* 1987;65: 189–195.
117. Bhamarapravati N, Yoksan S. Live attenuated tetravalent dengue vaccine. In: Gubler DJ, Kuno G (eds.) *Dengue and Dengue Hemorrhagic Fever*. CAB, New York, 1997, pp. 367–377.

118. Bhamarapravati N, Yoksan S. Live attenuated tetravalent dengue vaccine. *Vaccine* 2000;18(Suppl 2):44–47.
119. Vaughn DW, Hoke CH, Jr, Yoksan S *et al.* Testing of a dengue 2 live-attenuated vaccine (strain 16681 PDK 53) in ten American volunteers. *Vaccine* 1996;14:329–336.
120. Kanesa-thasan N, Sun W, Kim-Ahn G *et al.* Safety and immunogenicity of attenuated dengue virus vaccines (*Aventis Pasteur*) in human volunteers. *Vaccine* 2001;19(23–24):3179–3188.
121. Bhamarapravati N, Yoksan S. Study of bivalent dengue vaccine in volunteers. *Lancet* 1989;1:1077.
122. Sabchareon A, Lang J, Chanthavanich P *et al.* Safety and immunogenicity of tetravalent live-attenuated dengue vaccines in Thai adult volunteers: role of serotype concentration, ratio, and multiple doses. *Am J Trop Med Hyg* 2002;66(3):264–272.
123. Sabchareon A, Lang J, Chanthavanich P *et al.* Safety and immunogenicity of a three-dose regimen of two tetravalent live-attenuated dengue vaccines in five- to twelve-year-old Thai children. *Pediatr Infect Dis J* 2004;23(2):99–109.
124. Kitchener S, Nissen M, Nasveld P *et al.* Immunogenicity and safety of two live-attenuated tetravalent dengue vaccine formulations in healthy Australian adults. *Vaccine* 2006;24(9):1238–1241.
125. Sanchez V, Gimenez S, Tomlinson B *et al.* Innate and adaptive cellular immunity in flavivirus-naive human recipients of a live-attenuated dengue serotype 3 vaccine produced in Vero cells (VDV3). *Vaccine* 2006;24(23):4914–4926.
126. Mackowiak PA, Wasserman SS, Tacket CO *et al.* Quantitative relationship between oral temperature and severity of illness following inoculation with candidate attenuated dengue virus vaccines. *Clin Infect Dis* 1994; 19:948–950.
127. Kanesa-Thasan N, Edelman R, Tacket CO *et al.* Phase 1 studies of Walter Reed Army Institute of Research candidate attenuated dengue vaccines: selection of safe and immunogenic monovalent vaccines. *Am J Trop Med Hyg* 2003;69(6 Suppl):17–23.
128. Sun W, Edelman R, Kanesa-Thasan N *et al.* Vaccination of human volunteers with monovalent and tetravalent live-attenuated dengue vaccine candidates. *Am J Trop Med Hyg* 2003;69(6 Suppl):24–31.

129. Edelman R, Wasserman SS, Bodison SA *et al.* Phase I trial of 16 formulations of a tetravalent live-attenuated dengue vaccine. *Am J Trop Med Hyg* 2003;69(6 Suppl):48–60.
130. Sun W, Cunningham D, Wasserman SS, Perry J, Putnak JR, Eckels KH, Vaughn DW, Thomas, SJ, Kanesa-thasan N, Innis BL, Edelman R. Phase 2 clinical trial of three formulations of tetravalent live-attenuated dengue vaccine in flavivirus-naive adults. *Human Vaccines* 2008;4(6):(in press).
131. Simasathien S, Thomas SJ, Watanaveeradej V, Nisalak A, Barberousse C, Innis BL *et al.* Safety and immunogenicity of a tetravalent live-attenuated dengue vaccine in flavivirus naive children. *Am J Trop Med Hyg* 2008; 78(3):426–433.
132. Burton DR, Williamson RA, Parren PW. Antibody and virus: binding and neutralization. *Virology* 2000;270(1):1–3.
133. Murphy BR, Prince GA, Walsh EE *et al.* Dissociation between serum neutralizing and glycoprotein antibody responses of infants and children who received inactivated respiratory syncytial virus vaccine. *J Clin Microbiol* 1986;24(2):197–202.
134. Polack FP, Hoffman SJ, Crujeiras G, Griffin DE. A role for nonprotective complement-fixing antibodies with low avidity for measles virus in atypical measles. *Nat Med* 2003;9(9):1209–1213.
135. Griffin DE, Ward BJ, Esolen LM. Pathogenesis of measles virus infection: a hypothesis for altered immune responses. *J Infect Dis* 1994;170(Suppl 1): S24–31.
136. Graham BS, Henderson GS, Tang YW, Lu X, Neuzil KM, Colley DG. Priming immunization determines T helper cytokine mRNA expression patterns in lungs of mice challenged with respiratory syncytial virus. *J Immunol* 1993;151(4):2032–2040.
137. Guirakhoo F, Kitchener S, Morrison D, Forrat R, McCarthy K, Nichols R, Yoksan S, Duan X, Emak TH, Kanesa-Thasan N, Bedford P, Lang J, Quentin-Millet MJ, Monath TP. Live attenuated chimeric yellow fever dengue 2 (ChimeriVax-DEN2) vaccine Phase 1 clinical trial for safety and immunogenicity: effect of yellow fever pre-immunity in induction of cross neutralizing abtibody response to all 4 dengue serotypes. *Hum Vaccin* 2006;2(2):60–67.
138. Vaughn DW. Invited commentary: dengue lessons from Cuba. *Am J Epidemiol* 2000;152(9):800–803.

139. Buchy P, Vo VL, Bui KT *et al.* Secondary dengue virus type 4 infections in Vietnam. *Southeast Asian J Trop Med Public Health* 2005;36(1):178–185.
140. Burke DS, Nisalak A, Johnson DE, Scott RM. A prospective study of dengue infections in Bangkok. *Am J Trop Med Hyg* 1988;38(1):172–180.
141. Endy TP, Chunsuttiwat S, Nisalak A *et al.* Epidemiology of inapparent and symptomatic acute dengue virus infection: a prospective study of primary school children in Kamphaeng Phet, Thailand. *Am J Epidemiol* 2002;156(1):40–51.
142. Lai CJ, Zhao BT, Hori H, Bray M. Infectious RNA transcribed from stably cloned full-length cDNA of dengue type 4 virus. *Proc Natl Acad Sci USA* 1991;88:5139–5143.
143. Henchal EA, Putnak JR. The dengue viruses. *Clin Microbiol Rev* 1990;3(4):376–396.
144. Men R, Bray M, Clark D, Chanock RM, Lai CJ. Dengue type 4 virus mutants containing deletions in the 3′ noncoding region of the RNA genome: analysis of growth restriction in cell culture and altered viremia pattern and immunogenicity in rhesus monkeys. *J Virol* 1996;70:3930–3937.
145. Durbin AP, Karron RA, Sun W *et al.* Attenuation and immunogenicity in humans of a live dengue virus type-4 vaccine candidate with a 30-nucleotide deletion in its 3′-untranslated region. *Am J Trop Med Hyg* 2001;65(5):405–413.
146. Durbin AP, Whitehead SS, McArthur J *et al.* rDEN4 Delta 30, a live attenuated dengue virus type 4 vaccine candidate, is safe, immunogenic, and highly infectious in healthy adult volunteers. *J Infect Dis* 2005;191(5):710–718.
147. Whitehead SS, Falgout B, Hanley KA, Blaney JE, Jr, Markoff L, Murphy BR. A live, attenuated dengue virus type 1 vaccine candidate with a 30-nucleotide deletion in the 3′ untranslated region is highly attenuated and immunogenic in monkeys. *J Virol* 2003;77(2):1653–1657.
148. Hanley KA, Lee JJ, Blaney JE, Jr, Murphy BR, Whitehead SS. Paired charge-to-alanine mutagenesis of dengue virus type 4 NS5 generates mutants with temperature-sensitive, host range, and mouse attenuation phenotypes. *J Virol* 2002;76(2):525–531.
149. Blaney J, Johnson D, Manipon G *et al.* Genetic basis of attenuation of dengue virus type 4 small plaque mutants with restricted replication in

suckling mice and in SCID mice transplanted with human liver cells. *Virology* 2002;300(1):125.

150. Polo S, Ketner G, Levis R, Falgout B. Infectious RNA transcripts from full-length dengue virus type 2 cDNA clones made in yeast. *J Virol* 1997;71(7): 5366–5374.
151. Brinton MA, Fernandez AV, Dispoto JH. The 3′-nucleotides of flavivirus genomic RNA form a conserved secondary structure. *Virology* 1986;153(1): 113–121.
152. Zeng L, Falgout B, Markoff L. Identification of specific nucleotide sequences within the conserved 3′-SL in the dengue type 2 virus genome required for replication. *J Virol* 1998;72(9):7510–7522.
153. Blackwell JL, Brinton MA. Translation elongation factor-1 alpha interacts with the 3′ stem-loop region of West Nile virus genomic RNA. *J Virol* 1997;71(9):6433–6444.
154. Chen CJ, Kuo MD, Chien LJ, Hsu SL, Wang YM, Lin JH. RNA–protein interactions: involvement of NS3, NS5, and 3′ noncoding regions of Japanese encephalitis virus genomic RNA. *J Virol* 1997;71(5): 3466–3473.
155. Puri B, Nelson W, Porter KR, Henchal EA, Hayes CG. Complete nucleotide sequence analysis of a Western Pacific dengue-1 virus strain. *Virus Genes* 1998;17(1):85–88.
156. Markoff L, Pang X, Houng HS *et al.* Derivation and characterization of a dengue type 1 host range-restricted mutant virus that is attenuated and highly immunogenic in monkeys. *J Virol* 2002;76(7):3318–3328.
157. Innis BL, Eckels KH, Kraiselburd E *et al.* Virulence of a live dengue virus vaccine candidate: a possible new marker of dengue virus attenuation. *J Infect Dis* 1988;158(4):876–880.
158. Bray M, Lai CJ. Construction of intertypic chimeric dengue viruses by substitution of structural protein genes. *Proc Natl Acad Sci USA* 1991;88: 10342–10346.
159. Lai CJ, Bray M, Men R *et al.* Evaluation of molecular strategies to develop a live dengue vaccine. *Clin Diagn Virol* 1998;10(2-3):173–179.
160. Chen W, Kawano H, Men R, Clark D, Lai CJ. Construction of intertypic chimeric dengue viruses exhibiting type 3 antigenicity and neurovirulence for mice. *J Virol* 1995;69:5186–5190.

161. Blaney JE, Jr, Durbin AP, Murphy BR, Whitehead SS. Development of a live attenuated dengue virus vaccine using reverse genetics. *Viral Immunol* 2006;19(1):10–32.
162. Durbin AP, McArthur JH, Marron JA *et al.* rDEN2/4Delta30(ME), a live attenuated chimeric dengue serotype 2 vaccine, is safe and highly immunogenic in healthy dengue-naive adults. *Hum Vaccin* 2006; 2(6):255–260.
163. Kinney RM, Butrapet S, Chang GJ *et al.* Construction of infectious cDNA clones for dengue 2 virus: strain 16681 and its attenuated vaccine derivative, strain PDK-53. *Virology* 1997;230(2):300–308.
164. Butrapet S, Huang CY, Pierro DJ, Bhamarapravati N, Gubler DJ, Kinney RM. Attenuation markers of a candidate dengue type 2 vaccine virus, strain 16681 (PDK-53), are defined by mutations in the 5′ noncoding region and nonstructural proteins 1 and 3. *J Virol* 2000;74(7):3011–3019.
165. Butrapet S, Kinney RM, Huang CY. Determining genetic stabilities of chimeric dengue vaccine candidates based on dengue 2 PDK-53 virus by sequencing and quantitative TaqMAMA. *J Virol Methods* 2006; 131(1):1–9.
166. Huang CY, Butrapet S, Pierro DJ *et al.* Chimeric dengue type 2 (vaccine strain PDK-53)/dengue type 1 virus as a potential candidate dengue type 1 virus vaccine. *J Virol* 2000;74(7):3020–3028.
167. Huang CY, Butrapet S, Tsuchiya KR, Bhamarapravati N, Gubler DJ, Kinney RM. Dengue 2 PDK-53 virus as a chimeric carrier for tetravalent dengue vaccine development. *J Virol* 2003;77(21):11436–11447.
168. Rice CM, Grakoui A, Galler R, Chambers TJ. Transcription of infectious yellow fever RNA from full-length cDNA templates produced by *in vitro* ligation. *New Biol* 1989;1(3):285–296.
169. Guirakhoo F, Weltzin R, Chambers TJ *et al.* Recombinant chimeric yellow fever – dengue type 2 virus is immunogenic and protective in nonhuman primates. *J Virol* 2000;74(12):5477–5485.
170. Guirakhoo F, Pugachev K, Zhang Z *et al.* Safety and efficacy of chimeric yellow fever – dengue virus tetravalent vaccine formulations in nonhuman primates. *J Virol* 2004;78(9):4761–4775.
171. Guirakhoo F, Arroyo J, Pugachev KV *et al.* Construction, safety, and immunogenicity in nonhuman primates of a chimeric yellow fever — dengue virus tetravalent vaccine. *J Virol* 2001;75(16):7290–7304.

172. Monath TP, Myers GA, Beck RA *et al.* Safety testing for neurovirulence of novel live, attenuated flavivirus vaccines: infant mice provide an accurate surrogate for the test in monkeys. *Biologicals* 2005;33(3):131–144.
173. Higgs S, Vanlandingham DL, Klingler KA *et al.* Growth characteristics of ChimeriVax-Den vaccine viruses in *Aedes aegypti* and *Aedes albopictus* from Thailand. *Am J Trop Med Hyg* 2006;75(5):986–993.
174. Whalen RG. DNA vaccines for emerging infectious diseases: what if? *Emerg Infect Dis* 1996;2(3):168–175.
175. Kochel T, Wu SJ, Raviprakash K *et al.* Inoculation of plasmids expressing the dengue-2 envelope gene elicit neutralizing antibodies in mice. *Vaccine* 1997;15(5):547–552.
176. Wang B, Boyer J, Srikantan V *et al.* DNA inoculation induces neutralizing immune responses against human immunodeficiency virus type 1 in mice and nonhuman primates. *DNA Cell Biol* 1993;12(9):799–805.
177. Ulmer JB, Donnelly JJ, Parker SE *et al.* Heterologous protection against influenza by injection of DNA encoding a viral protein. *Science* 1993;259(5102):1745–1749.
178. Sedegah M, Hedstrom R, Hobart P, Hoffman SL. Protection against malaria by immunization with plasmid DNA encoding circumsporozoite protein. *Proc Natl Acad Sci USA* 1994;91(21):9866–9870.
179. Konishi E, Yamaoka M, Kurane I, Mason PW. A DNA vaccine expressing dengue type 2 virus premembrane and envelope genes induces neutralizing antibody and memory B cells in mice. *Vaccine* 2000;18(11–12):1133–1139.
180. Raviprakash K, Kochel TJ, Ewing D *et al.* Immunogenicity of dengue virus type 1 DNA vaccines expressing truncated and full length envelope protein. *Vaccine* 2000;18(22):2426–2434.
181. Raviprakash K, Porter KR, Kochel TJ *et al.* Dengue virus type 1 DNA vaccine induces protective immune responses in rhesus macaques. *J Gen Virol* 2000;81(Pt 7):1659–1667.
182. Raviprakash K, Apt D, Brinkman A *et al.* A chimeric tetravalent dengue DNA vaccine elicits neutralizing antibody to all four virus serotypes in rhesus macaques. *Virology* 2006;353(1):166–173.
183. Raviprakash K, Marques E, Ewing D *et al.* Synergistic neutralizing antibody response to a dengue virus type 2 DNA vaccine by incorporation of lysosome-associated membrane protein sequences and use of plasmid expressing GM-CSF. *Virology* 2001;290(1):74–82.

184. Eo SK, Gierynska M, Kamar AA, Rouse BT. Prime-boost immunization with DNA vaccine: mucosal route of administration changes the rules. *J Immunol* 2001;166(9):5473–5479.
185. Doolan DL, Hoffman SL. DNA-based vaccines against malaria: status and promise of the multi-stage malaria DNA vaccine operation. *Int J Parasitol* 2001;31(8):753–762.
186. Tellier MC, Pu R, Pollock D *et al.* Efficacy evaluation of prime-boost protocol: canarypoxvirus-based feline immunodeficiency virus (FIV) vaccine and inactivated FIV-infected cell vaccine against heterologous FIV challenge in cats. *AIDS* 1998;12(1):11–18.
187. Imoto J, Konishi E. Dengue tetravalent DNA vaccine increases its immunogenicity in mice when mixed with a dengue type 2 subunit vaccine or an inactivated Japanese encephalitis vaccine. *Vaccine* 2007;25(6): 1076–1084.
188. Gurunathan S, Klinman DM, Seder RA. DNA vaccines: immunology, application, and optimization. *Annu Rev Immunol* 2000;18:927–974.
189. Klinman DM, Takeno M, Ichino M *et al.* DNA vaccines: safety and efficacy issues. *Springer Semin Immunopathol* 1997;19(2):245–256.
190. Jechlinger W. Optimization and delivery of plasmid DNA for vaccination. *Expert Rev Vaccines* 2006;5(6):803–825.
191. Nichols WW, Ledwith BJ, Manam SV, Troilo PJ. Potential DNA vaccine integration into host cell genome. *Ann N Y Acad Sci* 1995;772:30–39.
192. Martin T, Parker SE, Hedstrom R *et al.* Plasmid DNA malaria vaccine: the potential for genomic integration after intramuscular injection. *Hum Gene Ther* 1999;10(5):759–768.
193. Parker SE, Borellini F, Wenk ML *et al.* Plasmid DNA malaria vaccine: tissue distribution and safety studies in mice and rabbits. *Hum Gene Ther* 1999;10(5):741–758.
194. Mor G, Singla M, Steinberg AD, Hoffman SL, Okuda K, Klinman DM. Do DNA vaccines induce autoimmune disease? *Hum Gene Ther* 1997;8(3): 293–300.
195. Prud'homme GJ, Lawson BR, Chang Y, Theofilopoulos AN. Immunotherapeutic gene transfer into muscle. *Trends Immunol* 2001;22(3):149–155.
196. von Herrath MG, Whitton JL. DNA vaccination to treat autoimmune diabetes. *Ann Med* 2000;32(5):285–292.

197. Karin N. Gene therapy for T cell-mediated autoimmunity: teaching the immune system how to restrain its own harmful activities by targeted DNA vaccines. *Isr Med Assoc J* 2000;(2 Suppl):63–68.
198. Mandl CW, Aberle JH, Aberle SW, Holzmann H, Allison SL, Heinz FX. *In vitro*-synthesized infectious RNA as an attenuated live vaccine in a flavivirus model [see comments]. *Nat Med* 1998;4(12):1438–1440.
199. Dubensky TW, Jr, Polo JM, Liu MA. Live virus vaccines: something old, something new, something borrowed... [news; comment]. *Nat Med* 1998; 4(12):1357–1358.
200. Simmons JS, St John JH, Reynolds FHK. Experimental studies of dengue. *Philippine J Sci* 1931;44:1–252.
201. Siler JF, Hall MW, Hitchens AP. Dengue: its history, epidemilogy, mechanism of transmission, etiology, clinical manifestations, immunity, and prevention. *Philippine J Sci* 1926;29:1–304.
202. Putnak R, Barvir DA, Burrous JM *et al.* Development of a purified, inactivated, dengue-2 virus vaccine prototype in Vero cells: immunogenicity and protection in mice and rhesus monkeys. *J Infect Dis* 1996;174:1176–1184.
203. Putnak R, Cassidy K, Conforti N *et al.* Immunogenic and protective response in mice immunized with a purified, inactivated, dengue-2 virus vaccine prototype made in fetal rhesus lung cells. *Am J Trop Med Hyg* 1996;55(5):504–510.
204. Aihara H, Takasaki T, Toyosaki-Maeda T, Suzuki R, Okuno Y, Kurane I. T-cell activation and induction of antibodies and memory T cells by immunization with inactivated Japanese encephalitis vaccine. *Viral Immunol* 2000;13(2):179–186.
205. Hoke CH, Jr., Nisalak A, Sangawhipa N *et al.* Protection against Japanese encephalitis by inactivated vaccines. *N Engl J Med* 1988;319:608–614.
206. Craig SC, Pittman PR, Lewis TE *et al.* An accelerated schedule for tick-borne encephalitis vaccine: the American military experience in Bosnia. *Am J Trop Med Hyg* 1999;61(6):874–878.
207. Brinton MA, Kurane I, Mathew A *et al.* Immune mediated and inherited defences against flaviviruses. *Clin Diagn Virol* 1998;10(2-3):129–139.
208. Rey FA, Heinz FX, Mandl C, Kunz C, Harrison SC. The envelope glycoprotein from tick-borne encephalitis virus at 2 angstrom resolution. *Nature* 1995;375:291–298.

209. Trent DW, Kinney RM, Huang CY. Recombinant dengue virus vaccines. In: Gubler DJ, Kuno G (eds.) Dengue and Dengue Hemorrhagic Fever. CAB, New York, 1997.
210. Fonseca BAL, Khoshnood K, Shope RE, Mason PW. Flavivirus type-specific antigens produced from fusions of a portion of the E protein gene with the Escherichia coli TRPE gene. *Am J Trop Med Hyg* 1991;44(5):500–508.
211. Srivastava AK, Putnak JR, Warren RL, Hoke CH, Jr. Mice immunized with a dengue type 2 virus E and NS1 fusion protein made in *Escherichia coli* are protected against lethal dengue virus infection. *Vaccine* 1995;13(13):1251–1258.
212. Simmons M, Nelson WM, Wu SJ, Hayes CG. Evaluation of the protective efficacy of a recombinant dengue envelope B domain fusion protein against dengue 2 virus infection in mice. *Am J Trop Med Hyg* 1998;58(5):655–662.
213. Kelly EP, Greene JJ, King AD, Innis BL. Purified dengue 2 virus envelope glycoprotein aggregates produced by baculovirus are immunogenic in mice. *Vaccine* 2000;18(23):2549–2559.
214. Velzing J, Groen J, Drouet MT *et al.* Induction of protective immunity against dengue virus type 2: comparison of candidate live attenuated and recombinant vaccines. *Vaccine* 1999;17(11-12):1312–1320.
215. Bielefeldt-Ohmann H, Beasley DW, Fitzpatrick DR, Aaskov JG. Analysis of a recombinant dengue-2 virus–dengue-3 virus hybrid envelope protein expressed in a secretory baculovirus system. *J Gen Virol* 1997;78(Pt 11): 2723–2733.
216. Sugrue RJ, Fu J, Howe J, Chan YC. Expression of the dengue virus structural proteins in *Pichia pastoris* leads to the generation of virus-like particles. *J Gen Virol* 1997;78(Pt 8):1861–1866.
217. Zhao BT, Prince G, Horswood R *et al.* Expression of dengue virus structural proteins and nonstructural protein NS1 by a recombinant vaccinia virus. *J Virol* 1987;61:4019–4022.
218. Deubel V, Kinney RM, Esposito JJ *et al.* Dengue 2 virus envelope protein expressed by a recombinant vaccinia virus fails to protect monkeys against dengue. *J Gen Virol* 1988;69:1921–1929.
219. Men R, Wyatt L, Tokimatsu I *et al.* Immunization of rhesus monkeys with a recombinant of modified vaccinia virus Ankara expressing a truncated envelope glycoprotein of dengue type 2 virus induced resistance to dengue type 2 virus challenge. *Vaccine* 2000;18(27):3113–3122.

220. Fonseca BAL, Pincus S, Shope RE, Paoletti E, Mason PW. Recombinant vaccinia viruses co-expressing dengue-1 glycoproteins prM and E induce neutralizing antibodies in mice. *Vaccine* 1994;12(3):279–285.
221. Konishi E, Fujii A. Dengue type 2 virus subviral extracellular particles produced by a stably transfected mammalian cell line and their evaluation for a subunit vaccine. *Vaccine* 2002;20(7-8):1058–1067.
222. Simmons M, Porter KR, Hayes CG, Vaughn DW, Putnak R. Characterization of antibody responses to combinations of a dengue virus type 2 DNA vaccine and two dengue virus type 2 protein vaccines in rhesus macaques. *J Virol* 2006;80(19):9577–9585.
223. Apt D, Raviprakash K, Brinkman A *et al.* Tetravalent neutralizing antibody response against four dengue serotypes by a single chimeric dengue envelope antigen. *Vaccine* 2006;24(3):335–344.
224. Roehrig JT, Johnson AJ, Hunt AR, Beaty BJ, Mathews JH. Enhancement of the antibody response to flavivirus B-cell epitopes by using homologous or heterologous T-cell epitopes. *J Virol* 1992;66:3385–3390.
225. Becker Y. Dengue fever virus and Japanese encephalitis virus synthetic peptides, with motifs to fit HLA class I haplotypes prevalent in human populations in endemic regions, can be used for application to skin Langerhans cells to prime antiviral CD8(+) cytotoxic T cells (CTLs) — a novel approach to the protection of humans. *Virus Genes* 1994;9: 33–45.
226. Huang JH, Wey JJ, Sun YC, Chin C, Chien LJ, Wu YC. Antibody responses to an immunodominant nonstructural 1 synthetic peptide in patients with dengue fever and dengue hemorrhagic fever. *J Med Virol* 1999;57(1):1–8.
227. Garcia G, Vaughn DW, Del Angel RM. Recognition of synthetic oligopeptides from nonstructural proteins NS1 and NS3 of dengue-4 virus by sera from dengue virus-infected children. *Am J Trop Med Hyg* 1997;56(4): 466–470.
228. Durbin AP, McArthur J, Marron JA, Blaney JE Jr, Thumar B, Wanionek K, Murphy BR, Whitehead SS. The live attenuated dengue serotype 1 vaccine rDEN1delta30 is safe and highly immunogenic in healthy adult volunteers. *Hum Vaccin* 2006;2(4):167–173.

5

Clinical Features of Dengue

Jeremy Farrar

Introduction

Millions of individuals across the tropical and subtropical world get infected with dengue viruses every year. A small percentage of them present with clinical disease, and an even smaller percentage present to a hospital, or a tertiary referral hospital. Especially outside the dengue-endemic countries, clinical descriptions published in the international literature are mostly from tertiary referral hospitals. With the enormous increase in tourism, business-related travel, global deployment of military and international non-governmental organizations in recent decades, dengue cases have been seen outside traditional endemic areas. The daytime biting habits of the *Aedes* mosquito and the urban habitat visited by most international travellers make it all but impossible to avoid exposure (bed nets probably offer little protection). There is no vaccine or prophylaxis available. Dengue infections in travellers are monitored by TropNetEurop, established in 1999,[1] to provide data on approximately 51,000 patients per year across 24 sites within Europe (wwwtropneteurope/dengue). The United States Centers for Disease Control provides a similar website to clinicians and the public (www.cdc.gov/dengue). During a three-year period 294 cases of dengue were reported to the TropNetEurop site. Most (78%) were returning from holidays or business-related travel (9%) to South and South-east Asia and the Americas.[2]

Of the many clinical features associated with dengue infections, from the standpoint of threat to life and clinical intervention the most important is increased vascular permeability leading to dengue shock syndrome (DSS). Children are particularly prone to the development of shock, in all likelihood because of age-related differences in capillary fragility that may make them more susceptible to the capillary leak syndrome than adults. As careful intervention at this stage may have a profound impact on outcome and prevent many of the secondary complications, such as haemorrhage, early clinical recognition of the syndrome is crucial (see Chap. 6, "Management of Dengue").

This chapter does not attempt to catalogue all symptoms or signs associated with dengue infection but attempts to place important features in a context of clinical and epidemiological decision-making.

Classification Schemes

Seminal work on dengue in the 1960s and 1970s, much of it reported from Thailand, has provided the classic clinical descriptions, definitions and treatment strategies of dengue haemorrhagic fever and dengue shock syndrome.[3] These studies have led to the World Health Organization case definition and classification of the disease and guidelines for treatment.[4] The case definitions and further classification of dengue fever (DF) and dengue haemorrhagic fever (DHF) are outlined in Table 1.

Clinical Features

In children, dengue is usually an asymptomatic infection. The ratio of asymptomatic to symptomatic infection is in favor of the former and depends on a multitude of epidemiological factors, but is approximately 1:40.[5] By the age of 6, 60% of children in southern Vietnam are circulating dengue antibodies. Similar results are reported from other endemic regions.[6–8] The vast majority of these dengue-immune individuals have suffered no apparent illness. However, in well-documented epidemics, particularly due to dengue 1 in island settings whose population had no known previous history of dengue, the clinical attack rate may have been as high as 40%.[9]

Table 1. WHO case definitions.

Case definition for dengue haemorrhagic fever

The following must all be present:

- Fever, or history of acute fever, lasting 2–7 days, occasionally biphasic.
- Haemorrhagic tendencies, evidenced by at least one of the following:
  - A positive tourniquet test;*
  - Petechiae, ecchymosis or purpura;
  - Bleeding from the mucosa, gastrointestinal tract, injection sites or other locations;
  - Haematemesis or melaena.
- Thrombocytopenia (100,000 cells per mm^3 or less).†
- Evidence of plasma leakage due to increased vascular permeability, manifested by at lest one of the following:
  - A rise in the haematocrit equal to or greater than 20% above the average for age, sex and population;
  - A drop in the haematocrit following volume replacement treatment equal to or greater than 20% of baseline;
  - Signs of plasma leakage such as pleural effusion, ascites and hypoproteinemia.

Case definition for dengue shock syndrome

All of the above four criteria for DHF must be present plus evidence of circulatory failure manifested by:

- Rapid and weak pulse, and
- Narrow pulse pressure [< 20 mm Hg (2.7 kPa)] or manifested by:
  - Hypotension for age, and
  - Cold, clammy skin and restlessness.

---

* The tourniquet test is performed by inflating a blood pressure cuff on the upper arm to a point midway between the systolic and diastolic pressures for 5 minutes. A test is considered positive when 20 or more petechiae per 2.5 cm (1 inch) square are observed. The test may be negative or mildly positive during the phase of profound shock. It usually becomes positive, sometimes strongly positive, if the test is conducted after recovery from shock.

† This number represents a direct count using a phase-contrast microscope (normal is 200,000–500,000 per mm^3). In practice, for outpatients, an approximate count from a peripheral blood smear is acceptable. In normal persons, 4–10 platelets per oil-immersion field (100×; the average of the readings from 10 oil-immersion fields is recommended) indicates an adequate platelet count. An average of ≤ 3 platelets per oil-immersion field is considered low (i.e., < 100,000 per mm^3).

Source: *Dengue Haemorrhagic Fever: Diagnosis, Treatment, Prevention and Control*, 2nd ed. World Health Organization, Geneva, 1997.

Dengue fever

Children

Primary infections with dengue virus types 2 and 4 are thought to be largely inapparent, regardless of age.[10,11–13] Primary infections with dengue types 1 and 3 are more often overt. Manifestations vary with age and from patient to patient. In infants and young children, the disease may be undifferentiated or characterized by a 1–5-day fever, pharyngeal inflammation, rhinitis and mild cough. Infants and children with high fevers may experience febrile convulsions. Distinctive mean incubation period, duration of illness or spectra of clinical findings could characterize disease with different dengue types, although this has as yet not been studied carefully. The DHF syndrome is differentiated from DF by its association with thrombocytopenia and capillary leakage.[4]

Adults

In classic DF, after an incubation period of 2–7 days, there is a sudden onset of fever, accompanied by a sensation of chilliness, which rapidly rises to 39.5–41.5°C (103–106°F), usually accompanied by frontal or retro-orbital headache. Occasionally, back pain precedes the fever. A transient, macular, generalized rash that blanches under pressure may be seen during the first 24–48 hours of fever. The pulse rate may be slow in proportion to the degree of fever. Myalgia or bone pain occurs soon after onset and increases in severity. During the second to the sixth day of fever, nausea and vomiting are apt to occur; during this phase, generalized lymphadenopathy, cutaneous hyperesthesia or hyperalgesia, taste aberrations and pronounced anorexia may develop.

Coincident with or 1 or 2 days after defervescence, a generalized, morbilliform, maculopapular rash may appear, sparing the palms and soles. It disappears in 1–5 days. In some cases, there is oedema of the palms and soles. Desquamation may occur. About the time of appearance of this second rash, the body temperature, which has fallen to normal, may become elevated slightly and establish the biphasic temperature curve.

Epistaxis, petechiae and purpuric lesions, although uncommon, may occur at any stage of the disease. Swallowed blood from epistaxis may be passed by the rectum or vomited, and can be interpreted as bleeding of gastrointestinal origin. Gastrointestinal bleeding, menorrhagia and bleeding from other organs have been observed in some DF outbreaks.[14–16] There is evidence that peptic ulcer predisposes to gastrointestinal haemorrhage; in some cases, patients may be in danger of exsanguinating, but without abnormal vascular permeability.[15] This syndrome can be confused with DSS and contributes to a misunderstanding of the pathogenesis of DHF/DSS. The mechanism of the haemorrhagic diathesis occurring commonly with dengue virus infections is not known, but speculation centres on platelet abnormalities. After the febrile stage, prolonged asthenia, mental depression, bradycardia and ventricular extrasystoles are common in adults.[17]

Dengue haemorrhagic fever (DHF grades I and II) is a severe form of dengue illness and is characterized by haemoconcentration, thrombocytopenia and coagulation abnormalities. When accompanied by narrow pulse pressure or hypotension, the illness is designated as *dengue shock syndrome (DHF grades III and IV)* (approximately 25% of hospitalized cases). Fatality rates of DHF vary between 1% and 5%, although much higher rates have been reported.

This is an acute vascular permeability syndrome accompanied by abnormal haemostasis. The incubation period of DHF/DSS is unknown but is presumed to be similar to that of DF. In children, the progression of the illness is characteristic.[4,18–20] A relatively mild first phase with abrupt onset of fever, malaise, vomiting, headache, anorexia and cough may be followed after 2–5 days by rapid deterioration and physical collapse. In Thailand, the median day of admission to hospital after onset of fever is day 4. In this second phase, the patient usually manifests cold, clammy extremities, a warm trunk, a flushed face and diaphoresis. Patients are restless and irritable and complain of midepigastric pain.

Dengue shock syndrome

Patients, particularly those in clinical shock, may show symptoms of mental obtundation with varying neurological reflex changes suggestive of encephalopathy. There frequently are scattered petechiae on the forehead and extremities, spontaneous ecchymoses may appear, and easy bruisability and bleeding at sites of venipuncture are common (black-and-white photos only). There may be circumoral and peripheral cyanosis. Respiration is rapid and often laboured. The pulse is weak, rapid and thready, and the heart sounds are faint. The pulse pressure is frequently narrow (20 mm Hg); the systolic and diastolic pressures may be low or unobtainable. The liver may become palpable two or three finger breadths below the costal margin and usually is firm and non-tender. A chest radiograph shows unilateral (*right*) or bilateral pleural effusions. Approximately 10% of patients manifest gross ecchymosis or gastrointestinal bleeding. After a 24- or 36-hour period of crisis, convalescence is fairly rapid in children who recover. The temperature may return to normal before or during the stage of shock. Higher mortality rates tend to occur in settings where there is less clinical experience in managing DSS, particularly the delicate fluid balance and avoidance of fluid overload. Complications include pleural effusions, shock, pneumonia, liver dysfunction or failure, encephalopathy, and severe bleeding, including cerebral and pulmonary haemorrhage.

Infants

Infants are at high risk of developing severe disease following dengue infection but are often omitted from clinical descriptions and research studies. In Bangkok in 1964, the incidence of DHF/DSS in infants under the age of one was 17/1000.[21] In a recent study in Vietnam, Thailand, Burma and Indonesia, infants accounted for approximately 5% of all DHF cases, with a median age at admission of seven months.[22] The age distribution of cases is consistent with the hypothesis of antibody-dependent immune enhancement resulting from antenatal transfer of dengue antibodies. When sub-protective levels of these antibodies are reached, residual antibodies result in enhanced dengue infection and disease. Clinically, infants often present with failure to feed, coryza, hepatomegaly, drowsiness and vomiting, Shock and bleeding are more common in infants than in children.[23–25]

Children

The tendency to shock in infants and children (or the protection of adults against developing DSS) may have a physiological explanation (see below — "Plasma Leak").

Adults

DHF and DSS have been generally thought to occur only in children. This impression derives from the extensive clinical experience in South-east Asia where, because of high transmission rates, the disease was typically restricted to childhood. By contrast, the early literature on DF mostly centred on adults.[26] With the global spread of dengue it is likely that there will be an increase in the number of adults presenting with severe dengue.

Dengue illnesses in adult patients may present differently to those in children. In an outbreak in New Delhi, India, in 1996, adults appear to have had a high rate of bleeding manifestations and less overt shock.[27,28] Rates of haematemesis (23–30%) and malaena (14%) were higher than usually reported from children, while DSS was low (2.5%) compared with children in the same epidemic where DSS occurred in approximately 48% of all hospitalized children in one series.[29,30] In a primary dengue 1 epidemic in Taiwan, underlying peptic ulcers led to gastrointestinal bleeding. Whether underlying disease, i.e. peptic ulcers in adults, or a different pathophysiological response contributes to such high rates is not known. No difference in the level of thrombocytopenia was reported in the Indian cases.

Differential Diagnosis

Classical dengue illness can be an easy diagnosis to make in endemic regions with experienced clinical staff and a high prior probability that a febrile illness with rash and thrombocytopenia is caused by dengue. Most of the symptoms and signs accompanying dengue infection are common to many febrile illnesses, with few features that reliably discriminate dengue especially early.[11] The differential diagnosis inevitably is very large. It is region- and even country- and season-specific. It includes measles, rubella, enterovirus, influenza, typhoid, chikungunya, scarlet fever, malaria, leptospirosis, hepatitis A, rickettsiosis, bacterial sepsis, Hantaan infection, viral haemorrhagic fevers (including Ebola, Lassa fever, etc.), West Nile fever, O'nyong-nyong fever and Rift Valley fever (usually without a rash). Because of the variation in clinical findings and the multiplicity of possible causative agents, the descriptive term "dengue-like disease" should be used until a specific aetiologic diagnosis is provided by the laboratory.

Specific Issues

Plasma leak

Clearly, there is an increase in microvascular permeability that leads to DSS. However, the underlying pathophysiological mechanism of haemodynamic shock in DSS remains tantalizingly unknown.

Using strain gauge plethysmography, increased microvascular permeability was demonstrated during the period around defervescence in DHF/DSS.[31] Using a related technique, Libraty *et al.* showed similar changes in the intravascular : extravascular fluid compartments.[32] Increased microvascular permeability was also observed in patients with the DF syndrome.[11] In addition, workers in Vietnam have recently demonstrated an increase in protein clearance in patients with DF, different in magnitude but qualitatively similar to the clearance seen in DSS. These studies suggest that there is a spectrum of capillary permeability and plasma leak across all individuals with overt dengue illnesses rather than distinct pathological mechanisms underlying DF and DHF.[33]

The above studies suggest that two separate mechanisms may exist to account for the elevated vascular permeability in patients with DHF. Some level of vascular permeability that lasts for at least several days occurs in all patients with clinically overt dengue. The second event is a rapid and marked but self-limiting increase in microvascular permeability, superimposed upon the first, which allows plasma water to flood out of the intravascular compartment and leads to sudden hypovolemic shock as the patients' compensatory mechanisms (increased lymphatic drainage, reabsorptive capacity) fail to cope. Gambel *et al.* also observed in a separate study that the capillary fragility of healthy children was twice that of healthy adults.[31,34] This higher microvascular permeability in childhood probably results from the greater density and surface area of growing microvessels in children than in adults and may help explain why children are more prone to develop DSS than adults, as was observed in Cuba, where children 3–14 years old with secondary DEN-2 infections had significantly higher rates of shock and much higher mortality rates (14.5-fold higher) than young adults aged 15–39 years.[35]

Dengue in pregnancy and neonates

There are scattered reports in the literature of dengue infection acquired during pregnancy causing problems for both mother and baby. However, the systematic prospective studies needed to calculate the true burden of dengue on pregnancy are not available. A threefold increase in neural tube defects was noted in babies conceived during a 1988 outbreak of dengue in northern India compared with historical controls.[36] By contrast, among nine Taiwanese women with confirmed dengue infection during their first trimester, no adverse effects on the pregnancy or baby were noted.[37] Among 22 pregnant women with a dengue like illness in French Guiana followed to term during 1992–1998, 3 babies died and 3 were born prematurely.[38] The same group reported even more severe consequences of presumptive dengue infections for the mothers, including increased risk of premature delivery, severe haemorrhaging during Caesarean section and abruptio placenta. In this series 8/38 babies were born prematurely. There were 5 foetal deaths, and 4 cases of acute foetal distress during labour and 2 cases of mother-to-child dengue transmission.[39] This high level of foetal wastage was not observed in Cuba, where children born to 59 women 5–9 months after the 1981 dengue 2 epidemic were followed for 5 years. There were 4 babies born who were anti-dengue IgM-positive. These newborns showed no apparent clinical damage at birth or at follow-up when they reached 5 years of age.[40]

Mother-to-baby transmission of the dengue virus has been the subject of a number of case reports. A case of severe respiratory distress, intracerebral haemorrhage and neonatal death has been reported.[41] More commonly, vertical transmission of the dengue virus has been associated only with a mild febrile illness and thrombocytopenia on the third-ninth day after delivery,[41,42] Reports from India and French Guiana on effects particularly of neural tube defects and foetal death are not borne out by studies in Cuba.

Large prospective studies will be needed to clarify the impact of dengue infections on gravid females.

The tourniquet test

The tourniquet test is performed by inflating a blood pressure cuff on the upper arm of the patient to a point midway between systolic and diastolic pressure for 5 minutes, and the number of resulting petechiae counted in a 2.5 cm square on the volar aspect of the forearm just distal to the antecubital fossa. A test is considered positive when 20 or more petechiae are observed in the 2.5 cm square.

The World Health Organization recommends use of the tourniquet test to document "haemorrhagic tendencies", one of the four elements in the clinical case definition of DHF (the others are fever, or history of acute fever; thrombocytopenia; and evidence of plasma leakage). Haemorrhagic tendencies also include petechiae, ecchymoses or purpura, bleeding from the mucosa, gastrointestinal tract, injection sites, haematemisis and or malaena. In some dengue-endemic countries a positive tourniquet test is thought to be required for the diagnosis of DHF; often, a positive tourniquet test supplants examination for the presence of other elements of the case definition.

Is the tourniquet test a valuable predictor of DHF? What does the evidence show? In a study of 240 children in Delhi in 1996 with WHO case definitions applied (all patients had confirmatory serology), the tourniquet test was positive in 40% of children with DF, 62% with DHF, and 64% with DSS.[29] In 110 adult patients hospitalized with DHF in north India in 1996, the tourniquet test was positive in 39.1% of all DHF cases. In 172 Thai children, 36% with DF and 52% with DHF had positive tourniquet tests. In addition, a positive tourniquet test was found in 21% of children with a nondengue viral infection. A positive tourniquet test was significantly more likely in children with dengue infection.[11] In 905 Vietnamese children (548 with dengue confirmed serologically and 357 in whom dengue was excluded serologically), the test had a sensitivity of 41.6%, a specificity of 94.4%, a positive predictive value of 98.3% and a negative predictive value of 17.3%. In this study other bleeding signs were invariably present and the tourniquet test provided additional information to aid diagnosis in only 5% of cases.[30]

Overall, a positive standard tourniquet test is reasonably specific for dengue infection, if performed on children suspected to have dengue in an endemic area where the probability of dengue is high. It should be remembered that a negative test does not exclude dengue infection. A careful inspection of the skin for petechiae or other bleeding can contribute importantly to the correct diagnosis.

The tourniquet test should be regarded as suggestive of dengue infection but not used as an absolute criterion for making the diagnosis. Nor is the test helpful in defining the severity of illness. It can be difficult to interpret in dark-skinned individuals. The fact that petechiae may be difficult to observe on dark skin may contribute to the differences seen in the results of the tourniquet test in different populations. The time taken to perform a tourniquet test may be better applied to an overall assessment of the patient with suspected dengue infection.

Coagulopathy

The exact mechanisms responsible for bleeding tendencies during dengue infections are not understood. Thrombocytopenia (platelet count <100,000/μl) forms one of the four criteria in the WHO case definition of DHF. In DHF, platelet function is abnormal and transfused platelets have a markedly shortened survival time.[43] Prothrombin and partial thromboplastin times are mildly prolonged and fibrinogen levels slightly reduced. Reductions in the blood levels of specific coagulation factors, including II, V, VII, VIII, IX, X, antithrombin and alpha-2-antiplasmin have been demonstrated in several studies on DHF. Levels of protein C, protein S and antithrombin III are normal or mildly reduced. Except when dengue is complicated with massive fluid overload and multiorgan dysfunction, bleeding is not caused by classical disseminated intravascular coagulation (DIC).[43]

There are abnormalities in all the major pathways of the coagulation cascade (low levels of the natural anticoagulant proteins, together with increased levels of the major procoagulants tissue factor and PAI-1). These findings are compatible with the possibility that dengue infection activates fibrinolysis in the absence of a thrombotic stimulus, degrading fibrinogen directly, and prompting a marked and effective secondary activation of various procoagulants' homeostatic pathways. Bleeding in dengue probably results from a combination of thrombocytopenia, poor platelet function and increased fibrinolysis. (What about damage to capillaries?)

In interpreting data from studies on coagulation factors in dengue, or indeed on other soluble mediators (cytokines, chemokines, histamine and other vasoactive peptides), many of which have a molecular weight in the range of 30,000–70,000 daltons, i.e. similar to that of albumin at 69,000 daltons, in the presence of increased vascular permeability many of these proteins will leak into extravascular spaces along with albumin. When children are resuscitated with intravascular fluids, there is dilutional effect that may result in reduced plasma levels of small proteins. Low levels of mediators have commonly and incorrectly been attributed to increased consumption or decreased production during the acute plasma leakage stage of dengue. Blood levels of soluble mediators should be corrected for the plasma leak and the dilutional effect of therapy.

Rarely does clinically significant bleeding in DHF lead to a drop in haemoglobin or require a blood transfusion (<1% of all dengue admissions in most series). Before the era of modern replacement therapy, severe gastrointestinal haemorrhages were reported at higher incidence than today. Inevitably, the attribution of disease severity to haemorrhagic manifestations of DHF may unintentionally distract clinicians from focusing on the major problem, i.e. vascular permeability and the management of the resulting shock.[43–47]

Severe bleeding typically develops once a patient gets into a vicious cycle of unresponsive shock, massive fluid replacement, fluid overload, worsening metabolic state. Finally, haemorrhage is part of a widespread multiorgan dysfunction. Careful and early fluid resuscitation when shock is first recognized may prevent this cycle and hence the development of severe bleeding.

Neurological manifestations

Dengue has been associated in a number of case reports and small patient series with many neurological phenomena. There have been few well-designed prospective clinical descriptive studies conducted to elucidate the incidence of neurological manifestations in dengue.

Retrospective studies of patients with suspected dengue from Cuba, India, Nicaragua, Thailand and Vietnam document an encephalopathic syndrome with severe dengue illnesses (0.5–3%) admitted to tertiary referral hospitals.[48–51] In Vietnam the question was reversed to ask how many patients admitted to hospital with clinically diagnosed encephalitis had evidence of concurrent dengue infection. The answer was that approximately 5% (16/378) of such patients had serological evidence of acute dengue infection compared to 1% of controls.[52] In 60% of these patients the CNS syndrome was the presenting complaint.

The commonest neurological manifestations associated with dengue infection are headache (approximately 40–70% of hospitalized patients), restlessness (10%) and muscle pains (10%). More severe neurological features include altered consciousness that may be accompanied by generalized convulsions. Patients may be in a coma (Glasgow Coma Score ≤ 11; Blantyre Coma Score ≤ 3), but a milder alteration of consciousness, lethargy, drowsiness and occasionally agitation are more common. Focal upper motor neuron signs, extrapyramidal features and transverse myelitis have been reported. The cerebrospinal fluid (CSF) pressure in patients with encephalopathic signs is moderately elevated (10–20 cms) with a normal CSF including the blood glucose ratio and a mildly raised protein. There may be no cells in the CSF or a mild lymphocyte pleocytosis (< 500 cells/μl). In patients with severe encephalopathy the mortality rate may be as high as 22%.[48] Survivors were left with persisting convulsions, personality changes and residual spasticity. Magnetic resonance imaging (MRI) performed on 18 patients with dengue-associated encephalopathy showed diffuse cerebral oedema and scattered focal lesions with no predilection for the basal ganglia as seen in Japanese encephalitis.[48] There are few autopsy studies in dengue. Most show cerebral oedema, vascular congestion, and focal haemorrhage.[53–56] There was no pathological evidence of viral encephalitis, nor have dengue viruses been isolated from brain tissue. In 18 autopsies conducted on children dying from dengue in Burma, Rosen *et al.* found no dengue virus in the brain by viral isolation or RT-PCR.[57]

Myalgia or back pain is a presenting feature in 10–50% of patients. The creatinine phosphokinase is usually normal or moderately elevated. Muscle biopsy shows a mild perivascular mononuclear infiltrate and lipid accumulation.[58] Anecdotally, myalgia and back pain can be debilitating and can continue for many weeks after the end of the febrile period. Mental depression in the weeks and months after dengue may be relatively common. Suicides have been reported during the recovery period after DF.[59] Acute polyradiculitis with muscle weakness and areflexia (compatible with the Guillain–Barre syndrome) and presenting 4–6 weeks after an acute dengue illness has been reported.[60]

The available evidence suggests that the most common neurological manifestations in dengue are secondary to severe systemic dengue illness including hepatic failure and circulatory collapse. However, it seems probable from the growing literature that dengue can directly cause an as-yet-poorly-defined neurological syndrome characterized by reduced consciousness, convulsions and occasionally long term sequellae in the absence of the classical clinical picture of dengue.

Liver involvement

The liver may be enlarged in a high percentage of all children admitted with serologically confirmed severe dengue, although the incidence in a South American study is lower.[29,50,61] Hepatomegaly was significantly associated with DHF compared to the dengue fever syndrome (55% vs. 18%; $p < 0.01$).[62] Jaundice is rare, even in severe DSS.[49,62]

Aspartate aminotransferase (AST) and alanine aminotransferase (ALT) blood levels were elevated in 60–90% of children with DHF.[63] The serum bilirubin, alkaline phosphatase, and gamma-glutamyl transpeptidase (G-GT) were elevated in 7%, 16% and 83% of patients. In the majority of cases the elevation in the transaminases was mild to moderate, but in a small group (7–10%) it can be 10× the upper limit of normal.[64]

Changes in the liver enzymes do not appear to be affected by coinfection with hepatitis B or hepatitis C or between primary and secondary dengue infection, but were significantly higher in patients with bleeding episodes. It is likely that liver involvement worsens the propensity for haemorrhage in dengue.

Liver biopsies, mostly at postmortem, have been reported in only a handful of patients. In a series of five children who died of DHF, Huerre reported severe, diffuse hepatitis with midzonal necrosis and steatosis and focal areas of necrosis in two patients. In one of the patients Kupffer cells were destroyed in areas with focal or severe necrosis. Dengue viral antigen was detected in hepatocytes and Kupffer cells.[65]

Hepatic failure associated with encephalopathy was reported in 18 patients with severe shock, metabolic acidosis, DIC and multiorgan failure, including both hepatic and brain dysfunction.[66] Lum reported 8 children (4% of the total dengue admissions) with hepatic failure and encephalopathy from a retrospective analysis of 20 severe admissions to a hospital in Kuala Lumpur, Malaysia, in a 2-year period, 1990–1991. Interestingly 4 of the 8 children were under 1 year of age; the others were 3, 6, 10 and 11 years old. Two of the 8 had Grade 4 DHF and 6 had Grade 3. Serum ammonia was normal in all patients, making Reye's syndrome unlikely.[67]

Dermatology

The skin is involved at some stage in the majority of patients with dengue. During the first few days of illness there may be a generalized macular blush and the skin is sensitive to touch. Later, spontaneous petechiae develop in up to 80% of patients and bleeding at injection sites and into the conjunctival and mucous membranes is common. An irritating haemorrhagic erythema of the palms and soles of the hands and feet, which may desquamate, has been described in adults. During the recovery period a macular, discrete and occasionally itchy rash can develop. This has a characteristic appearance with extensive erythematous areas surrounding discrete islands of apparently normal skin. Histology of the affected skin shows a non-specific minor predominantly T-cell lymphocytic dermal vasculitis.[68] *In vitro* experiments have confirmed that dermal Langerhan's cells are capable of supporting viral replication.[69]

Renal, bone marrow, myocardium, and other diseases (thalassaemia, G6PD, etc.)

Renal involvement in dengue is rare except as part of severe multiorgan failure.[70,71] Percutaneous needle biopsies were performed on the kidneys of 20 patients with DHF and renal involvement. Immune complex and C3 were localized in the glomeruli and electron microscopy showed focal thickening of the glomerular basement membrane, with hypertrophy of mesangial cells suggestive of a degree of glomerulonephritis.

The bone marrow may be involved in dengue infection and may be a site of viral replication. There is bone marrow hypocellularity, increased megakaryocytes, and characteristic but still poorly described atypical lymphocytes during the acute infection.[72–79]

References

1. Jelinek T. Dengue fever in international travelers. *Clin Infect Dis* 2000;31(1):144–147.
2. Jelinek T *et al.* Epidemiology and clinical features of imported dengue fever in Europe: sentinel surveillance data from TropNetEurop. *Clin Infect Dis* 2002;35(9):1047–1052.
3. Cohen SN, Halstead SB. Shock associated with dengue infection. I. Clinical and physiologic manifestations of dengue hemorrhagic fever in Thailand, 1964. *J Pediatr* 1966;68(3):448–456.
4. WHO. *Dengue Haemorrhagic Fever: Diagnosis, Treatment, Prevention and Control.* World Heath Organization, 1997.
5. Halstead SB. Epidemiology of dengue and dengue haemorrhagic fever. *Dengue and Dengue Hemorrhagic Fever*, 1st ed. CAB, Wallingford, Oxford.
6. Bartley LM *et al.* Assessment of the factors associated with flavivirus seroprevalence in a population in southern Vietnam. *Epidemiol Infect* 2002;128(2):213–220.
7. Ooi EE *et al.* Dengue seroepidemiology in Singapore. *Lancet* 2001;357(9257):685–686.
8. Guzman MG *et al.* Epidemiologic studies on dengue in Santiago de Cuba, 1997. *Am J Epidemiol* 2000;152(9):793–799; discussion 804.

9. Gubler DJ. Dengue and dengue hemorrhagic fever; its history and resurgence as a global public health problem. In: DJ Gulber, G Kuno (eds.) *Dengue and Dengue Hemorrhagic Fever*. CAB International, Wallingford, Oxford, 1997.
10. Guzman MG *et al.* Dengue hemorrhagic fever in Cuba, 1981: a retrospective seroepidemiologic study. *Am J Trop Med Hyg* 1990;42(2):179–184.
11. Kalayanarooj S *et al.* Early clinical and laboratory indicators of acute dengue illness. *J Infect Dis* 1997;176(2):313–321.
12. Vaughn DW *et al.* Dengue viremia titer, antibody response pattern, and virus serotype correlate with disease severity. *J Infect Dis* 2000;181(1):2–9.
13. Vaughn DW, Invited commentary: dengue lessons from Cuba. *Am J Epidemiol* 2000;152(9):800–803.
14. Rice L, Dengue fever: a clinical report of the Galveston epidemic of 1922. *Am J Trop Med Hyg* 1923;3:73–90.
15. Tsai JC, KCH, Chen PC. Upper gastrointestinal bleeding in dengue fever. *Am J Gastroenterol* 1991;86:33–35.
16. Wen KH, Sheu MM, Chung CB *et al.* The ocular fundus findings in dengue fever. *Kaohsiung J Med Sci* 1989;5:24–30 (in Chinese).
17. Lumley GF, Taylor FH. *Dengue.* School of Public Health and Tropical Medicine Service Publication Number 3, Australasian Medical Publishing Company, 1943.
18. Clarke JL. Privatization of mosquito control services in urban areas. *Kaohsiung J Med Sci* 1994;10:S74–S77 (in Chinese).
19. Nimmannitya S, Halstead SB, Cohen SN *et al.* Dengue and chikungunya virus infections in man in Thailand, 1962–1964. *Am J Trop Med Hyg* 1969;18:954–971.
20. Nimmannitya S. Clinical spectrum and management of dengue haemorrhagic fever. *Southeast Asian J Trop Med Public Health* 1987;18(3):392–397.
21. Halstead SB *et al.* Dengue chikungunya virus infection in man in Thailand, 1962–1964. 3. Clinical, epidemiologic, and virologic observations on disease in non-indigenous white persons. *Am J Trop Med Hyg* 1969;18(6):984–996.
22. Halstead SB *et al.* Dengue hemorrhagic fever in infants: research opportunities ignored. *Emerg Infect Dis* 2002;8(12):1474–1479.
23. Dietz V *et al.* The 1986 dengue and dengue hemorrhagic fever epidemic in Puerto Rico: epidemiologic and clinical observations. *P R Health Sci J* 1996;15(3):201–210.

24. Hongsiriwon S. Dengue hemorrhagic fever in infants. *Southeast Asian J Trop Med Public Health* 2002;33(1):49–55.
25. Martinez E *et al.* Dengue fever and hemorrhagic dengue in infants with a primary infection. *Rev Cubana Med Trop* 1993;45(2):97–101.
26. Ibrahim NM, I Cheong. Adult dengue haemorrhagic fever at Kuala Lumpur Hospital: retrospective study of 102 cases. *Br J Clin Pract* 1995;49(4):189–191.
27. Tripathi BK *et al.* Experience in adult population in dengue outbreak in Delhi. *J Assoc Physicians India* 1998;46(3):273–276.
28. Anuradha S *et al.* The 1996 outbreak of dengue hemorrhagic fever in Delhi, India. *Southeast Asian J Trop Med Public Health* 1998;29(3):503–506.
29. Kabra SK *et al.* Dengue haemorrhagic fever in children in the 1996 Delhi epidemic. *Trans R Soc Trop Med Hyg* 1999;93(3):294–298.
30. Cao XT *et al.* Evaluation of the World Health Organization standard tourniquet test and a modified tourniquet test in the diagnosis of dengue infection in Vietnam. *Trop Med Int Health* 2002;7(2):125–132.
31. Bethell DB *et al.* Noninvasive measurement of microvascular leakage in patients with dengue hemorrhagic fever. *Clin Infect Dis* 2001;32(2): 243–253.
32. Libraty DH *et al.* Assessment of body fluid compartment volumes by multifrequency bioelectrical impedance spectroscopy in children with dengue. *Trans R Soc Trop Med Hyg* 2002;96(3):295–299.
33. Wills BA, Oragui EE, Dung NM, Loan HT, Chau NV, Farrar JJ, Levin M. Dengue shock syndrome evidence for selective loss of plasma protein by a charge dependent molecular sieving process. *J Infect Dis* 2004;190(4):810–818.
34. Gamble J *et al.* Age-related changes in microvascular permeability: a significant factor in the susceptibility of children to shock? *Clin Sci (Lond)* 2000;98(2):211–216.
35. Guzman MG *et al.* Effect of age on outcome of secondary dengue 2 infections. *Int J Infect Dis* 2002;6(2):118–124.
36. Sharma JB, Gulati N. Potential relationship between dengue fever and neural tube defects in a northern district of India. *Int J Gynaecol Obstet* 1992;39(4):291–295.
37. Chong KY, Lin KC. A preliminary report of the fetal effects of dengue infection in pregnancy. *Kaohsiung J Med Sci* 1989;5(1):31–34 (in Chinese).
38. Carles G, Peiffer H, Talarmin A. Effects of dengue fever during pregnancy in French Guiana. *Clin Infect Dis* 1999;28(3):637–640.

39. Carles G *et al.* Dengue fever and pregnancy: a study of 38 cases in French Guiana. *J Gynecol Obstet Biol Reprod (Paris)* 2000;29(8):758–762.
40. Fernandez R *et al.* Study of the relationship dengue–pregnancy in a group of Cuban mothers. *Rev Cubana Med Trop* 1994;46(2):76–78.
41. Chye JK *et al.* Vertical transmission of dengue. *Clin Infect Dis* 1997;25(6):1374–1377.
42. Boussemart T *et al.* Prenatal transmission of dengue: two new cases. *J Perinatol* 2001;21(4):255–257.
43. Wills BA *et al.* Coagulation abnormalities in dengue hemorrhagic fever: serial investigations in 167 Vietnamese children with dengue shock syndrome. *Clin Infect Dis* 2002;35(3):277–285.
44. Krishnamurti C *et al.* Mechanisms of hemorrhage in dengue without circulatory collapse. *Am J Trop Med Hyg* 2001;65(6):840–847.
45. Mairuhu AT *et al.* Is clinical outcome of dengue-virus infections influenced by coagulation and fibrinolysis? A critical review of the evidence. *Lancet Infect Dis* 2003;3(1):33–41.
46. Bhamarapravati N. Hemostatic defects in dengue hemorrhagic fever. *Rev Infect Dis* 1989;11(Suppl 4):S826–S829.
47. Srichaikul T *et al.* Fibrinogen metabolism and disseminated intravascular coagulation in dengue hemorrhagic fever. *Am J Trop Med Hyg* 1977;26(3): 525–532.
48. Cam BV *et al.* Prospective case-control study of encephalopathy in children with dengue hemorrhagic fever. *Am J Trop Med Hyg* 2001;65(6): 848–851.
49. Kabra SK *et al.* Dengue hemorrhagic fever: clinical manifestations and management. *Indian J Pediatr* 1999;66(1):93–101.
50. Harris E *et al.* Clinical, epidemiologic, and virologic features of dengue in the 1998 epidemic in Nicaragua. *Am J Trop Med Hyg* 2000;63(1–2):5–11.
51. Pancharoen C, Mekmullica J, Thisyakorn U. Primary dengue infection: what are the clinical distinctions from secondary infection? *Southeast Asian J Trop Med Public Health* 2001;32(3):476–480.
52. Solomon T *et al.* Neurological manifestations of dengue infection. *Lancet* 2000;355(9209):1053–1059.
53. Bhamarapravati N, Tuchinda P, Boonyapaknavik V. Pathology of Thailand haemorrhagic fever: a study of 100 autopsy cases. *Ann Trop Med Parasitol* 1967;61(4):500–510.

54. Janssen HL *et al.* Fatal cerebral oedema associated with primary dengue infection. *J Infect* 1998;36(3):344–346.
55. Fresh JW *et al.* Philippine hemorrhagic fever: a clinical, laboratory, and necropsy study. *J Lab Clin Med* 1969;73(3):451–458.
56. Couvelard A *et al.* Report of a fatal case of dengue infection with hepatitis: demonstration of dengue antigens in hepatocytes and liver apoptosis. *Hum Pathol* 1999;30(9):1106–1110.
57. Rosen L, Drouet MT, Deubel V. Detection of dengue virus RNA by reverse transcription-polymerase chain reaction in the liver and lymphoid organs but not in the brain in fatal human infection. *Am J Trop Med Hyg* 1999;61(5): 720–724.
58. Malheiros SM *et al.* Dengue: Muscle biopsy findings in 15 patients. *Arq Neuropsiquiatr* 1993;51(2):159–164.
59. Lopez-Velez R *et al.* Dengue in Spanish travelers returning from the tropics. *Eur J Clin Microbiol Infect Dis* 1996;15(10):823–826.
60. Esack A, Teelucksingh S, Singh N. The Guillain-Barre syndrome following dengue fever. W*est Indian Med J* 1999;48(1):36–37.
61. Chairulfatah A *et al.* Clinical manifestations of dengue haemorrhagic fever in children in Bandung, Indonesia. *Ann Soc Belg Med Trop* 1995;75(4):291–295.
62. Phuong CX, Nhan NT, Kneen R, Thuy PT, van Thien C, Nga NT, Thuy TT, Solomon T, Stepniewska K, Wills B; Dong Nai Study Group. Clinical diagnosis and assessment of severity of confirmed dengue infections in Vietnamese children: is the WHO classification system helpful? *Am J Trop Med Hyg* 2004;70(2):172–179.
63. Kuo CH *et al.* Liver biochemical tests and dengue fever. *Am J Trop Med Hyg* 1992;47(3):265–270.
64. Nguyen TL, Nguyen TH, Tieu NT. The impact of dengue haemorrhagic fever on liver function. *Res Virol* 1997;148(4):273–277.
65. Huerre MR *et al.* Liver histopathology and biological correlates in five cases of fatal dengue fever in Vietnamese children. *Virchows Arch* 2001;438(2): 107–115.
66. Nimmannitya S, Thisyakorn U, Hemsrichart V. Dengue haemorrhagic fever with unusual manifestations. *Southeast Asian J Trop Med Public Health* 1987;18(3):398–406.
67. Lum LC *et al.* Fulminant hepatitis in dengue infection. *Southeast Asian J Trop Med Public Health* 1993;24(3):467–471.

68. Desruelles F *et al.* Cutaneo-mucous manifestations of dengue. *Ann Dermatol Venereol* 1997;124(3):237–241.
69. Wu SJ *et al.* Human skin Langerhans cells are targets of dengue virus infection. *Nat Med* 2000;6(7):816–820.
70. Boonpucknavig V *et al.* Glomerular changes in dengue hemorrhagic fever. *Arch Pathol Lab Med* 1976;100(4):206–212.
71. Guzman MG *et al.* Fatal dengue hemorrhagic fever in Cuba, 1997. *Int J Infect Dis* 1999;3(3):130–135.
72. Gasem MH *et al.* Evaluation of a simple and rapid dipstick assay for the diagnosis of typhoid fever in Indonesia. *J Med Microbiol* 2002;51(2): 173–177.
73. La Russa VF, Innis BL. Mechanisms of dengue virus-induced bone marrow suppression. *Baillieres Clin Haematol* 1995;8(1):249–270.
74. Teruel-Lopez E. Dengue: a review. *Invest Clin* 1991;32(4):201–217.
75. Aung-Khin M, Ma-Ma K, Thant Z. Changes in the tissues of the immune system in dengue haemorrhagic fever. *J Trop Med Hyg* 1975;78(12): 256–261.
76. Halstead SB. Antibody, macrophages, dengue virus infection, shock, and hemorrhage: a pathogenetic cascade. *Rev Infect Dis* 1989;11(Suppl 4): S830–S839.
77. Miagostovich MP *et al.* Retrospective study on dengue fatal cases. *Clin Neuropathol* 1997;16(4):204–208.
78. Hathirat P *et al.* Abnormal hemostasis in dengue hemorrhagic fever. *Southeast Asian J Trop Med Public Health* 1993;24(Suppl 1):80–85.
79. Burke T. Dengue haemorrhagic fever: a pathological study. *Trans R Soc Trop Med Hyg* 1968;62(5):682–692.

6

Management of Dengue

Bridget Wills

Introduction

No specific therapy is available at the present time for symptomatic dengue infections. Effective treatment relies on good supportive care, with particular emphasis on fluid therapy and management of bleeding complications. In this chapter the basic principles of management for the common presentations of dengue will be reviewed. Management of unusual manifestations such as dengue encephalopathy or fulminant hepatitis is similar to the standard treatment of these disorders and will not be addressed.

Prompt resuscitation to restore circulating plasma volume is the cornerstone of therapy for patients with dengue shock syndrome (DSS), whilst for the less severe syndromes of dengue fever (DF) and dengue haemorrhagic fever (DHF) without shock, less aggressive parenteral fluid therapy is frequently indicated. In dengue-endemic areas established DSS is usually a clearly identifiable clinical entity. However, in the initial stages of infection before the development of haemoconcentration or shock, it is often difficult to differentiate between patients with DF and those with DHF. Therefore, it is convenient to consider management for two distinct clinical situations: first, supportive treatment for the early stages of symptomatic disease, whether DF or DHF, in the hope that complications may be prevented; and second, effective volume resuscitation for established shock. It is critical to understand that all parenteral fluids have specific indications and side effects in the same way as any other medication, and that significant complications may arise as a direct result of fluid therapy; this is particularly relevant in patients with increased vascular permeability, the central pathophysiological feature of DHF. In the following section the physiology of normal fluid balance will be reviewed briefly, since a basic understanding of this field together with some knowledge of the characteristics of the intravenous fluids commonly available is helpful in making decisions regarding appropriate fluid regimens for individual patients.

The physiology of normal fluid balance

Water accounts for about 60% of total body weight and is distributed between three main fluid compartments.[1] The intracellular fluid (ICF) compartment contains approximately two thirds of total body water, while the remaining one third is contained within the extracellular fluid (ECF) compartment, distributed between the intravascular compartment or plasma (25%) and the interstitial compartment (75%) (Fig. 1). Thus the normal circulating plasma volume constitutes only 25% of one third of total body fluid, i.e. 8% of total body water.

The ICF is separated from the ECF by a selective cell membrane that is freely permeable to water but not to most solutes, specific transport systems being required for the movement of ions and proteins. Sodium and potassium are asymmetrically distributed across the membrane, with sodium the predominant cation and major determinant of osmolality of the ECF, and potassium the predominant cation of the ICF. A change in the osmolality of either compartment results in the rapid movement of water across the membrane to restore osmotic equilibrium. In contrast, within the extracellular compartment, plasma and interstitial fluid communicate freely through the capillary wall, which is highly permeable both to water and to small ions, but not to proteins. Thus the ionic composition of the two ECFs is very similar, but the protein content of plasma is considerably greater than that of interstitial

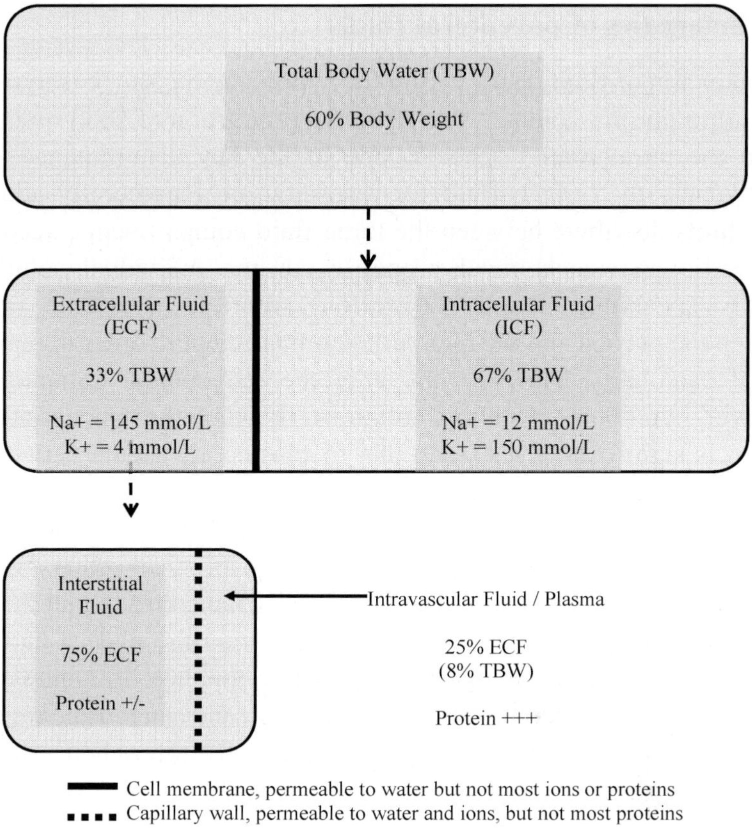

Fig. 1. Schematic diagram illustrating the approximate volumes and biochemical constituents of the major body fluid compartments.

fluid. Fluid movement across the capillary wall is determined by a combination of the hydrostatic pressure generated by the heart, and the oncotic pressure created by the excess protein present in plasma.[2] In patients with increased vascular permeability, fluid, electrolytes and proteins leak from the plasma to the interstitial compartment, altering the usual dynamic balance such that 80–90% of the ECF may be located in the interstitial compartment, with only 10–20% remaining in the circulation.

Characteristics of parenteral fluids

The main aim of fluid resuscitation for hypovolaemic shock is to restore circulating plasma volume. Therefore the proportion of fluid remaining within the intravascular space is one of the key determinants of the effectiveness of a given fluid for resuscitation. Parenterally administered fluids distribute between the three fluid compartments, according to specific physicochemical properties of the individual solutions.[1] Broadly speaking, isotonic crystalloid solutions distribute rapidly between the plasma and the interstitial compartment in order to maintain normal osmolality. They do not enter the intracellular compartment. However, hypotonic crystalloid solutions, in which the concentration of sodium is significantly less than that of plasma, distribute between all three body fluid compartments, the proportion of fluid entering the intracellular compartment being directly dependent upon the sodium concentration of the solution. One litre of physiological saline administered parenterally to a healthy person will distribute throughout the ECF space only, with 25% (250 ml) remaining within the plasma and 75% passing to the interstitial space within a few minutes. In contrast, one litre of 5% dextrose, which contains no sodium and can thus be regarded effectively as free water, will distribute equally throughout all three fluid compartments. Only 25% of one third of the total volume, i.e. approximately 80 ml of fluid, will remain within the plasma. In patients with capillary leak syndrome the volume of parenterally administered fluid remaining in the plasma is reduced even further and reflects whatever new balance has been established between the two ECF compartments, e.g. 80%/20% or 90%/10% instead of the usual 75%/25%. Thus it follows that in situations where restoration of circulating plasma volume is a priority, only isotonic crystalloid solutions should be considered. Any crystalloid solution with a final sodium concentration of less than around 130 mmol/L is contraindicated; physiological (0.9%) saline, lactated Ringer's solution etc. are acceptable but half-normal (0.45%) saline or 5% dextrose in 0.18% saline should not be used in these circumstances.

Theoretically, colloid solutions offer advantages over crystalloid solutions for emergency resuscitation of hypovolaemic shock. First, immediate distribution is primarily within the intravascular compartment, limited by the permeability of the capillary wall to the particular constituent molecules. Second, the colloid molecules increase plasma oncotic pressure, thereby altering the balance of fluid flux across the endothelium and drawing fluid back into the intravascular compartment from the interstitial space. Thus, in contrast to crystalloid solutions, a bolus of a colloid solution provides temporary volume expansion in excess of the actual volume of fluid infused (Table 1). All synthetic colloids are polydisperse, with a range of molecules of different molecular weights in one solution. One of the major determinants of the magnitude of the effect on plasma oncotic pressure is the average molecular weight of the colloid molecules; in general, small molecules exert a greater osmotic effect than larger molecules at the same concentration. However, large molecules remain within the circulation for a longer time period as the small molecules are rapidly excreted by the kidneys or are lost from the circulation by leakage into the interstitium. In patients with increased vascular permeability, leakage of these small colloid molecules may be marked and this may exacerbate the severity of fluid overload since the leaked molecules continue to exert an osmotic effect from within the interstitial space.[3]

Another significant determinant of the effectiveness of a particular colloid preparation for volume replacement is the concentration of the solution. Hyperoncotic solutions have considerably greater effect on plasma oncotic pressure than iso-oncotic preparations of the same molecule and are thus able to draw more interstitial fluid back into the intravascular compartment. However, there are concerns about the safety of hyperoncotic solutions in clinical use. The development of acute renal failure is a well-recognised complication, particularly in hypovolaemic patients, and in many countries preparations such as 10% Dextran 40 are no longer thought to be suitable for use as volume expanders.[4,5] Other concerns relating to the use of colloid solutions include the possibility of adverse effects on blood coagulation, and the well-recognised potential to cause allergic reactions.[6] Haemostatic competence is significantly reduced by dextrans and high molecular weight starch preparations (MW 400,000) but as yet there is little evidence for clinically relevant effects on coagulation by medium molecular weight

Table 1. Characteristics of various different colloids used for plasma volume support.

	Initial volume expansion (%)*	Duration of volume effect (hours)	Adverse effect on coagulation	Allergic potential	Other significant side effects
3% Gelatin (MW = 35,000)	60–80	3–4	+/–	++	
10% Dextran 40 (MW = 40,000)	170–180	4–6	++	+	Renal failure in hypovolaemic patients
6% Dextran 70 (MW = 70,000)	100–140	6–8	++	+	
6% Hydroxyethylstarch (MW = 200,000/0.5)	100–140	6–8	+	+/–	
6% Hydroxyethylstarch (MW = 400,000)	80–100	12–24	++	+	

* Refers to the infused volume.

starch solutions (MW 200,000) or gelatins. Gelatins are associated with the highest incidence of allergic reactions. Dextrans are also well known to cause hypersensitivity reactions, but prophylaxis with a dextran hapten effectively prevents such problems. In addition, dextran solutions are manufactured by a process involving bacterial culture and degradation, and residual pyrogens may be sufficient to induce significant febrile reactions (as distinct from true allergic reactions) in some recipients.[7,8] Starch solutions rarely cause acute allergic reactions, although late onset pruritus is a well-recognised complication following treatment with high molecular weight starches.

The relevant characteristics of a range of commonly available colloid fluids are detailed in Table 1. By comparison, equal volumes of isotonic crystalloid solutions are thought to provide about one quarter of the plasma volume support but are generally safe, reaction-free, and have only dilutional effects on coagulation. For many years there has been vigorous debate in the medical literature regarding the use of colloids compared with crystalloids for volume replacement in critically ill patients, irrespective of the underlying disease process.[9–11] Recently, the Saline versus Albumin Fluid Evaluation (SAFE) Study, a randomized and blinded multi-centre trial that recruited 7000 patients, established that albumin and normal saline were equally effective for fluid resuscitation in a heterogeneous population of patients in intensive care units, but suggested that there was a treatment effect favouring albumin in patients with severe sepsis.[12]

Clinical studies of specific fluid therapy in patients with dengue

Although volume replacement is recognised as the single critical therapeutic intervention for most patients with DSS, very few formal trials have been carried out to determine the optimal management regimen, in part because of the difficulties inherent in performing such studies on critically ill patients. Thus current recommendations for fluid therapy in patients with dengue remain largely empirical rather than evidence-based.[13,14] To date only three randomised and blinded clinical trials that investigate the impact of different regimens have been published. An initial pilot study involved 50 children with DSS; all the subjects did well but minor differences were detected in the immediate clinical responses to different fluids, with significantly greater improvements in surrogate markers of recovery such as cardiac index, haematocrit, and blood pressure among children who received a colloid rather than a crystalloid at first presentation.[15] The second study recruited 230 children with DSS over one year in a single hospital, and again all recovered well. Although the results of this study suggested that early treatment with colloid solutions might influence overall recovery in the more severely ill patients, the study was statistically under-powered.[16] The third study established, in a large group of children, that for those with moderately severe shock Ringer's lactate was as effective as colloid therapy, and that for those with severe shock 6% Dextran 70 and 6% HES performed equally well although the dextran preparation was associated with significantly more side effects.[8] Use of Ringer's lactate was not evaluated in the severely shocked group, because of concerns about the potential for development of severe respiratory compromise if large volumes of fluid were required.

From these studies and knowledge of the intrinsic properties of different fluid solutions, we can conclude that the majority of children with DSS can be treated successfully with isotonic crystalloid solutions. If a colloid is judged to be necessary (as discussed in the following section) a medium molecular weight preparation which combines good initial plasma volume support with good intravascular persistence and an acceptable side effect profile, particularly in relation to haemostatis, is probably the preparation of choice. Further research is needed to determine whether early treatment with a colloid confers a true advantage in those with severe shock.

Management

A practical guide to management will be presented in this section. Worldwide, the majority of dengue infections occur in school age children. The principles of management outlined here are directed primarily towards this group. The specific problems encountered in treating infants and adults will be addressed only briefly where relevant.

Outpatient management

For most patients hospital admission is unnecessary during the first 2–3 days of illness. Differentiation between dengue infections (DF or DHF) and other common infections of childhood is often not possible at this stage. Reasons for admission include the following:

- Persistent vomiting or severe abdominal pain
- A febrile convulsion, or high fever in a child with a history of febrile convulsions
- Mucosal bleeding (epistaxis, gum bleeding, gastrointestinal bleeding) or unusual skin bleeding/bruising
- Restlessness, lethargy or exhaustion
- Any signs of cardiovascular compromise (rarely, DSS may occur on day 3)
- Reduced urine output

In the absence of any of these features it is safe to manage the patient at home, provided that the parents are competent, have access to reasonable transport facilities and are given clear instructions to return promptly for review if any of the above complications arise.

Oral rehydration should be encouraged with ORS or similar preparations, together with a light diet. The parent or guardian should be instructed in fever control with fans, tepid sponging etc. Paracetamol (10–15 mg/kg/4–6-hourly) is the preferred antipyretic agent. Aspirin and non-steroidal anti-inflammatory drugs such as ibuprofen are contraindicated because of the potential to cause gastritis and bleeding, as well as anti-platelet effects in patients likely to have significant thrombocytopenia. In addition, aspirin should not be prescribed for children because of the risk of Reye's syndrome.[17]

Ideally the haematocrit and platelet count should be checked at the first visit, to provide a baseline for comparison with subsequent values, and the child should be reviewed daily until the fever has settled. During this period, daily haematocrit measurements may reveal a progressive increase in the degree of haemoconcentration signalling the development of DHF — if the increase in haemoconcentration is rapid or severe, this should prompt referral to a suitable inpatient facility. However, if the increase in haematocrit is mild (10–15%) the child may continue to be managed at home provided that he or she is cardiovascularly stable, tolerates oral fluids well and exhibits no bleeding manifestations. At each review the possibility of an alternative diagnosis (typhoid, malaria, measles etc.) must be borne in mind and suitable investigations or interventions arranged as necessary.

Inpatient management of DF/DHF without shock

Careful attention to fluid balance is the most important aspect of care for children without established shock who require hospital admission. In the majority of cases significant vascular leakage becomes apparent around the fourth or fifth day of illness, with reabsorption of fluid commencing between the sixth and the seventh day. Occasionally, however, capillary leak becomes apparent as early as day 3 of illness and in such cases is often severe. Judicious use of parenteral fluids from an early stage when haemoconcentration is first noted may modify the course of the disease and prevent the development of shock, but it is critical to appreciate that parenteral fluid given early on may prejudice subsequent management should the child progress to shock after some hours or days. If significant fluid overload with large pleural effusions and ascites has already developed by the time shock supervenes, the administration of further fluid to counteract the cardiovascular compromise may be singularly problematic. To minimise the risk of such iatrogenic complications, very careful consideration needs to be given to the nature and volume of all parenteral fluid therapy. This, and other aspects of inpatient care, are discussed below.

Symptomatic care

The principles of basic symptomatic care for inpatients without shock are similar to those for outpatient management. High fever should be controlled with paracetamol and simple measures such as fans or tepid sponging. Oral fluids and a light diet should be encouraged in all children able to tolerate them.

Monitoring

Patients must be observed closely, including regular pulse and blood pressure monitoring, until they have been afebrile for at least 24 hours without antipyretics. Detailed fluid intake and output should be documented, particularly for those patients requiring intravenous fluid therapy. Initially the haematocrit should be checked daily or twice daily, but if it starts to rise or the patient's clinical condition deteriorates, more frequent estimations are required. A platelet count should be measured on admission and then every one to two days, but if the patient develops significant mucosal bleeding or severe thrombocytopenia is documented (platelet count $< 30{,}000\ \text{mm}^3$), more frequent checks are advisable.

Intravenous fluid therapy

Intravenous fluid therapy is indicated for those with repeated vomiting, any signs of cardiovascular compromise or a very high or rapidly rising haematocrit. Only isotonic crystalloid solutions, which distribute primarily within the extracellular fluid compartment (see above) and thus provide the greatest intravascular support, should be used. Suitable preparations include physiological (0.9%) saline, Ringer's lactate and Ringer's acetate solutions, but pure 5% dextrose and preparations including 5% dextrose diluted in hypotonic electrolyte solutions should be avoided. Hypoglycaemia is uncommon in school age children and most patients are able to tolerate some oral fluid. In the event that a dextrose-containing preparation is considered necessary, 5% dextrose in full strength physiological saline should be used, or small volumes (1 ml/kg) of 10–20% dextrose may be given in addition to the basic fluid regimen.

The minimum volume of fluid possible to maintain cardiovascular stability and good urine output should be prescribed. It may be necessary to commence with a volume of 5–6 ml/kg/hour for the first one to two hours, but ideally the infusion rate should be reduced to maintenance levels of 2–3 ml/kg/h as soon as possible, depending on the clinical response. Changes in the haematocrit are helpful in guiding fluid therapy but it is important to recognise that it is the clinical response that matters, rather than normalisation of the haematocrit. If the patient is warm, well perfused and has a good urine output, then a relatively high but stable haematocrit is acceptable. In most patients parenteral fluid therapy is required only for 24–48 hours, since the capillary leak resolves spontaneously after this time.

Management of haemorrhagic complications

Mucosal bleeding in the form of epistaxis, gum bleeding, gastrointestinal or vaginal bleeding may occur, but is usually minor in children with DF or DHF without shock. Rarely, however, massive gastrointestinal bleeding requiring transfusion may occur. For most patients routine inpatient observation with regular monitoring of the platelet count is all that is necessary. However, if overt bleeding is persistent or major gastrointestinal bleeding is suspected, the patient should be transferred to a high dependency unit with facilities for close observation, coagulation screening tests should be performed, and blood should be grouped, in readiness for cross-matching. Nasal packing may be required for persistent epistaxis but should only be performed by experienced staff. Transfusion is very rarely necessary, but if so, fresh blood must be used, since such patients commonly have profound thrombocytopenia. In these circumstances platelet concentrates and/or fresh frozen plasma may be helpful. However, platelet concentrates are of no value for the treatment of thrombocytopenia in the absence of major bleeding and may be harmful;[18] thrombocytopenia improves spontaneously within a few days, the half-life of transfused platelets is markedly reduced such that they are effective only for a few hours,[19–21] and the patient may develop fluid overload. Finally, the patient may be exposed unnecessarily to the risks of transmissible agents in blood products. Children with dengue infections and profound thrombocytopenia (< 20,000 platelets/mm^3) but without major bleeding should be managed expectantly with bed rest and protection from trauma, particularly protection from invasive procedures such as intramuscular injections.

Management of established dengue shock syndrome

Treatment of established DSS is a medical emergency. Children admitted with established shock and those who develop cardiovascular compromise while in hospital should be looked after on a designated ward, preferably a high dependency or intensive care unit, staffed by medical and nursing personnel experienced in the management of such cases. Volume replacement with parenteral fluid therapy is the cornerstone of treatment. No other specific treatments have been shown to be beneficial. Steroid therapy, previously employed regularly for the treatment of DHF, has shown no convincing benefit in several randomised controlled trials and is no longer recommended.[22,23] Similarly, carabazochrome, an agent thought to influence haemostasis and vascular function, did not improve the severity of plasma leakage in a randomised controlled trial and should not be given.[24]

Initial assessment

Rapid clinical assessment of cardiovascular status (pulse, blood pressure, peripheral perfusion, urine output and mental state) determines initial management. A diagnosis of DSS is made when the pulse pressure narrows to 20 mm Hg or less accompanied by signs of impaired peripheral perfusion, but even at this stage the child may appear disconcertingly well and complain only of vague abdominal pain and tiredness. Paradoxically, many children do not develop a significant tachycardia despite profound hypovolaemia. The results of basic laboratory investigations including a haematocrit and platelet count are also useful in assessing severity but initiation of treatment must not be delayed pending their availability. Detailed clinical examination should be carried out once resuscitation is in progress. The following features are commonly associated with severe disease and a complicated clinical course:

- Unrecordable pulse and blood pressure (grade IV DHF).
- Very narrow pulse pressure, $\leq$ 10 mm Hg (severe grade III DHF).[16]
- Any evidence of compromised cerebral perfusion (lethargy, irritability, drowsiness or restlessness).
- Presentation with shock early in the course of the disease (before day 4 of fever).[25]
- Marked elevation of the haematocrit ($\geq$ 50% in children).[25] However, it should be remembered that it is the proportionate increase in

haematocrit which indicates severity rather than any absolute value. A child with mild chronic anaemia and a baseline haematocrit of 30%, presenting with DSS and a haematocrit of 40%, is relatively more haemoconcentrated than another child with a baseline value of 42% and a haematocrit of 50% at the time of shock.

- Clinically apparent pleural effusions or ascites at the time of presentation with shock. Large volumes of fluid must be present to be clinically detectable, implying either recent onset of catastrophic leak or a steady loss of fluid over a longer period of time before the development of haemodynamic compromise.

After the initial rapid assessment, resuscitation with parenteral fluids should be commenced immediately. Reliable intravenous access must be secured as soon as possible and, rarely, in patients with profound shock a venous cut-down or insertion of an intraosseous line may be necessary. All patients with shock or respiratory compromise should receive oxygen by facemask or nasal cannulae. A regular schedule of half- to one-hourly clinical observations should be instituted together with a detailed record of all fluid intake and output. The haematocrit should be measured every two hours for the first six hours and thereafter every four to six hours until the patient is stable. Immediate access to haematocrit results is extremely important, since decisions about fluid management need to be reviewed frequently and a delay of several hours in obtaining essential results from a distant laboratory is likely to compromise the standard of care. A simple micro-haematocrit centrifuge should be located on or immediately adjacent to the DSS ward and all staff should be trained to use it properly.

Intravenous fluid therapy

Immediate restoration of a stable and effective circulation with parenteral fluid therapy is the primary aim of treatment. Subsequently the focus of management shifts towards ongoing support of the circulation for the limited time period during which capillary leak persists. Extreme care is needed to balance the requirement for intravenous fluid to maintain plasma volume against the inherent risk of leakage of the administered fluid into the interstitial space. The leaked fluid may contribute to the

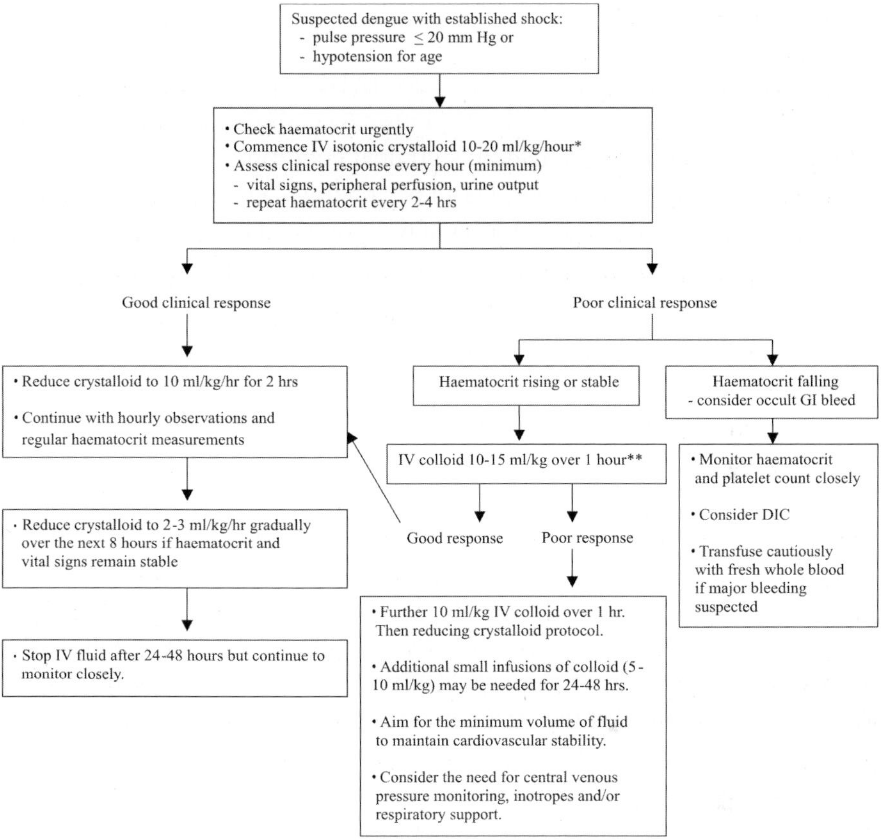

Fig. 2. Flow chart for suggested volume replacement for patients with dengue shock syndrome.

* Use only isotonic crystalloid solutions such as Ringer's lactate or normal saline.

** Use a medium molecular weight iso-oncotic colloid preparation, e.g. 6% Dextran 70 or 6% Hydroxyethylstarch MW 200,000.

development of pleural effusions, ascites and respiratory compromise, and the potential downward spiral towards multi-organ failure, disseminated intravascular coagulation and death. Inevitably those patients with severe capillary leak syndrome, most at risk of these complications, are also those most in need of aggressive circulatory support. A flow chart detailing a suggested regimen for fluid replacement is shown in Fig. 2.

Grade III DHF

For the majority of patients with grade III DHF resuscitation should be started with an isotonic crystalloid solution (see above) at a rate of 10–20 ml/kg over one hour. If the patient's clinical condition has stabilised after this time (wider pulse pressure, warm peripheries, a reduction in the heart rate), the rate of fluid administration may be reduced to 10 ml/kg/h for 1–2 hours, and then gradually reduced to maintenance levels over the next 6–8 hours. A suitable schedule for children up to 10–12 years of age might be as follows: 10 ml/kg/h for 2 hours: 7.5 ml/kg/h for 2 hours: 5 ml/kg/h for 4 hours: 2–3 ml/kg/h for 24–36 hours. Older children/teenagers and children who are overweight require proportionately less fluid per kilogram body weight and the volumes given after the initial resuscitation should be adjusted downwards. For most patients intravenous therapy can be stopped after this time, provided the clinical condition has been stable for 24 hours.

If there is evidence of ongoing cardiovascular compromise after the first hour of treatment (no improvement in the pulse pressure or pulse rate, persisting peripheral shutdown, a rising haematocrit), a colloid solution should be substituted for the crystalloid solution, at an initial rate of 10–15 ml/kg over one hour. At present there is no conclusive evidence (see earlier discussion and Table 1) as to which colloid solution is most effective for resuscitation, but many authorities consider an iso-oncotic, medium molecular weight preparation to be the best choice currently available; examples are 6% Dextran 70 and 6% HES (MW 200,000). Clinicians should become familiar with the use of one or two of the suitable locally available preparations.

After one hour of colloid infusion, treatment should revert to the reducing schedule of isotonic crystalloid as detailed above, provided that the patient's condition has improved. If not, a further infusion of colloid at a rate of 10 ml/kg may be given over one hour. However, it must be remembered that all colloids influence coagulation to some degree, although clinically significant effects are unlikely with infused volumes of less than 20–25 ml/kg/day. In treating a disease with an intrinsic effect on coagulation it is preferable to limit the use of any colloid preparation to the minimum necessary to support the circulation at any one time, in the expectation that further boluses may be required in the ensuing 24–48 hours during which capillary leak persists.

Frequent observation of vital signs, mental state and urine output, together with serial haematocrit measurements, are used to assess the response to treatment. After initial resuscitation most patients can be successfully managed with the reducing schedule of isotonic crystalloid fluid until the re-absorptive phase of the illness begins. However, some may experience further episodes of cardiovascular decompensation after the initial episode and may require supplementary treatment with small infusions of 5–10 ml/kg of colloid. In deciding whether a further infusion of colloid is necessary for a particular patient, all clinical and laboratory information should be taken into account and a conservative approach adopted. Thus a child who is warm, well perfused, passing good volumes of urine and who has a stable haematocrit of 48% on day 6 of illness probably does not require a colloid infusion even though the pulse pressure may be only 20 mm Hg. A child with a similar clinical picture, but in whom the haematocrit has risen from 44% to 48% on day 4 of illness, is likely to deteriorate over the next few hours and require intervention. In all cases the key to successful management lies in frequent and careful reassessment by experienced personnel.

Grade IV DHF

Children with no recordable pulse or blood pressure (DHF grade IV) should be managed more vigorously. However, the spectrum of shock severity is continuous, with an arbitrary demarcation line between DHF grade III and grade IV, and there is some evidence to suggest that patients with severe grade III disease (pulse pressure $\leq$ 10 mm Hg) also merit aggressive intervention.[16]

There is general agreement that profoundly shocked patients usually require resuscitation with colloid therapy, but debate continues as to whether this should be administered immediately, or after initial treatment with a rapid bolus of a crystalloid solution. One strategy is to treat with 20 ml/kg of an isotonic crystalloid as a bolus over 15 minutes, followed by a colloid infusion of 10–20 ml/kg given more or less rapidly, depending on the patient's response to the initial crystalloid infusion. Other experienced clinicians consider that it is preferable to start immediately with a bolus of colloid to try to limit the overall volume of fluid given, in the expectation that these severely compromised patients are likely to require considerable ongoing fluid support in the ensuing 24–48 hours. Notwithstanding their initial severity, even most of these patients do improve with aggressive initial volume replacement, and can be managed subsequently, as suggested above for children with grade III DHF. Supplementary treatment with one or two further small boluses of colloid is quite likely to prove necessary but a conservative approach to decision-making is still essential.

The small number of children who fail to improve require high-level intensive care, and arrangements should be made for urgent transfer to a suitable facility. Management is complex and difficult, as with all critically ill patients, and will only be touched on here. Central venous pressure (CVP) monitoring may provide useful information to direct fluid therapy. However, central venous cannulation is not without risk to the patient and should only be carried out by experienced personnel with facilities for immediate treatment of complications if they arise. Inotropic support is often required in addition to volume support. Significant pleural effusions and respiratory compromise are likely to develop and pleural and ascitic drainage, together with respiratory support in the form of continuous positive airway pressure (CPAP) or assisted ventilation, may all prove to be necessary. Metabolic and electrolyte derangements are common and should be actively sought and corrected.

Blood transfusion

Transfusion is indicated only for those with life-threatening bleeding and should be undertaken with extreme care because of the problem of fluid overload. In children with DSS major bleeding is almost always associated with very severe or prolonged shock, and is usually from the gastrointestinal tract.[26] Underlying causes include profound thrombocytopenia, hypofibrinogenaemia and deranged fibrinolysis, together with evidence of disseminated intravascular coagulation in cases with marked tissue hypoxia and acidosis. Internal bleeding may not become apparent for many hours until the first melaena stool is passed. It should be considered in all those who fail to improve clinically after appropriate fluid resuscitation, particularly if the haematocrit is falling and the abdomen is distended and tender. In the event that major internal bleeding is suspected in a child with shock, a small volume of fresh whole blood (5–10 ml/kg) should be given over 1–2 hours and the response observed. Further small transfusions may be given subsequently if there is a good clinical response and significant bleeding is confirmed. Platelet concentrates and fresh frozen plasma may also be helpful but are effective only for a few hours and should be reserved for those with life-threatening bleeding, for the reasons previously noted. Usage should be guided by the platelet count and coagulation profile. Two small recent studies indicate that recombinant activated factor VII, by enhancing thrombin generation locally at sites of vascular injury, may be helpful in controlling life-threatening bleeding, but further research is needed.[27,28]

Significant mucosal bleeding, particularly gastrointestinal bleeding, is more common in adults with dengue infection and may contribute to shock.[29,30] This probably reflects the greater frequency of gastritis and peptic ulcer disease in this age group.[30,31] Capillary leak syndrome does occur but is generally less severe than in children. More aggressive transfusion may be necessary in adult patients, although care is still required to avoid the pitfalls of fluid overload.

Fluid overload

Clinically significant fluid overload develops in several situations. Most commonly, it follows administration of inappropriate intravenous fluid in excessive amounts and/or too rapidly to patients with moderate capillary leak, or else continued parenteral fluid therapy once the re-absorptive phase of the disease has commenced. Rarely, it may be seen in patients with catastrophic leak for whom support of the circulation is not possible without administration of large volumes of fluid. In addition, it may occur in patients with underlying chronic diseases, particularly cardiac or renal disorders. Careful attention to management guidelines together with frequent reassessment of the patient by experienced staff should help to limit the occurrence of iatrogenic fluid overload, while early identification of the rare patients with catastrophic leak or severe underlying disease may allow pre-emptive intervention before significant respiratory compromise occurs.

Early respiratory signs of fluid overload include tachypnoea and recession, with pleural effusions and ascites. More severe overload has secondary effects on cardiac function, increasing pre- and afterload, altering cardiovascular efficiency, and eventually resulting in hypotension and circulatory failure. Pulmonary oedema, cyanosis and respiratory failure are late manifestations. Differentiation between haemodynamic instability resulting from fluid overload and that caused by inadequate treatment of the underlying hypovolaemia is critically important, since management of these two situations differs markedly. Measurement of CVP may be helpful; those patients with pure fluid overload have high CVP readings, whilst those with mixed hypovolaemia and fluid overload usually have low filling pressures. Two-dimensional echocardiography can also provide useful information on cardiac function and filling pressures, and, together with the results of chest radiography and/or ultrasound assessment of the severity of pleural effusions, regular CVP measurements, pulse oximetry and blood gas analysis, should be used to guide management.

For children in whom shock has resolved and the cardiovascular indices (pulse and blood pressure) are stable, those with mild to moderate overload but without significant respiratory symptoms should be observed on bed rest for 24–48 hours and all intravenous fluid therapy should be stopped. In most cases spontaneous re-absorption of fluid will occur with a concomitant diuresis, and no further treatment proves to be necessary. Symptomatic patients with tachypnoea or breathlessness and large effusions should receive a diuretic (e.g. frusemide 1 mg/kg/dose oral or IV) once or twice daily for 24 hours, together with facial oxygen and strict bed rest. Continuous positive airway pressure in the form of nasal or facial CPAP may also help to reduce the work of breathing.[32] Those children with severe overload and secondary haemodynamic instability (i.e. high CVP readings in the presence of narrow pulse pressure or hypotension) should receive oxygen and diuretics, and support with an inotropic agent such as dopamine or dobutamine should be considered. Inotrope infusions of 5–10 μg/kg/min may be necessary for 24 hours or so whilst the diuretics take effect, but even in these circumstances the majority of patients improve with appropriate supportive care.

The rare patients, usually with catastrophic capillary leak, who remain hypovolaemic (i.e. low CVP readings with cardiovascular compromise and peripheral vasoconstriction) when significant fluid overload becomes apparent are extremely difficult to manage and have a high mortality. Repeated small boluses of colloid are frequently necessary to support the circulation, combined with inotropic agents, often in high doses. Maintenance crystalloid therapy should be reduced to a minimum and diuretics are contraindicated since their effect will be to deplete the intravascular compartment further. Aspiration of large pleural effusions or drainage of ascites is helpful in relieving respiratory symptoms, and early intervention with positive pressure ventilation, preferably before the development of frank pulmonary oedema, may be life-saving. Once pulmonary oedema is established the technical difficulties of mechanical ventilation become increasingly complex and the outlook is grave. Metabolic derangements, disseminated intravascular coagulation and multi-organ failure are common complications and contribute to the high mortality in this group.

Management of DHF–DSS in infancy

DHF–DSS occurs infrequently in infants under one year of age but special care must be taken over fluid management in this age group. Fluid accounts for a greater proportion of body weight in infants,[33] and minimum daily requirements are correspondingly greater; cardiovascular and renal function are still developing and there is a smaller reserve for coping with disturbance; finally, capillary beds are intrinsically more permeable than those of older children or adults.[34] Both early cardiovascular compromise and significant fluid overload are more likely to occur in infants with capillary leak syndrome. Thus, all infants must be treated as high-risk patients and merit early intervention with colloids, similar to older children with grade IV disease.

Conclusions

Over the last 40 years the incidence of dengue infections, particularly the more severe manifestations of DHF and DSS, has increased dramatically, and dengue is now one of the most common reasons for hospital admission among children in Asia.[35] Despite the fact that the mechanisms underlying both the coagulopathy and the transient increase in vascular permeability remain poorly understood, considerable improvements have been made in management over the years. The WHO-sponsored promotion of national treatment guidelines, the increased availability of parenteral fluids, and the development of excellent networks for support and education of health care personnel at all levels have contributed to a marked reduction in mortality in many South-east Asian countries.[13,14] The majority of children now admitted to hospitals within the region respond to appropriate supportive care and are discharged home well after a brief hospital stay. As dengue emerges as a significant contributor to morbidity and mortality across the Americas and in the Pacific region, similar efforts will be required to develop appropriate national guidelines and educational programmes that address the particular problems encountered. For example, in areas of lower endemicity adults often make up a larger proportion of the symptomatic patients, underlying disorders such as diabetes, atherosclerosis and chronic liver disease are common, and bleeding tends to be more of a problem than overt vascular leakage — specific management protocols need to be designed that reflect these differences.

It may be many years before biological control measures and/or vaccination have a significant impact on the global burden of dengue disease. In the interim, as well as supporting and improving individual case management through local dengue control initiatives, research is needed to try to identify risk factors that might predict the development of DHF/DSS and to investigate novel treatments for those with severe disease. Useful insights may be gained from consideration of the management of other diseases with similar problems of capillary leak and/or haemorrhage, e.g. the use of recombinant factor VIIa for uncontrollable bleeding. In addition, as new and different parenteral fluid therapies become available, management guidelines must be reappraised at regular intervals, and careful clinical research encouraged in order to provide the necessary evidence on which to base future recommendations.

References

1. Griffel MI, Kaufman BS. Pharmacology of colloids and crystalloids. *Crit Care Clin* 1992;8(2):235–253.
2. Michel CC, Curry FE. Microvascular permeability. *Physiol Rev* 1999;79(3):703–761.
3. Haupt MT, Kaufman BS, Carlson RW. Fluid resuscitation in patients with increased vascular permeability. *Crit Care Clin* 1992;8(2):341–353.
4. Mailloux L, Swartz CD, Capizzi R *et al.* Acute renal failure after administration of low-molecular weight dextran. *N Engl J Med* 1967;277(21): 1113–1118.
5. Moran M, Kapsner C. Acute renal failure associated with elevated plasma oncotic pressure. *N Engl J Med* 1987;317(3):150–153.
6. Haljamae H. Albumin: to use or not to use? Contemporary alternatives? In: Baron J-F, Treib J (eds.) *Volume Replacement.* Springer-Verlag, Berlin, 1998.
7. Martis L, Patel M, Giertych J *et al.* Aseptic peritonitis due to peptidoglycan contamination of pharmacopoeia standard dialysis solution. *Lancet* 2005;365(9459):588–594.
8. Wills BA, Nguyen MD, Ha TL *et al.* Comparison of three fluid solutions for resuscitation in dengue shock syndrome. *N Engl J Med* 2005;353(9): 877–889.
9. Choi PT, Yip G, Quinonez LG, Cook DJ. Crystalloids vs. colloids in fluid resuscitation: a systematic review. *Crit Care Med* 1999;27(1):200–210.
10. Alderson P, Schierhout G, Roberts I, Bunn F. Colloids versus crystalloids for fluid resuscitation in critically ill patients. *Cochrane Database Syst Rev* 2000(2):CD000567.
11. Bunn F, Alderson P, Hawkins V. Colloid solutions for fluid resuscitation. *Cochrane Database Syst Rev* 2003(1):CD001319.
12. Finfer S, Bellomo R, Boyce N, French J, Myburgh J, Norton R. A comparison of albumin and saline for fluid resuscitation in the intensive care unit. *N Engl J Med* 2004;350(22):2247–2256.
13. WHO. *Dengue Haemorrhagic Fever: Diagnosis, Treatment, Prevention and Control.* World Health Organization, Geneva, 1997.

14. WHO. *Guidelines for Treatment of Dengue Fever/Dengue Haemorrhagic Fever in Small Hospitals*. WHO Regional Office for Southeast Asia, New Delhi, 1999.
15. Dung NM, Day NP, Tam DT *et al*. Fluid replacement in dengue shock syndrome: a randomized, double-blind comparison of four intravenous-fluid regimens. *Clin Infect Dis* 1999;29(4):787–794.
16. Ngo NT, Cao XT, Kneen R *et al*. Acute management of dengue shock syndrome: a randomized double-blind comparison of 4 intravenous fluid regimens in the first hour. *Clin Infect Dis* 2001;32(2):204–213.
17. Glasgow JF, Middleton B. Reye syndrome — insights on causation and prognosis. *Arch Dis Child* 2001;85(5):351–353.
18. Kumar ND, Tomar V, Singh B, Kela K. Platelet transfusion practice during dengue fever epidemic. *Indian J Pathol Microbiol* 2000;43(1):55–60.
19. Mitrakul C, Poshyachinda M, Futrakul P, Sangkawibha N, Ahandrik S. Hemostatic and platelet kinetic studies in dengue hemorrhagic fever. *Am J Trop Med Hyg* 1977;26(5 Pt 1):975–984.
20. Isarangkura P, Tuchinda S. The behavior of transfused platelets in dengue hemorrhagic fever. *Southeast Asian J Trop Med Public Health* 1993; 24(Suppl 1):222–224.
21. Wang S, He R, Patarapotikul J, Innis BL, Anderson R. Antibody-enhanced binding of dengue-2 virus to human platelets. *Virology* 1995;213(1):254–257.
22. Sumarmo, Talogo W, Asrin A, Isnuhandojo B, Sahudi A. Failure of hydrocortisone to affect outcome in dengue shock syndrome. *Pediatrics* 1982;69(1):45–49.
23. Tassniyom S, Vasanawathana S, Chirawatkul A, Rojanasuphot S. Failure of high-dose methylprednisolone in established dengue shock syndrome: a placebo-controlled, double-blind study. *Pediatrics* 1993;92(1):111–115.
24. Tassniyom S, Vasanawathana S, Dhiensiri T, Nisalak A, Chirawatkul A. Failure of carbazochrome sodium sulfonate (AC-17) to prevent dengue vascular permeability or shock: a randomized, controlled trial. *J Pediatr* 1997; 131(4):525–528.
25. Phuong CX, Nhan NT, Kneen R *et al*. Clinical diagnosis and assessment of severity of confirmed dengue infections in Vietnamese children: is the World Health Organization classification system helpful? *Am J Trop Med Hyg* 2004;70(2):172–179.

26. Wills BA, Oragui EE, Stephens AC *et al.* Coagulation abnormalities in dengue hemorrhagic fever: serial investigations in 167 Vietnamese children with dengue shock syndrome. *Clin Infect Dis* 2002;35(3):277–285.
27. Chuansumrit A, Tangnararatchakit K, Lektakul Y *et al.* The use of recombinant activated factor VII for controlling life-threatening bleeding in dengue shock syndrome. *Blood Coagul Fibrinolysis* 2004;15(4):335–342.
28. Chuansumrit A, Wangruangsatid S, Lektrakul Y, Chua MN, Zeta Capeding MR, Bech OM. Control of bleeding in children with dengue hemorrhagic fever using recombinant activated factor VII: a randomized, double-blind, placebo-controlled study. *Blood Coagul Fibrinolysis* 2005;16(8):549–555.
29. Wali JP, Biswas A, Handa R, Aggarwal P, Wig N, Dwivedi SN. Dengue haemorrhagic fever in adults: a prospective study of 110 cases. *Trop Doct* 1999;29(1):27–30.
30. Chiu YC, Wu KL, Kuo CH *et al.* Endoscopic findings and management of dengue patients with upper gastrointestinal bleeding. *Am J Trop Med Hyg* 2005;73(2):441–444.
31. Tsai CJ, Kuo CH, Chen PC, Changcheng CS. Upper gastrointestinal bleeding in dengue fever. *Am J Gastroenterol* 1991;86(1):33–35.
32. Cam BV, Tuan DT, Fonsmark L *et al.* Randomized comparison of oxygen mask treatment vs. nasal continuous positive airway pressure in dengue shock syndrome with acute respiratory failure. *J Trop Pediatr* 2002;48(6):335–339.
33. De Bruin WJ, Greenwald BM, Notterman DA. Fluid resuscitation in pediatrics. *Crit Care Clin* 1992;8(2):423–438.
34. Gamble J, Bethell D, Day NP *et al.* Age-related changes in microvascular permeability: a significant factor in the susceptibility of children to shock? *Clin Sci (Lond)* 2000;98(2):211–216.
35. Monath TP. Dengue: the risk to developed and developing countries. *Proc Natl Acad Sci USA* 1994;91(7):2395–2400.

7

Pathogenesis: Risk Factors Prior to Infection

Scott B. Halstead

Introduction

The pathogenesis of all infectious diseases is a product of host as well as microbial factors. For the host, extrinsic factors such as immune status and intrinsic factors such as age, sex, innate immune system and individual or species level genetics may operate singly or in combination. Organisms possess a variety of virulence factors. In most instances, homologous or heterologous immunity to pathogens increases host defenses. Dengue viruses appear unique among human pathogens in that the immune status of the host may modify disease in two directions — toward increased severity or in the direction of infections accompanied by mild or no disease or complete protection. This regulatory mechanism appears to derive from attributes unique to the dengue virus group. First, four types have evolved from a common ancestor[1] endowing the dengue viruses with many common antigens but sufficient critical difference(s) to permit sequential infection with different types. Second, a principal organ supporting dengue virus infection is the mononuclear phagocytic cell system. These cells are capable of supporting antibody-dependent enhancement of infection (ADE). For evidence of the growth of dengue viruses in mononuclear phagocytes and the role ADE plays in controlling the severity of dengue disease, the reader is referred to reviews of the epidemiological and experimental evidence supporting ADE during dengue infections, to the extensive *in vitro* experience with ADE caused by viruses from a wide variety of taxons and with *in vivo* ADE in many species of domestic animals.[2–9]

Building on those earlier reviews, this chapter identifies and discusses viral and host factors present prior to onset of infection that play a role in controlling the severity of dengue infections.

Host Factors

Cells that support dengue infection

Only a few histopathologic studies describe cellular and organ damage done by dengue virus infection in human hosts, and none of them were published within the past 25 years.[10–15] To learn more about the tissues that support dengue virus infections, conventional pathologic studies have been supplemented by efforts to isolate the dengue virus or localize dengue viral antigens in peripheral blood leukocytes, biopsies or organs at autopsy using various virus isolation methods, immunohistological tests, PCR on tissue suspensions or *in situ* hybridization. These studies are summarized in Table 1. As is evident, cells of macrophage morphology in skin, liver, renal glomerulus, thymus, spleen, blood and lymph nodes are the most common sites where dengue virus antigens have been found. Dengue viruses have been isolated from cells identified as monocytes or lymphocytes in peripheral blood.[16] Techniques that visualize dengue viral antigens cannot discriminate between attachment of antigens to the surface of cells and phagocytosis of viral antigens, *in vivo*. For this reason *in situ* hybridization is needed to identify sites of viral replication. Very few such studies have been published. In these, blood monocytes and macrophages in liver, skin, spleen and thymus were probe-positive.[17–19] Three studies have isolated the dengue virus or observed viral antigen in the human central nervous system.[20–22] In the brain, the virus was localized to neurons, endothelial or microglial cells. However, when a sensitive viral RNA amplification method was applied to a large number

Table 1. Studies on the localization of dengue viruses, antigens or viral RNA to tissues or cells in surgical biopsies or autopsies from human beings experiencing dengue infections.

Method	Observation	Patients (pos/studied)	References
Virus isolation from autopsy tissues, DSS, 2nd infection	Recovery of virus from liver, heart, lymph node, lung and bone marrow	13/132	90–92, 93–95
Virus isolation from autopsy tissues, DSS, 1st infection, infants	Recovery of virus from liver, spleen, thymus and lung	4/4	92, 94
Virus isolation from PBL, DF/DHF 1st and 2nd infections	Recovery of virus from washed, adherent monocytes and lymphocytes	76/322	16
Virus isolation from biopsies, DSS, 2nd infection	Recovery of DENV-3 from liver	1/1	96
Immunohistology on skin biopsies, DSS, 2nd infection, DF 1st infection	Dengue antigen in dermal macrophages; Langerhans cells in epidermis	15/54	97, 98
Immunohistology on autopsy tissues, DSS/2nd infection	Liver (hepatocytes, Kupffer cells), macrophages in spleen, thymus, lung, intestines; brain parenchyma	6/26	20, 22, 24, 99
Immunohistology on autopsy tissues, 1st infection	DENV-1 antigen in Kupffer cells; splenic, thymic, pulmonary macrophages; endothelial cells, brain	6/6	21, 100
Immunohistology on autopsy, biopsy/PBL, DF/DHF	Dengue antigen in liver, spleen, lymph node macrophages, blood monocytes and lymphocytes and endothelial cells	9/24	17

(*Continued*)

Table 1. (*Continued*).

Method	Observation	Patients (pos/studied)	References
Electron microscopy on kidney biopsies, DHF, 2nd infection	Flavivirus-like particles in glomerular macrophages; none in endothelial cells	12/53	101
Electron microscopy on skin biopsies, DHF, 2nd infection	No necrosis or virus-like particles in endothelial cells	0/44	102
PCR on autopsy tissues, DSS, 2nd infection	Dengue viral RNA in liver, spleen, lymph node, but not brain	31/40	23, 93, 103
RNA hybridization on autopsy tissues, DSS, 2nd infection	DENV-2 in macrophages, liver, spleen, thymus	2/2	18, 19
RNA hybridization on autopsies, biopsies, PBL, DF/DHF	Dengue RNA in splenic macrophages, blood monocytes, lymphocytes	3/24	17

of human tissues, no evidence could be found of dengue virus replication in the CNS.[23] Growth of dengue viruses has been demonstrated in immature skin dendritic cells, and once infected these cells promptly migrate out of skin toward regional lymph nodes.[3] Dengue 1 antigen was found in hepatocytes obtained from a patient dying during what may have been a primary dengue infection; these cells also showed apoptosis.[24] Studies on autopsy tissues complemented by efforts to grow dengue viruses in human tissue explants, primary cells and human cell lines suggest an inability of dengue 1 virus to replicate in mature Kupffer cells, while hepatocyte cells and cell lines do become infected but then promptly die as the result of apoptosis.[25–28] Hepatocytes may be an important reservoir of dengue infection. Of interest, apoptotic dengue-infected hepatocytes engulfed by Kupffer cells form Councilman bodies, the classical stigmata of dengue and yellow fever liver pathology. While these studies are provocative, in order to understand dengue pathogenesis a much greater effort must be made to study the natural history and tissue involvement in dengue infections.

Immune status

Antibodies derived from prior dengue infection

The first evidence that a prior dengue infection somehow modifies a subsequent dengue infection from mild to severe was inferred from a highly significant association found between occurrence of secondary type antibody responses and severe illness.[29] These data are summarized in Table 2. Some authors have critiqued this evidence, suggesting that when multiple dengue serotypes circulate secondary (including tertiary and quaternary) dengue infections must be more common than primary infections.[30–33] With respect to a second as opposed to a third or fourth infection, this assumption cannot be true as second infections can only follow a first infection and not every individual will experience a second infection. Neutralizing antibody profiles on sera from older residents of dengue-hyperendemic areas provide evidence that infections with third and fourth dengue types do occur. But the age of DHF/DSS cases plus numerous epidemiological studies point to severe disease occurring during second heterotypic dengue infections (see below). A small number of third dengue infections result in mild and severe disease (Mammen MP, personal

Table 2. Incidence of secondary dengue infections in the open population and among dengue-infected outpatients and dengue-infected inpatients with syndromes varying in severity.

Severity group	Primary	Secondary	% Secondary
All children	165,794	126,728	29*
FUO/OPDss	33	61	65†
FUO/Hosp.	13	21	64‡
DHF/no shock	65	262	80‡
DSS	2	160	99‡

* Ref. 4, † Ref. 104, ‡ Ref. 105.

Table 3. Summary results of prospective cohort studies of dengue infections. DHF/DSS occurred in children whose preinfection serological status was known.

Place	Year	Serol. test*	Ages (years)	Cohort total	Infections 1st	Infections 2nd	DHF cases 1st	DHF cases 2nd	Reference
Koh Samui	1966	HI	2–12	336	26	83	0	3	106
Rayong	1980	PRNT	4–10	881	93	112	0	4	49
Bangkok	1980	PRNT†	4–15	1757	31	59	0	7	44
Jogjakarta	1995–96	PRNT	4–9	1837	162	120	0	7	52

* HI — Hemagglutination-inhibition; PRNT — plaque reduction neutralization test.

† Paired sera screened by HI and then tested by PRNT.

communication). Fourth dengue infections are largely inapparent. For an expanded discussion on this controversy, see Chap. 14.

The role of second dengue infections has been documented in four prospective seroepidemiologic studies in which hospital DHF cases occurred among a prebled cohort (Table 3). In three of these studies, sera from children prior to hospitalization for DHF had neutralizing antibody profiles consistent with a single antecedent dengue infection. A total of 21 children were hospitalized with DHF/DSS among 379 who experienced a second dengue infection, while no cases were observed among 312 children experiencing a primary infection ($p < 0.0001$). The pathogenic role of second dengue infections was definitively proven in three epidemics in Cuba where DHF/DSS occurred in persons who were infected with DENV-1 in 1977–79 and then infected with DENV-2 in 1981 and 1997[34,35] and following the introduction of DENV-3 in 2001–2.[36] These data are discussed in greater detail below. Mathematical models of sequential dengue infections also show that the age-specific DHF hospitalization curves in dengue-endemic countries are consistent with second but not third or fourth sequential infections.[37] The age distribution of children admitted to the Children's Hospital, Yangon, Myanmar, is similar to 1962–64 Bangkok data used in the analysis (Fig. 1). Remarkably, the mathematical model predicted 58.5 DHF/DSS cases per 1000 secondary dengue infections — a ratio very similar to those documented in prospective studies (see below).

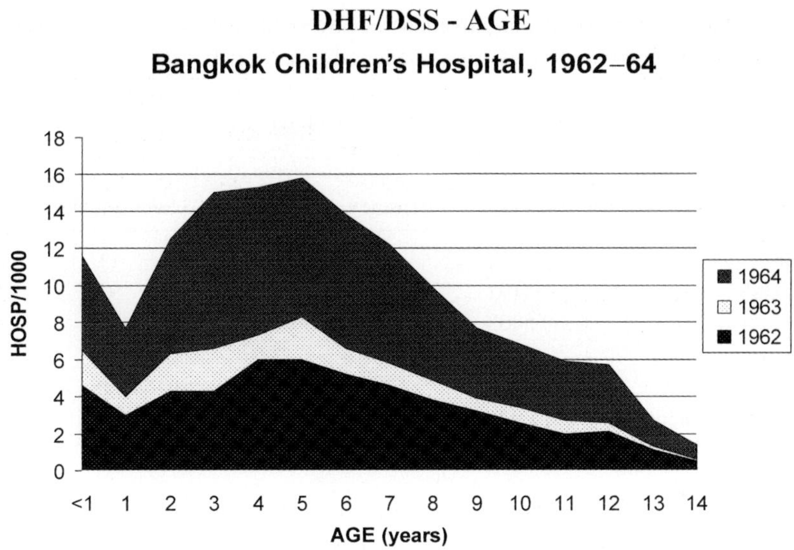

Fig. 1. Yearly age-specific hospitalization rates per 1000 for Bangkok and Thonburi, 1962–64. (Ref. 89, cited with permission.)

The role of immune status (antibody) in changing the clinical expression of dengue can be dramatic. Growing evidence suggests that primary DENV-2 and -4 infections, at least in children, are silent while overt disease accompanies second dengue infections with these viruses.[35,38,39] Table 4 summarizes evidence from the 1997 epidemic in Santiago de Cuba, in which an Asian genotype DENV-2 resulted in silent infections in susceptible adults and children. In the same outbreak, a very high proportion of DENV-1–immune adults infected with DENV-2 developed either dengue fever or DHF/DSS. Individuals old enough to have acquired DENV-1 infections in 1977–79 were all adults in 1997.

Antibodies acquired passively

The bimodal distribution of hospitalized DHF/DSS cases by age suggested the possibility that severe disease was caused by two different etiological mechanisms (Fig. 1). Remarkably, DHF in infants was almost exclusively associated with primary dengue infections.[40] These cases

Table 4. High rate of inapparent infections in susceptible persons and overt disease in those circulating antibody derived from previous dengue 1 infection. Primary infections included persons of all ages. Secondary infections occurred only in persons aged 18 and above. [Data from 1997 dengue 2 epidemic in Santiago de Cuba, Ref. 35, cited with permission.]

Cases	Total	Primary	%	Secondary	%
Deaths	12	1	0.008	11	0.2
DHF/DSS	193	2	0.015	91	1.9
Dengue fever	5003	395	3.0	4608	95.8
Total	5208	398	3.0	4810	100
Inapparent	12,718	12,718	97.0	0	0
Total dengue infections	17,926	13,116		4810	

occurred in an unusual age distribution (Fig. 2).[41] A role for maternal antbodies is illustrated in Fig. 3. The infant DHF hospitalization curve resembles the titration curve for enhancing antibodies *in vitro*.[42] When added to monocytes in the presence of dengue viruses, maternal antibodies exhibit two biological properties: neutralization and enhancement. Initially they protect the infant but later they are capable of enhancing disease. The phenomenon was studied in 13 infants hospitalized for DHF accompanied by a primary DENV-2 infection.[43] Infants were hospitalized between 4 and 12 months of age. All were born to mothers whose dengue serum neutralizing antibody profiles suggested multiple previous dengue infections. With maternal sera serving as surrogates for cord blood, each mother transferred to their infants IgG_1 antibodies that neutralized DENV-2 (Table 5). As antibodies of different titers were catabolized (diluted) to nonprotective levels, infants became susceptible to enhanced DENV-2 virus infections; the higher the PRNT of maternal anti–DENV-2 received at birth, the older the infant at hospitalization for a DENV-2 infection.

Incidence rates of DHF and/or DSS during second dengue infections

The hospitalization rate for severe dengue infections among children experiencing secondary dengue infections has been measured repeatedly

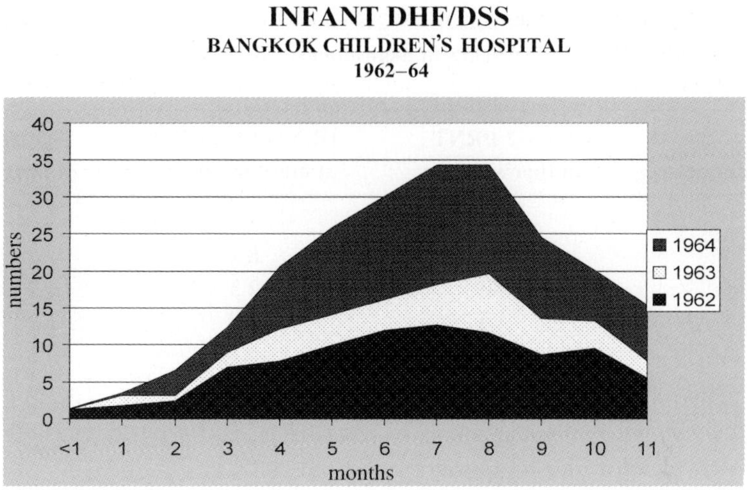

Fig. 2. Age distribution of infants admitted with DHF/DSS to Bangkok Children's Hospital, 1962–64. (Ref. 89, cited with permission.)

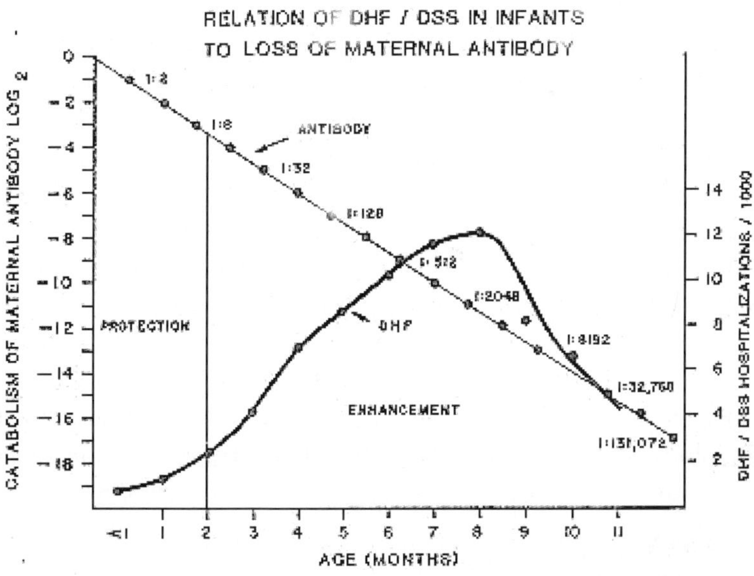

Fig. 3. Schema demonstrating the relationship between the protecting and enhancing effect of passively acquired maternal dengue antibodies and the age distribution of primary infection DHF/DSS cases. (Ref. 4, cited with permission.)

Table 5. Temporal relationship of DENV-2 neutralizing antibodies in mothers' sera compared with age at onset of DHF/DSS in their infants. (Modified from Kliks *et al.*[43]; cited with permission.)

Case number	DENV-2 $PRNT_{50}$ mother's serum	DENV-2 $PRNT_{50}$ at onset of DHF†	Age at onset of DHF (mos)
1	30	1.9	4
2	50	6.2	3
3	80	2.5	6
4	90	0.7	8
5	200	3.1	7
6	290	4.5	7
7	350	2.7	8
8	360	45.0	4
9	420	3.3	8
10	500	15.2	6
11	720	5.6	8
12	2000	3.9	11
13	8200	4.0	12

* Reciprocal titer as determined by the method of Ref. 107.

† Predicted assuming half-life of maternal IgG dengue antibody of 35 days.

over the past 40 years, most being in the range of 2–3% (Tables 6 and 7). Eight published studies provide data on serologically characterized DHF cases admitted to hospital(s) in communities in which primary and secondary dengue infection rates were measured by testing antibodies in pre- and post-rainy-season serum pairs (Table 6). The most dramatic and persuasive of these studies are from Cuba, where SE Asian genotype DENV-2 viruses were introduced into an island population previously exposed only to the DENV-1 virus.[34,35] In the first DHF/DSS outbreak, the introduction of DENV-2 in 1981 followed a DENV-1 outbreak lasting from 1977 to 1979. No dengue transmission had occurred in Cuba since the end of World War II.[34] After years of strict mosquito abatement, control measures lapsed and DENV-2 was again introduced in 1997, this time in an outbreak restricted to the eastern city Santiago de Cuba. There, only individuals who were 18 years or older were immune to DENV-1 from the islandwide epidemic of 1977–79. DHF/DSS cases were restricted to

Table 6. Primary and secondary dengue infections in selected open populations and among DHF/DSS cases in children aged one year or older.

Place	Year	Serol test	Age years	Total population	Infections 1st	Infections 2nd	DHF/DSS 1st	DHF/DSS 2nd	References
Bangkok	1962	HI	1–15	2097	412	313	n.d.	n.d.	4, 46, 105
				842,451	165,794	125,728	297* (DHF)	2528* (DHF)	
							12* (DSS)	1428* (DSS)	
Koh Samui	1966	HI	1–15	13,975	1900	2700	1	33[†]	106
Koh Samui	1967	HI	1–15	13,975	n.d.	n.d.	0	15[‡]	108
Rayong	1980	PRNT	<1–10	8885	1717	920	0	18[§]	49
Yangon, Myanmar	1984–88	PRNT	2–6	3579	472	448	7	138[¶]	51
Havana	1981	PRNT	3–14	359,879	20,393	59,875	18	1213[//]	34
Santiago de Cuba	1997	PRNT	20+	325,310	13,116	4810	3	202	35

Viruses isolated:

* Children's Hospital study only: DENV-1,[10] DENV-2,[10] DENV-3.[4]

[†] DENV-1,[1] DENV-2,[16] DENV-3.[7]

[‡] Only from DF cases: DENV-4.[7]

[§] DENV-2.[7]

[¶] DENV-1,[3] DENV-2,[13] DENV-3,[4] DENV-4.[1]

[//] DENV-2 isolated only.

Table 7. DHF/DSS or DSS hospitalization rates per 1000 secondary dengue infections calculated from data derived from prospective cohort studies or from epidemiologically defined populations.

Place	Year	Ages (years)	Secondary dengue infections		DHF/1000 2nd infections	DSS/1000 2nd Infections	Reference
			DHF/DSS				
Bangkok	1962	1–15	2528/1428	125,728	20.1*	11.4*	4
Koh Samui	1966	2–12	3	83	36.1		106
Koh Samui	1966	1–15	33	2700	12.2		106
Rayong	1980	4–10	4	112	–	35.7*	49
Rayong	1980	<1–10	18	920	–	19.6*	49
Rayong	1980	<1–10	10	48	–	208.0*,†	49
Bangkok	1980	4–15	7	59	118.6*,‡		44
Havana	1981	3–14	1213	59,875	20.3		34
Yangon	1984–88	2–6	138	4181	33.0*		51
Jogjakarta	1995–96	4–9	7	120	58.3*		52
Santiago de Cuba	1997	18+	202	4810	42.0*		35

* Cases graded by WHO criteria.

† DSS cases among secondary DENV-2 infections only.

‡ Denominator likely to be too small as secondary infections determined by insensitive HI test.

individuals 18 years old and older. In 2001–02, DENV-3 circulated in the Havana area, producing a small DHF outbreak. Only individuals who were infected in the sequence of DENV-1 then DENV-3 experienced severe disease.[36] No DHF was observed in tertiary infections, i.e. persons infected first with DENV-1 and -2 then with DENV-3 (Guzman MG, personal communication).

Hospitalization rates during secondary dengue infection have also been measured in the dengue-endemic countries Thailand and Myanmar (Burma). These data permitted the calculation of DHF/DSS or DSS hospitalization rates per 1000 secondary dengue infections (Table 7). Among seven population-based studies on dengue in children, DHF rates ranged from 12.2 to 118.6/1000 secondary dengue infections, with a mean of 42.7/1000. Four independent studies in Thailand, Myanmar and Cuba produced tightly clustered values of 20.1–36.1/1000 secondary dengue infections. An eighth measurement was made on Cuban adults infected with DENV-1 then DENV-2 at an interval of 18–20 years. Four studies estimated attack rates per 1000 secondary dengue infections separately for dengue shock syndrome (Table 7). The highest DSS rate, 208.0/1000 secondary dengue infections, was calculated for infections occurring in the specific infection sequence of DENV-1 then DENV-2. The same infection sequence was thought to cause most DHF cases in the 1980 Bangkok school study and also yielded high infection- ratios.[44]

In a large Bangkok study, Scott *et al.* found records of five children with severe disease accompanied by antibody response classified as primary.[45] An earlier large study in Bangkok had included similar cases (Table 6). In large data sets from Bangkok (1962), Yangon (1984–88), Havana (1981) and Santiago (1997), primary infections in children constituted 7.3%, 5.0%, 1.5% and 1.5% of all DHF cases, respectively. DHF hospitalizations per 1000 primary infections in children are calculated from data provided in Table 6. For Bangkok, Yangon, Havana and Santiago, the rates were 1.9, 0.9, 0.9 and 0.23/1000, respectively (mean of 1.0). When means of primary and secondary infection population-based hospitalization rates are compared, DHF/DSS occurs about 40 times more frequently during secondary than primary infections. Because observations from large studies cannot be of research quality, there is always uncertainty as to whether clinical cases or antibody responses were correctly classified. When data are of high quality, primary infections in children are nearly always associated with cases classified as dengue fever or DHF without shock.[38] These comparisons, of course, exclude infants under the age of one year.

Incidence rates of DHF and/or DSS in infants with passively acquired dengue antibodies

Hospitalization rates have been estimated for DHF/DSS in infants based on a random sample of infant DHF/DSS cases studied virologically at Bangkok Children's Hospital in 1962.[46] The number of infant DHF cases with primary antibody responses was used to estimate total admissions for the year and then by ratio to all Bangkok and Thonburi hospitals. An age-stratified serological survey and data from the 1960 census were used to estimate susceptibles and population size for 3–11-month-old infants. Using the 1962 dengue infection rate for Bangkok, infant DHF hospitalizations per 1000 primary infections were calculated (Table 8). This rate is about one-half the secondary infection DHF rate in older children (Table 7). In calculating this rate it is assumed that infants are at risk of primary infection DHF for 3–11 months. However, passively acquired maternal dengue antibodies degrade in such a way that each infant is at risk for a period of only about one month.[43] Recalculated for a one-month period, the rate becomes 110.6/1000 (Table 8), or four times the estimated hospitalization rate for Bangkok children during secondary dengue infections. This rate is higher than the 42.7/1000 mean rate of childhood DHF

Table 8. Hospitalization rates for DHF in infants less than one year old per 1000 primary dengue infections, estimated using at-risk periods of one or nine months.

Place, year	At-risk period	At-risk	DHF cases	DHF/ 1000	Reference
Bangkok, 1962	9 months	26,519*	326	12.3	4
	1 month	2947†	326	110.6	

* Estimated dengue infections in infants aged 3–11 months (see Fig. 1) based upon 1962 infection rate of 41%.

† Estimated dengue infections assuming infant is at risk of DHF/DSS for period of four weeks (see text).

hospitalizations/secondary infections (see above), suggesting that passive antibody is more effective than infection acquired immunity in enhancing dengue disease.

Factors that control illness severity during secondary infections

Many factors have been identified that influence the clinical expression of secondary dengue infections. Factors intrinsic to the host are presented below. Here we discuss antibody and viral factors.

Heterotypic neutralizing antibodies

Infections with dengue viruses raise antibodies to a wide range of antigens. Indeed, there is evidence from human cross-challenge experiments performed by Sabin that following DENV-1 infection adults were not susceptible to DENV-2 infection for three months.[47] Varying amounts of heterotypic neutralizing antibodies are raised following a primary dengue infection. The natural histories of antibody responses following infection by DENV-1, -2, -3 or -4 viruses including heterotypic antibodies are essentially unknown. These antibodies are critically important, as shown in Table 9. Low dilutions of preinfection sera from children who had inapparent secondary DENV-2 infections almost invariably neutralized DENV-2, while preinfection sera from children who developed an illness requiring hospitalization had little or no detectable heterotypic

Table 9. The moderating effect of heterotypic neutralizing antibodies during secondary dengue infections; the relationship between preillness neutralizing antibodies derived from previous dengue infection and either inapparent infection or hospitalization among a cohort of 40 Bangkok schoolchildren, June–November 1980. (Cited with permission from Ref. 44)

Dengue 2 NT antibodies in undiluted sera, June 1980	Admitted to hospital with DHF	Number of school absences
Yes	1	29
No	6	4
	7	33

neutralizing antibodies.[48] This observation suggests that low levels of heterotypic neutralizing antibodies (most were anti–DENV-1) did not prevent but down-regulated DENV-2 infections. Around one-fifth of monotypically dengue-immune schoolchildren lacked heterotypic DENV-2 antibodies and developed DHF when infected by DENV-2. This is virtually the same ratio for DSS during sequential DENV-1 then DENV-2 in Rayong, Thailand (see Table 7).

It is not known whether heterotypic antibodies are raised to antigens common to two or more dengue viruses and/or as a result of a specific antibody response repertoire controlled by the host. However, there is evidence that DENV-2 viruses may have evolved to escape from heterotypic neutralization (see below).

Sequence of viral infection

The sequence of infection may be highly determinative of disease severity. Only secondary DENV-2 infections were pathogenic in the 1980 cohort study at Rayong, Thailand.[49] The specific infection sequences associated with DSS cases were known from virus isolations in acute phase sera and antibodies in preillness sera, or by applying the original antigenic sin phenomenon to paired sera.[50] Although secondary DENV-1 infections were most common that year, DSS occurred only during secondary DENV-2 infections (Table 10). Burmese workers came to a similar conclusion in their 1984–88 longitudinal seroepidemiological study in Yangon, Myanmar.[51] By contrast, in an Indonesian study (Table 10), DSS was associated with sequences ending in DENV-1, -3 and -4, but not DENV-2, even though secondary DENV-2 infections were common.[52] DENV-3 was associated with an outbreak of DHF/DSS on Tahiti in a population that had prior infection experience with DENV-1 and DENV-2.[53] This DENV-3 infection seems to have sensitized children less than 12 years of age to DHF/DSS during DENV-1 infections that occurred in 2001 (Hubert, personal communication, 2002). Clearly, DHF/DSS is not universally associated with secondary DENV-2 infection. Furthermore, over time pathogenic sequences may change. This conclusion is implied by the constant

Table 10. Incidence of DHF/DSS cases among secondary dengue infections by specific sequence in childhood cohorts studied in Thailand and Indonesia.[49,52]

Place, year, ages	First dengue infection	Second dengue infection: Dengue 1	Dengue 2	Dengue 3	Dengue 4
Rayong (Thailand)	Dengue 1	–	10/48	0/9	0/20
1980, ages <1–10	Dengue 2	0/257	–	0/34	0/82
	Dengue 3	0/125	6/92	–	0/40
	Dengue 4	0/114	2/84	0/15	–
Jogjakarta	Dengue 1	–	0/31	1/6	0/6
(Indonesia),	Dengue 2	3/41	–	0/7	1/8
1995–96, ages	Dengue 3	0/9	0/2	–	0/4
4–9 years	Dengue 4	0/4	0/5	0/1	–

changes in dengue viruses recovered from DHF/DSS over many years at the Bangkok Children's Hospital (see Chap. 3).

Interval between the first and second dengue infections

DHF/DSS epidemics occurred in Cuba in 1981 and 1997. In both instances, severe disease was caused by Asian genotype DENV-2 viruses infecting DENV-1–immune individuals.[54] DENV-1 was transmitted in 1977–79 and is the only known introduction of this virus into Cuba. As soon as epidemics were recognized, comprehensive vector control was started and dengue transmission promptly terminated. Complete control of new dengue infections made it possible to measure infection rates associated with these epidemics from the prevalence of neutralizing antibodies in randomly collected sera.[34,35] Age-specific secondary DENV-2 infection rates were estimated for outbreaks in each city. DHF cases and deaths reported by age from Havana in 1981 and Santiago in 1997 were then used to calculate DHF and death rates per 10,000 secondary DENV-2 infections.[55] When DHF and death rates are compared in identical age groups, 15–65+ years, the 1997 epidemic can be seen to be 8 and 25 times more severe than that of 1981, respectively (Table 11). Correspondingly, case fatality rates among 15–39-year-olds in Havana and Santiago were 0.6% and 2.6%, respectively, differing by a factor of 4.7-fold.

Table 11. Increased severity of clinical disease in 15–65+-year-olds experiencing dengue 2 infections 4 and 20 years following dengue 1 infection. (Cited from Ref. 55 with permission.)

Place, year	Secondary DENV-2 infections	DHF cases	Deaths	DHF/10^4 secondary DENV-2	Deaths/10^4 secondary DENV-2	Case fatality rate (%)
Havana, 1981	242,070	1329	21	54.9	0.9	1.58
Santiago, 1997	4810	202	11	419.9	22.8	5.4

Factors that control illness severity during primary infections in the presence of passively acquired antibody

It was noted (Table 5) that sera from mothers whose infants acquired primary infection DHF had variable amounts of DENV-2 neutralizing antibodies. Figure 3 demonstrates the positive correlation between the DENV-2 $PRNT_{50}$ titer of maternal antibodies (presumably the titer in infant sera at birth) and the age of the infant at hospitalization for DHF. Most infants were admitted within three weeks of the time that maternal antibodies were expected to fall to a titer of 1:10.[43] Infants are at risk when neutralizing antibodies have fallen below a protective threshold but titers of ADE antibodies are maximal.

Other host factors

Race

Dengue researchers have noted that during each DHF/DSS outbreak in Cuba, blacks were hospitalized at rates lower than expected based upon their distribution in the general population.[34] The phenomenon was more precisely documented in a retrospective seroepidemiological study in which black and white residents of a district of Havana were shown to have had similar DENV-1 and DENV-2 infection rates. Reduced dengue hospitalization rates in blacks in the face of high infection rates and in a society in which all ethnic groups have equal access to health care point to the existence of a resistance gene.

Table 12. Dengue antibody prevalence in a cohort of schoolchildren with residence in Port-au-Prince, Haiti, 1997. (Cited from Ref. 58.)

	Age (years)								
Infection status	**6**	**7**	**8**	**9**	**10**	**11**	**12**	**13**	**Total (%)**
No antibody	1	0	2	0	1	0	0	0	4 (1.9)
Monotypic	4	8	9	4	1	1	0	0	27 (12.7)
Multitypic	41	32	25	37	25	12	5	2	181 (84.9)
Total	46	40	36	41	27	13	5	2	210

It can be predicted that relative epidemiological silence will accompany transmission of multiple dengue serotypes in black populations. This is the situation in Haiti, where simultaneous transmission of at least three dengue serotypes was documented.[56,57] The prevalence of dengue-neutralizing antibodies in Port au Prince schoolchildren implies extremely high rates of sequential dengue infection (Table 12), but until now there have been no documented cases of DHF/DSS.[58] All four dengue virus serotypes have been recovered from Africa, where dengue fever epidemics are rare and outbreaks of DHF/DSS have not been reported.[59]

Genetic markers

Several studies have searched for individual genetic markers that might be correlated with severe vs mild disease outcome. Ideally, such studies should examine patients matched for age, sex and race who had precisely the same dengue infection exposure but a different clinical outcome. CD8 T cells are likely to contribute to successful elimination of infection without serious symptoms and to severe dengue disease. Inflammatory cytokines secreted by interactions of CD8 T cells and virus-infected cells, particularly in the monocyte/macrophage lineage, may damage endothelial cells, resulting in a cascade of pathological events. Polymorphisms at the HLA class I loci was significantly associated with DHF susceptibility but polymorphism in the HLA-DRB1 or TNF genes was not.[60] Further analysis of the HLA class I association was confined to the HLA-A and not the HLA-B gene and that two alleles were relevant. Children with

HLA-A*33 were less likely to develop DHF and children with HLA-A*24 allele were at increased risk. These data on HLA class I associations with DHF support the class I variation described in two earlier studies:

(1) HLA frequencies were measured in 82 shock syndrome patients from the 1981 Cuban epidemic and compared with 80 persons with no history of DHF but whose dengue infection experience was not measured.[61] The distribution of HLA-A2, HLA-B blank, HLA Cw1 and HLA-A29 antigens differed between the groups.
(2) The distribution of 15 A loci and of 20 B loci on lymphocytes from 87 Thai DHF/DSS patients were compared with cells from 138 adult blood donors.[62] All donors had dengue HI antibody, but specific infection experience was unknown. Significant differences were observed in HLA loci between DSS patients and controls (higher or lower: A2, B13, B blank).

Age

During the Cuban DHF/DSS outbreak, all persons aged 2 through 40 years had the same lifetime dengue virus experience — DENV-1 in 1977–79 and DENV-2 in 1981.[34] This unique epidemiology, the result of years of limited trade and intercountry travel in the region, provided a unique opportunity to assess the effect of age on the clinical expression of secondary DENV-2 infections.[63] DENV-1 and -2 infection rates measured in the Cerro District in central Havana were applied to the entire population of the city. Age-specific hospitalization rates were estimated for individuals experiencing secondary DENV-2 infections. As shown in Fig. 4, the youngest infants are at highest risk of DHF/DSS. Rates reach a nadir among teenagers and then rise again during the third and fourth decades of life. These are the first data to compare risk of DHF/DSS for children and adults. Children were at a 40-times-greater risk than adults from severe disease, due to vascular permeability during a secondary DENV-2 infection. The application of data from Fig. 4 corrects an error made in a mathematical model relating dengue transmission to age-specific hospitalizations of DHF/DSS. In the original publication it was necessary to postulate that DHF/DSS would result only when first and second dengue infections occurred within a five-year period.[37] This time limitation was imposed under the assumption that all age groups were equally susceptible to DHF/DSS during secondary dengue infections.

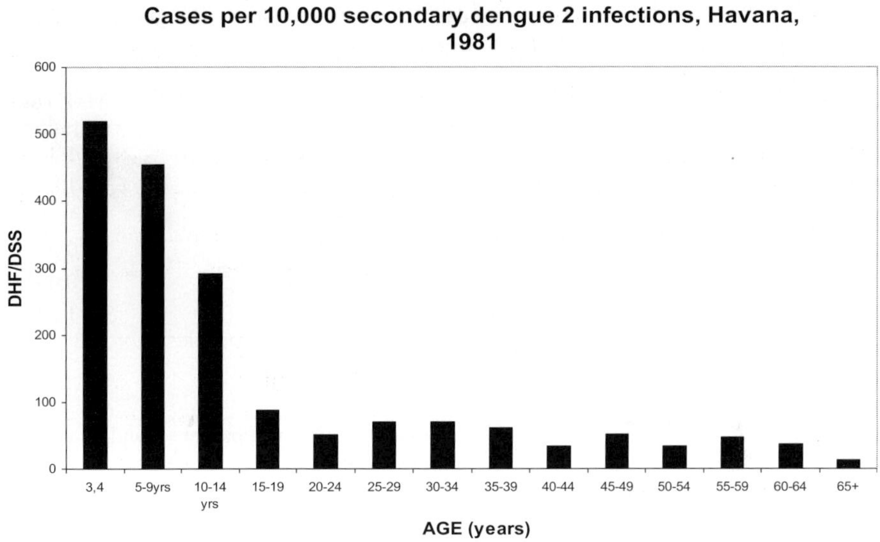

Fig. 4. Age-specific DHF/DSS hospitalization rates per 10,000 secondary DENV-2 infections for children and adults during the 1981 DENV-2 outbreak in Havana, Cuba. (Ref. 55 cited with permission.)

Sex

Many authors have found no significant difference in the hospitalization rates of males and females with DHF/DSS. Indeed, sex is not a factor among milder secondary dengue infections (Table 13). DHF grades I and II were observed at normal ratios, but DSS was more common in girls after the age of four years.[40] More DSS deaths have been reported in females than males in several series.[11,15]

Nutritional status

Although there is considerable anecdotal evidence that clinically significant malnutrition has a sparing effect on the severity of secondary dengue infections, this has been documented only by a single study.[64] It is difficult to select a group matched for children who are exposed to dengue infections. In the published study normal outpatients were chosen as controls (Table 14). The implication of the sparing effect of malnutrition on DSS is that as economies and nutrition improve in

Table 13. Sex and severity of secondary dengue infections, Bangkok Children's Hospital, 1962–64. (Cited from Ref. 109 with permission.)

Syndrome	Male	Female	M–F ratio
DF, outpatient	50	44	1.14
OPD control	106	100	1.06
DHF, no shock	136	115	1.18
DSS, all ages	72	123	0.59
DSS, > 4 years	(45)	(92)	(0.49)
Hospital control	42	34	1.24

Table 14. Effect of nutritional status on severity of clinically apparent secondary dengue infection syndromes. (Cited from Ref. 64 with permission.)

	Percent distribution		
		Controls	
Malutrition status	DHF/DSS $n = 100$	Infectious disease $n = 100$	Healthy children $n = 185$
None	87	28.8	76.4
1st degree	9	29.6	21.0
2nd degree	4	28.8	1.6
3rd degree	0	12.8	0

dengue-endemic areas without mosquito control or vaccination, DHF/DSS attack rates will increase. Recently, there has been some evidence that DHF in obese children is complicated, resulting in a poorer prognosis.[65]

Viral Factors

Genotypes

The concept that DHF/DSS is caused by intrinsic differences in dengue viruses has been a reflexive explanation for the "emergence" of this syndrome in SE Asia. When DENV-3 and -4 were associated with DHF/DSS

Table 15. Summary of studies looking for genetic sequences that correlate with dengue disease severity.

Serotype	Gene sequenced	DF	DHF, I&II	DSS	Observations	Reference
2	E	4	2	2	No differences	110
3	E (aa25-89)	9	11	1	"	111
3	PrM-E	1	2	2	"	112
2	E	1	1	1	"	"
2	E/NS-1 (240nt)	1	2	1	"	"
2	E/NS-1	0	2	1	"	113
2	C-PrM-E	1	2	1	1-2 aa	114
2	Entire	4	3	1	No differences	115
2	E/NS-1 (240nt)	30	462	–	"	116

in the Philippines in 1956, it was speculated that these serotypes were responsible for the "new" disease.[66] When serological studies suggested the existence of DENV-5 and -6, again they were thought to be DHF-associated.[67] When DENV-1 and -2 viruses were shown to cause dengue fever as well as DHF,[68] the search for virulent and nonvirulent dengue viruses focused on biological differences, e.g. *Aedes aegypti*- vs *Aedes albopictus*-transmitted viruses.[69,70] It seems logical that intrinsic virulence would have a genetic basis. Many dengue serotypes recovered from mild and severe dengue infections have been partially sequenced and compared (Table 15), but as yet no reproducible genetic differences have been found within DHF-endemic areas.

Antigenic structure

A simple observable correlate of genetic differences among viruses is antigenic changes. When Peruvian anti–DENV-1 sera were tested for their ability to neutralize American vs Asian genotype DENV-2 viruses, the former strains were highly neutralized while the latter were not (Table 16).[71] This suggests the existence on American genotype DENV-2 of envelope structure(s) analogous to structure(s) on DENV-1 viruses with the loss or modification of this structure on Asian genotype DENV-2 viruses.

Table 16. Geometric mean neutralizing antibody titers in 44 sera from DENV-1–immune residents of Iquitos, Peru, when tested by homologous dengue virus (DENV-1) or DENV-2 strains representing either American or Asian genotypes. (Cited from Ref. 81 with permission.)

Genotype	Asian	American	Asian	Asian
Virus	DENV-1	DENV-2	DENV-2	DENV-2
Strain	16007	IQT 2913	16881	OBS 8041
Isolated, place, date	Thailand, 1964	Peru, 1996	Thailand, 1964	Venezuela, 2000
GMT, $N = 44$	578	549	56	29

Rapid selection of mutants

The possibility of the rapid selection of neutralization escape mutants of dengue viruses was suggested by an observation made during the 1981 as well as the 1997 DHF/DSS epidemics in Cuba. In each outbreak month-to-month increases were observed in the fraction of severe compared with mild cases and in case fatality rates (Table 17).[72] It was hypothesized by the authors that successive transfers of DENV-2 viruses in DENV-1–immune hosts might result in rapid selection of mutants that lack the DENV-1–like antigenic structure(s). These viruses might be enhanced by antibodies directed against noncritical structural sites. This hypothesis was investigated studying the small number of dengue strains recovered from the 1997 epidemic.[73] Instead of changes in the envelope, systematic nucleotide changes were introduced into the nonstructural gene, NS-5.[74]

Neutralization escape mutants as a mechanism of emergence

Disease-severity-related viral genetic differences are well established. The first dengue virus recovered in the American hemisphere was DENV-2 TR 1751.[75] Outbreaks of dengue prior to World War II were attributed to DENV-2 by retrospective serological studies in Panama and Cuba.[34,76] In 1963, DENV-3 was introduced into the hemisphere and recognized in Puerto Rico.[77] It is not known how widely these two viruses were transmitted, but conditions for sequential infection existed and were

Table 17. Monthly DHF/DSS attack rates, case fatality rates and death rates during DENV-2 epidemics in all of Cuba, 1981, and Santiago de Cuba, 1997. (Cited from Ref. 72 with permission.)

	Cuba 1981				**Santiago 1997**			
Months	**June**	**July**	**August**	***P****	**May**	**June**	**July**	***P****
DHF/DSS	1881	6223	2120		37	132	29	
Deaths	38	77	40		1	6	5	
Total dengue	96,664	183,443	43,315		705	1785	244	
DHF/DSS/ dengue (%)	1.9	3.4	4.9	$< 10^{-7}$	5.2	7.4	11.9	< 0.01
Deaths/DHF/ DSS (%)	2.0	1.2	1.9	> 0.05	2.7	4.5	17.2	< 0.05
Deaths/ dengue (%)	0.04	0.04	009	< 0.01	0.14	0.34	2.05	< 0.01

* First month versus third month.

not accompanied by DHF/DSS. In 1977, DENV-1 was introduced into the Caribbean and quickly spread throughout the region.[78] Again, sequential infections (DENV-2–DENV-1 or DENV-2–DENV-3) were possible but there were no reports of DHF/DSS. Genomic studies on DENV-2 strains found differences between the American (non-DHF-related) and Asian (DHF-related) viruses.[79] DHF/DSS immediately accompanied an Asian genotype DENV-2 in Cuba in 1981 in individuals who had been infected during the 1977–79 virgin soil introduction of DENV-1.[54,80] But similar sequential introductions in of DENV-1 in 1990 and an American genotype DENV-2 in 1995 did not result in DHF/DSS in Peru.[81] Fortuitously, this event occurred in the Amazonian city of Iquitos, population 344,686, where a prospective seroepidemiological fever study was ongoing. This study permitted an estimate of the number of secondary DENV-2 infections. Forty-nine thousand secondary DENV-2 infections could have resulted in up to 10,000 cases of DHF/DSS. Careful study of hospital records found no DHF/DSS-like disease. In fact, secondary infections were accompanied by mild disease at attack rates far below the estimated dengue infections. Full-length sequences of the Asian and American DENV-2 genomes from viremic sera revealed a total of six encoded amino acid charge differences in the prM, E, NS4b, NS 5 genes along with structural changes in the 5′ and 3′ nontranslated regions.[82]

An explanatory model for a mechanism driving the emergence of DHF in dengue-hyperendemic regions can now be posited. The American DENV-2 virus is closer than Asian genotypes to sylvatic DENV-2 strains, the majority of which have been recovered in Africa.[1] This more "primitive" American genotype DENV-2 circulated for a long time alone and in relative silence in the American tropics. Because no other dengue types had been introduced, this strain was not under neutralization pressure from heterotypic dengue antibodies. SE Asian DENV-2 viruses evolved later than the American DENV-2 genotype and, at least since World War II, have cocirculated with other serotypes. Under these conditions it would not be surprising to learn that Asian genotype DENV-2 had lost surface structure(s) highly reactive with DENV-1 antibodies (neutralization escape). It is intriguing to speculate that this may be a generalized phenomenon affecting other dengue viruses circulating in SE Asia.

Discussion

Mechanisms regulating severity of dengue illnesses

Infected cell mass

Although this chapter has focussed on factors intrinsic to the virus or to the host or to the host's previous immunological experience, the implications of these data for mechanisms of disease severity should not be ignored. In the ADE hypothesis, the most important mechanism controlling disease severity of dengue is the total infected cell mass. It is important to understand that available evidence suggests that infected cell mass peaks *after* the cessation of viremia, based upon studies on rhesus monkeys.[83] That nonneutralizing dengue antibodies can enhance dengue infection was demonstrated by infecting rhesus monkeys which had received dengue antibodies by passive transfer.[84] It is plausible that antibodies alone were responsible for enhancing viremias in sequentially infected rhesus monkeys.[85] These monkeys were not ill. Perhaps this is to be expected since peak DENV-2 viremias in rhesus monkeys are five logs lower than in humans (Tables 18 and 20). An often-cited

Table 18. Dengue 2 viremia and infection parity in monkeys. (Cited from Ref. 85 with permission.)

Infection parity	n	Viremia onset day (mean)	P	Duration (days) (mean)	P	Mean peak viremia (PFU $\log_{10}$/1.0 ml)	P
Primary	24	3.0		4.0		2.7	
Secondary	44	3.6	0.05	3.4	0.5	3.7	< 0.001

Table 19. Virus load in acute phase sera, disease syndrome and infection parity. (Cited from Ref. 86 with permission.)

		Primary infections			Secondary infections		
	Syndrome	n	**Mean ($\log_{10}$)**	**Range**	n	**Mean ($\log_{10}$)**	**Range**
Dengue 3	Dengue fever	5	5.5	4.3–7.0	5	6.6	4.2–8.0
	DHF I & II	1	7.3	7.3	3	6.0	2.8–8.2
	DSS	2	5.4	4.6–6.4	9	6.1	3.8–8.2

Table 20. "Peak" virus load in acute phase sera from secondary dengue infections in relation to infection severity. (Cited from Ref. 38 with permission.)

	Dengue 1		Dengue 2	
Syndrome	n	**Peak mean ($\log_{10}$)**	n	**Peak mean ($\log_{10}$)**
Dengue fever	13	5.5*	16	5.5†
DHF I & II	10	7.3*	26	7.6†
DSS	3	8.3*	5	8.5†

* $R = 0.5$; $p = 0.01$.
† $R = 0.5$; $p < 0.01$.

study on children with DENV-3 infections in Indonesia documented mean viremias which were higher during certain secondary compared with primary DENV-3 infections (Table 19).[86] Because these were not serial bleedings begun early in the febrile period, but single point determinations on blood taken at various intervals after the time when peak viremias occur, only a range of titers was observed and these do not differ significantly.

All techniques needed to prove the ADE hypothesis are now available. Peak viremia and NS1 proteinemia should be the relevant surrogate for infected cell mass in dengue already demonstrated to predict oncoming disease severity during secondary DENV-1 and -2 infections (Table 20).[87,88] These data were obtained from children admitted to study within 48 hours after onset of fever and bled for five consecutive days, thus making it possible to identify the period of peak viremia. Peak viremia titers were carefully determined by assaying sera through the inoculation of mosquitoes via the intrathoracic route.[38] It seems likely that antibody and T cell responses will also be proportional to infected cell mass. A similar effort is required to measure peak viremias and immune responses in infants with DHF during primary dengue infections. To test the ADE hypothesis it will be crucial to compare peak viremias in individuals who are sick versus those who are not. For the inapparent infections or mild disease known to accompany primary DENV-2 and -4 infections, it is unlikely that sequential bleedings can ever be obtained to measure viremia kinetics. But, can there be any doubt that peak viremias during silent primary infections will be far lower than those accompanying clinically overt secondary dengue infections?

References

1. Wang E, Ni H, Xu R, Barrett AD, Watowich SJ, Gubler DJ *et al.* Evolutionary relationships of endemic/epidemic and sylvatic dengue viruses. *J Virol* 2000;74(7):3227–3234.
2. Halstead SB. Pathogenesis of dengue: challenges to molecular biology. *Science* 1988;239:476–481.
3. Wu SJ, Grouard-Vogel G, Sun W, Mascola JR, Brachtel E, Putvatana R *et al.* Human skin Langerhans cells are targets of dengue virus infection [see comments]. *Nat Med* 2000;6(7):816–820.
4. Halstead SB. Immunological parameters of togavirus disease syndromes. In: Schlesinger RW (ed.) *The Togaviruses: Biology, Structure, Replication.* Academic, New York, 1980, pp. 107–173.
5. Halstead SB. Dengue: hematologic aspects. *Seminhematol* 1982;19(2): 116–131.

6. Porterfield JS. Antibody-dependent enhancement of viral infectivity. *Adv Virus Res* 1986;31:335–355.
7. Halstead SB. Antibody-dependent enhancement of infection: a mechanism for indirect virus entry into cells. In: Wimmer E (ed.) *Cellular Receptors for Animal Viruses*. Cold Spring Harbor Laboratory Press, 1994, pp. 493–515.
8. Mascola JR, Matthieson, PM, Walker, MC, Halstead, SB, Burke, DS. Summary report: workshop on the potenital risks of antibody-dependent enhancement in human HIV vaccine trials. *AIDS Res Hum Retroviruses* 1993;9:1175–1184.
9. Halstead SB. Neutralization and antibody-dependent enhancement of dengue viruses. In: Chambers TJ, Monath, TP (eds.) *The Flaviviruses: Pathogenesis and Immunity*. Elsevier, New York, 2003, pp. 422–467.
10. Nelson ER. Hemorrhagic fever in children in Thailand: report of 69 cases. *J Pediatr* 1960;56:101–108.
11. Piyaratn P. Pathology of Thailand epidemic hemorrhagic fever. *Am J Trop Med Hyg* 1961;10:767–772.
12. Aikat BK. Pathology of mosquito-borne haemorrhagic fever in the Calcutta outbreak. *Bull World Health Organ* 1966;35:48–49.
13. Fresh JW, Reyes V, Clarke EJ, Uylangco CV. Philippine hemorrhagic fever: a clinical, laboratory, and necropsy study. *J Lab Clin Med* 1969:451–458.
14. Bhamarapravati N, Tuchinda P, Boonyapaknavik V. Pathology of Thailand haemorrhagic fever: a study of 100 autopsy cases. *Ann Trop Med Parasitol* 1967;61:500–510.
15. Burke T. Dengue haemorrhagic fever: a pathological study. *Trans R Soc Trop Med Hyg* 1968;625(5):682–692.
16. Scott RM, Nisalak A, Cheamudon U, Seridhoranakul S, Nimmannitya S. Isolation of dengue viruses peripheral blood leukocytes of patients with hemorrhagic fever. *J Infect Dis* 1980;141(1):1–6.
17. Jessie K, Fong MY, Devi S, Lam SK, Wong KT. Localization of dengue virus in naturally infected human tissues, by immunohistochemistry and *in situ* hybridization. *J Infect Dis* 2004;189:1411–1418.
18. Killen H, Osullivan MA. Detection of dengue virus by *in situ* hybridization. *J Virol Methods* 1993;41:135–146.
19. Kangwanpong D, Bhamarapravati N, Lucia HL. Diagnosing dengue virus infection in archived autopsy tissues by means of the *in situ* PCR method: a case report. *Clin Diagn Virol* 1995;3:165–172.

20. Bhoopat L, Bhamarapravati N, Attasiri C, Yoksarn S, Chaiwun B, Khunamornpong S *et al.* Immunohistochemical characterization of a new monoclonal antibody reactive with dengue virus-infected cells in frozen tissue using immunoperoxidase technique. *Asian Pac J Allergy Immunol* 1996;14(2):107–113.
21. Miagostovich MP, Ramos RG, Nicol AF, Nogueira RM, Cuzzi Maya T, Oliveira AV *et al.* Retrospective study on dengue fatal cases. *Clin Neuropathol* 1997;16(4):204–208.
22. Ramos C, Sanchez G, Pando RH, Baquera J, Hernandez D, Mota J *et al.* Dengue virus in the brain of a fatal case of hemorrhagic dengue fever. *J Neurovirol* 1998;4(4):465–468.
23. Rosen L, Drouet MT, Deubel V. Detection of dengue virus RNA by reverse transcription-polymerase chain reaction in the liver and lymphoid organs but not in the brain in fatal human infection. *Am J Trop Med Hyg* 1999;61(5):720–724.
24. Couvelard A, Marianneau P, Bedel C, Drouet MT, Vachon F, Henin D *et al.* Report of a fatal case of dengue infection with hepatitis: demonstration of dengue antigens in hepatocytes and liver apoptosis. *Hum Pathol* 1999;30(9):1106–1110.
25. Marianneau P, Cardona A, Edelman L, Deubel V, Despres P. Dengue virus replication in human hepatoma cells activates NF-kappaB which in turn induces apoptotic cell death. *J Virol* 1997;71(4):3244–3249.
26. Marianneau P, Flamand M, Deubel V, Despres P. Apoptotic cell death in response to dengue virus infection: the pathogenesis of dengue haemorrhagic fever revisited. *Clin Diagn Virol* 1998;10(2–3):113–119.
27. Marianneau P, Steffan AM, Royer C, Drouet MT, Jaeck D, Kirn A *et al.* Infection of primary cultures of human Kupffer cells by dengue virus: no viral progeny synthesis, but cytokine production is evident. *J Virol* 1999;73(6):5201–5206.
28. Marianneau P, Steffan AM, Royer C, Drouet MT, Kirn A, Deubel V. Differing infection patterns of dengue and yellow fever viruses in a human hepatoma cell line. *J Infect Dis* 1998;178(5):1270–1278.
29. Halstead SB, Nimmannitya S, Yamarat C, Russell PK. Hemorrhagic fever in Thailand; recent knowledge regarding etiology. *Jpn J Med Sci Biol* 1967;20s:96–103.

30. Rosen L. The pathogenesis of dengue haemorrhagic fever: a critical appraisal of current hypotheses. *S Afr Med J* 1986;Suppl:40–42.
31. Rosen L. Disease exacerbation caused by sequential dengue infections: myth or reality? *Rev Infect Dis* 1989;11(Suppl 4):S840–S842.
32. Rosen L. Comments on the epidemiology, pathogenesis, and control of dengue. *Med Trop* 1999;59(4):495–498.
33. Rosen L. The emperor's new clothes revisited, or reflections on the pathogenesis of dengue hemorrhagic fever. *Am J Trop Med Hyg* 1977;26:337–343.
34. Guzman MG, Kouri GP, Bravo J, Soler M, Vazquez S, Morier L. Dengue hemorrhagic fever in Cuba, 1981: a retrospective seroepidemiologic study. *Am J Trop Med Hyg* 1990;42:179–184.
35. Guzman MG, Kouri G, Valdes L, Bravo J, Alvarez M, Vazquez S *et al.* Epidemiologic studies on dengue in Santiago de Cuba, 1997. *Am J Epidemiol* 2000;152(9):793–799.
36. Gonzalez D, Castro OE, Kouri G, Perez JM, Martinez E, Vazquez S *et al.* Classical dengue hemorrhagic fever resulting from two dengue infections spaced 20 years or more apart: Havana, Dengue 3 epidemic, 2001–02. *Int J Infect Dis* 2005;9:280–285.
37. Fischer DB, Halstead SB. Observations related to pathogenesis of dengue hemorrhagic fever. V. Examination of age specific sequential infection rates using a mathematical model. *Yale J Biol Med* 1970;42:329–349.
38. Vaughn DW, Green S, Kalayanarooj S, Innis BL, Nimmannitya S, Suntayakorn S *et al.* Dengue viremia titer, antibody response pattern, and virus serotype correlate with disease severity. *J Infect Dis* 2000;181(1):2–9.
39. Buchy P, Vo VL, Trinh TX, Glaziou P, Le TT, Le VL *et al.* Secondary dengue virus type 4 infections in Vietnam. *Southeast Asian J Trop Med Public Health* 2005;36:178–185.
40. Halstead SB, Nimmannitya S, Cohen SN. Observations related to pathogenesis of dengue hemorrhagic fever. IV. Relation of disease severity to antibody response and virus recovered. *Yale J Biol Med* 1970;42:311–328.
41. Halstead SB, Lan NT, Myint TT, Shwe TN, Nisalak A, Soegijanto S *et al.* Infant dengue hemorrhagic fever: research opportunities ignored. *Emerg Infect Dis* 2002;12:1474–1479.

42. Morens DM, Halstead SB, Marchette NJ. Profiles of antibody-dependent enhancement of dengue virus type 2 infection. *Microb Pathog* 1987;3:231–237.
43. Kliks SC, Nimmannitya S, Nisalak A, Burke DS. Evidence that maternal dengue antibodies are important in the development of dengue hemorrhagic fever in infants. *Am J Trop Med Hyg* 1988;38(2):411–419.
44. Burke DS, Nisalak A, Johnson DE, Scott RM. A prospective study of dengue infections in Bangkok. *Am J Trop Med Hyg* 1988;38(1):172–180.
45. Scott RM, Nimmannitya S, Bancroft WH, Mansuwan P. Shock syndrome in primary dengue infections. *Am J Trop Med Hyg* 1976;25(6):866–874.
46. Halstead SB. The rubella problem in Hawaii. *Hawaii Med J* 1969;28: 387–388.
47. Sabin AB. Research on dengue during World War II. *Am J Trop Med Hyg* 1952;1:30–50.
48. Kliks SC, Nisalak A, Brandt WE, Wahl L, Burke DS. Antibody-dependent enhancement of dengue virus growth in human monocytes as a risk factor for dengue hemorrhagic fever. *Am J Trop Med Hyg* 1989;40(4):444–451.
49. Sangkawibha N, Rojanasuphot S, Ahandrik S, Viriyapongse S, Jatanasen S, Salitul V *et al.* Risk factors in dengue shock syndrome: a prospective epidemiologic study in Rayong, Thailand. I. The 1980 outbreak. *Am J Epidemiol* 1984;120:653–669.
50. Halstead SB, Rojanasuphot S, Sangkawibha N. Original antigenic sin in dengue. *Am J Trop Med Hyg* 1983;32(1):154–156.
51. Thein S, Aung MM, Shwe TN, Aye M, Zaw A, Aye K *et al.* Risk factors in dengue shock syndrome. *Am J Trop Med Hyg* 1997;56(5):566–572.
52. Graham RR, Juffrie M, Tan R, Hayes CG, Laksono I, Ma'roef C *et al.* A prospective seroepidemiologic study on dengue in children four to nine years of age in Yogyakarta, Indonesia. I. Studies in 1995–1996. *Am J Trop Med Hyg* 1999;61(3):412–419.
53. Chungue E, Spiegel A, Roux J, Laudon F, Cardines R. Dengue-3 in French Polynesia: preliminary data. *Med J Aust* 1990;152:557–558.
54. Guzman MG, Deubel V, Pelegrino JL, Rosario D, Marrero M, Sariol C *et al.* Partial nucleotide and amino acid sequences of the envelope and the envelope/nonstructural protein-1 gene junction of four dengue-2 virus strains isolated during the 1981 Cuban epidemic. *Am J Trop Med Hyg* 1995;52:241–246.

55. Guzman MG, Kouri G, Bravo J, Valdes L, Vazquez S, Halstead SB. Effect of age on outcome of secondary dengue 2 infections. *Int J Infect Dis* 2002;6:118–124.
56. Trofa AF, DeFraites RF, Smoak BL, Kanesa-thasan N, King AD, Burrous JM *et al*. Dengue fever in US military personnel in Haiti. *J Am Med Assoc* 1997;277(19):1546–1548.
57. Rossi CA, Drabick JJ, Gambel JM, Sun W, Lewis TE, Henchal EA. Laboratory diagnosis of acute dengue fever during the United Nations Mission in Haiti, 1995–1996. *Am J Trop Med Hyg* 1998;59(2):275–278.
58. Halstead SB, Streit TG, Lafontant JG, Putvatana R, Russell K, Sun W *et al*. Haiti: absence of dengue hemorrhagic fever despite hyperendemic dengue virus transmission. *Am J Trop Med Hyg* 2001;65:180–183.
59. Gubler DJ. Dengue and dengue hemorrhagic fever: its history and resurgence as a global public health problem. In: Gubler DJ, Kuno G (eds.) *Dengue and Dengue Hemorrhagic Fever*. CAB International, New York, 1997, pp. 1–22.
60. Loke H, Bethell DB, Phuong CX, Dung M, Schneider J, White NJ *et al*. Strong HLA class I–restricted T cell responses in dengue hemorrhagic fever: a double-edged sword? *J Infect Dis* 2001;184(11):1369–1373.
61. Paradoa Perez ML, Trujillo Y, Basanta P. Association of dengue hemorrhagic fever with the HLA system. *Haematologia* (*Budap*). 1987;20:83–87.
62. Chiewsilp P, Scott RM, Bhamarapravati N. Histocompatibility antigens and dengue hemorrhagic fever. *Am J Trop Med Hyg* 1981;30(5):1100–1105.
63. Alvarez M, Guzman MG, Pupo M, Morier L, Bravo J, Rodriguez R. Study of biologic attributes of Cuban dengue 2 Virus after serial passage in primary dog kidney cells. *Int J Infect Dis* 2001;5(1):35–39.
64. Thisyakorn U, Nimmannitya S. Nutritional status of children with dengue hemorrhagic fever. *Clin Infect Dis* 1993;16:295–297.
65. Kalayanarooj S, Nimmannitya S. Is dengue severity related to nutritional status? *Southeast Asian J Trop Med Public Health* 2005;36(2):378–384.
66. Hammon WM, Rudnick A, Sather GE. Viruses associated with epidemic hemorrhagic fevers of the Philippines and Thailand. *Science* 1960;131: 1102–1103.
67. Hammon WM, Sather, G.E. Virological findings in the 1960 hemorrhagic fever epidemic (dengue) in Thailand. *Am J Trop Med Hyg* 1964;13: 629–641.

68. Halstead SB, Udomsakdi S, Singharaj P, Nisalak A. Dengue and chikungunya virus infection in man in Thailand, 1962–1964. III. Clinical, epidemiology, and virologic observations on disease in non-indigenous white persons. *Am J Trop Med Hyg* 1969;18(6):984–996.
69. Rudnick A. Studies of the ecology of dengue in Malaysia: a preliminary report. *J Med Entomol* 1965;2:203–208.
70. Rudnick A. *Aedes aegypti* and haemorrhagic fever. *Bull World Health Organ* 1967;36:528–532.
71. Kochel TJ, Watts DM, Halstead SB, Hayes CG, Espinosa A, Felices V, Caceda R, Bautista T, Montoya Y, Douglas S, Russell KL. Effect of dengue-1 antibodies on American dengue-2 viral infection and dengue haemorrhagic fever. *Lancet* 2002;360:310–312.
72. Guzman MG, Kouri G, Halstead SB. Do escape mutants explain rapid increases in dengue case-fatality rates within epidemics? *Lancet* 2000;355(9218):1902–1903.
73. Rodriguez-Roche R, Alvarez M, Gritsun T, Rosario D, Halstead SB, Kouri G *et al.* Dengue virus type 2 in Cuba, 1997: conservation of E gene sequence in isolates obtained at different times during the epidemic. *Arch Virol* 2005;150:415–425.
74. Rodriguez-Roche R, Alvarez M, Gritsun T, Halstead SB, Kouri G, Gould EA *et al.* Virus evolution during a severe dengue epidemic in Cuba, 1997. *Virology* 2005;334:154–157.
75. Anderson CR, Downs WG. Isolation of dengue virus from a human being in Trinidad. *Science* 1956;124:224–225.
76. Rosen L. Observations on the epidemiology of dengue in Panama. *Am J Hyg* 1958;68:45–58.
77. Neff JM, Morris L, Gonzalez-Alcover R, Coleman PH, Lyss SB, Negron H. Dengue fever in a Puerto Rican community. *Am J Epidemiol* 1967;86:162–184.
78. PAHO. *Dengue in the Caribbean, 1977; 1978–1979*. PAHO, Montego Bay, Jamaica, 1978, pp. 1–186.
79. Rico-Hesse R. Molecular evolution and distribution of dengue viruses type 1 and 2 in nature. *Virology* 1990;174:479–493.
80. Kouri GP, Guzman MG, Bravo JR, Triana C. Dengue haemorrhagic fever / dengue shock syndrome: lessons from the Cuban epidemic, 1981. *Bull World Health Organ* 1989;67:375–380.

81. Watts DM, Porter KR, Putvatana P, Vasquez B, Calampa C, Hayes CG *et al.* Failure of secondary infection with American genotype dengue 2 to cause dengue haemorrhagic fever [see comments]. *Lancet* 1999;354(9188): 1431–1434.
82. Leitmeyer KC, Vaughn DW, Watts DM, Salas R, Villalobos I, de C *et al.* Dengue virus structural differences that correlate with pathogenesis. *J Virol* 1999;73(6):4738–4747.
83. Marchette NJ, Halstead SB, Falkler WA, Jr, Stenhouse A, Nash D. Studies on the pathogenesis of dengue infection in monkeys. 3. Sequential distribution of virus in primary and heterologous infections. *J Infect Dis* 1973;128(1):23–30.
84. Halstead SB. *In vivo* enhancement of dengue virus infection in Rhesus monkeys by passively transferred antibody. *J Infect Dis* 1979;140(4): 527–533.
85. Halstead SB, Shotwell H, Casals J. Studies on the pathogenesis of dengue infection in monkeys. II. Clinical laboratory responses to heterologous infection. *J Infect Dis* 1973;128(1):15–22.
86. Gubler DJ, Suharyono W, Lubis I, Eram S, Suliantisaroso J. Epidemic dengue hemorrhagic fever in rural Indonesia. I. Virological and epidemiological studies. *Am J Trop Med Hyg* 1979;28(4):701–710.
87. Libraty DH, Endy TP, Houng HS, Green S, Kalayanarooj S, Suntayakorn S *et al.* Differing influences of virus burden and immune activation on disease severity in secondary dengue-3 virus infections. *J Infect Dis* 2002;185(9):1213–1221.
88. Libraty DH, Young PR, Pickering D, Endy TP, Kalayanarooj S, Green S *et al.* High circulating levels of the dengue virus nonstructural protein NS1 early in dengue illness correlate with the development of dengue hemorrhagic fever. *J Infect Dis* 2002;186:1165–1168.
89. Halstead SB, Scanlon J, Umpaivit P, Udomsakdi S. Dengue and chikungunya virus infection in man in Thailand, 1962–1964: IV. Epidemiologic studies in the Bangkok metropolitan area. *Am J Trop Med Hyg* 1969;18(6):997–1021.
90. Dasaneyavaja A. First known isolation of a dengue virus from other human source than blood. In: Felsenfeld O (ed.) *Symposium on Haemorrhagic Fever* (1961). SEATO Medical Research Laboratory, Bangkok, Thailand, 1961, p. 61.

91. Nisalak A, Halstead SB, Singharaj P, Udomsakdi S, Nye SW, Vinijchaikul K. Observations related to pathogenesis of dengue hemorrhagic fever. 3. Virologic studies of fatal disease. *Yale J Biol Med* 1970;42(5):293–310.
92. Guzman MG, Kouri G, Morier L, Soler M, Fernandez A. A study of fatal hemorrhagic dengue cases in Cuba, 1981. *Bull Pan Am Health Organ* 1984;18(3):213–220.
93. Guzman MG, Alvarez M, Rodriguez R, Rosario D, Vazquez S, Valdes L *et al.* Fatal dengue hemorrhagic fever in Cuba, 1997. *Int J Infect Dis* 1999;3(3):130–135.
94. Yoksan S, Bhamarapravati N. Localization of dengue antigens in tissue specimens from fatal cases of dengue haemorrhagic fever. In: Pang T, Pathmanathan R (eds.) *International Conference on DHF* (1984). University of Malaya, Kuala Lumpur, 1984, pp. 406–410.
95. Chan YC, Lim KA, Ho BC. Recent epidemics of haemorrhagic fever in Singapore. *Jpn J Med Sci Biol* 1967;20S:81–88.
96. Sumarmo, Wulur H, Jahja E, Gubler DJ, Suharyono W, Sorensen K. Clinical observations on virologically confirmed fatal dengue infections in Jakarta, Indonesia. *Bull World Health Organ* 1983;61(4):693–701.
97. Boonpucknavig S, Boonpucknavig V, Bhamarapravati N, Nimmannitya S. Immunofluorescence study of skin rash in patients with dengue hemorrhagic fever. *Arch Pathol Lab Med* 1979;103:463–466.
98. Marovich J, Grouard-Vogel G, Louder J, Eller J, Sun W, Wu, SJ, Putvatana R *et al.* Human dendritic cells as targets of dengue virus infection. *J Investig Dermatol Symp Proc* 2001;6:219–224.
99. Bhamarapravati N, Boonpucknavig V. Pathogenetic studies on Thai hemorrhagic fever: immunofluorescent localization of dengue virus in human tissue. *Bull World Health Organ* 1966;35:50–51.
100. Hall WC, Crowell TP, Watts DM, Barros VL, Kruger H, Pinheiro F *et al.* Demonstration of yellow fever and dengue antigens in formalin-fixed paraffin-embedded human liver by immunohistochemical analysis. *Am J Trop Med Hyg* 1991;45(4):408–417.
101. Boonpucknavig V, Bhamarapravati N, Boonpucknavig S, Futrakul P, Tanpaichtr S. Glomerular changes in dengue hemorrhagic fever. *Arch Pathol Lab Med* 1976;100:206–212.

102. Sahaphong S, Riengrojpitak S, Bhamarapravati N, Chirachariyavej T. Electron microscopic study of the vascular endothelial cell in dengue hemorrhagic fever. *Southeast Asian J Trop Med Public Health* 1980;11: 194–204.
103. Sariol CA, Pelegrino JL, Martinez A, Arteaga E, Kouri G, Guzman MG. Detection and genetic relationship of dengue virus sequences in seventeen-year-old paraffin-embedded samples from Cuba. *Am J Trop Med Hyg* 1999;61(6):994–1000.
104. Halstead SB, Nimmannitya S, Margiotta MR. Dengue and chikungunya virus infection in man in Thailand, 1962–1964: II. Observations on disease in outpatients. *Am J Trop Med Hyg* 1969;18(6):972–983.
105. Nimmannitya S, Halstead SB, Cohen S, Margiotta MR. Dengue and chikungunya virus infection in man in Thailand, 1962–1964. I. Observations on hospitalized patients with hemorrhagic fever. *Am J Trop Med Hyg* 1969;18(6):954–971.
106. Russell PK, Yuill TM, Nisalak A, Udomsakdi S, Gould D, Winter PE. An insular outbreak of dengue hemorrhagic fever. II. Virologic and serologic studies. *Am J Trop Med Hyg* 1968;17(4):600–608.
107. Russell PK, Nisalak A. Dengue virus identification by the plaque reduction neutralization test. *J Immunol* 1967;99(2):291–296.
108. Winter PE, Nantapanich S, Nisalak A, Udomsakdi S, Dewey RW, Russell PK. Recurrence of epidemic dengue hemorrhagic fever in an insular setting. *Am J Trop Med Hyg* 1969;18(4):573–579.
109. Nimmannitya S, Halstead SB, Cohen SN, Margiotta MR. Dengue and chikungunya virus infection in man in Thailand, 1962–1964. I. Observations on hospitalized patients with hemorrhagic fever. *Am J Trop Med Hyg* 1969;18(6):954–971.
110. Blok J, Samuel S, Gibbs AJ, Vitarana UT. Variation of the nucleotide and encoded amino acid sequences of the envelope gene from eight dengue-2 viruses. *Arch Virol* 1989;105:39–53.
111. Chungue E, Deubel V, Cassar O, Laille M, Martin PMV. Molecular epidemiology of dengue-3 viruses and genetic relatedness among dengue-3 strains isolated from patients with mild or severe form of dengue fever in French Polynesia. *J Gen Virol* 1993;74:2765–2770.

112. Lee E, Gubler DJ, Weir RC, Dalgarno L. Genetic and biological differentiation of dengue-3 isolates obtained from clinical cases in Java, Indonesia, 1976–1978. *Arch Virol* 1993;133:113–125.
113. Thant KZ, Morita K, Igarashi A. Sequences of E/NS1 gene junction from four dengue-2 viruses of northeastern Thailand and their evolutionary relationships with other dengue-2 viruses. *Microbiol Immunol* 1995;39:581–590.
114. Thant KZ, Morita K, Igarashi A. Detection of the disease severity-related molecular differences among new Thai dengue-2 isolates in 1993, based on their structural proteins and major non-structural protein NS1 sequences. *Microbiol Immunol* 1996;40:205–216.
115. Mangada MN, Igarashi A. Molecular and *in vitro* analysis of eight dengue type 2 viruses isolated from patients exhibiting different disease severities. *Virology* 1998;244(2):458–466.
116. Rico-Hesse R, Harrison LM, Nisalak A, Vaughn DW, Kalayanarooj S, Green S *et al.* Molecular evolution of dengue type 2 virus in Thailand. *Am J Trop Med Hyg* 1998;58(1):96–101.

8

Elimination of Infection

Alan L. Rothman and Francis A. Ennis

Introduction

Resistance to viral infection is rarely absolute. Once a single viral particle has successfully infected a cell, mechanisms are required to limit further spread of the pathogen. The previous chapter has described the humoral immune responses to dengue virus infection, which prevent virus infection of cells at the earliest stage, i.e. viral entry. This chapter will focus on those immune mechanisms that eliminate viral infection once it has been established.

Innate and Adaptive Cellular Immune Responses

Animals confront a variety of pathogenic viruses and other microorganisms. These in turn use a variety of approaches to circumvent host defenses. Therefore, it is not surprising that animals have developed multiple systems to eliminate viral infection once it has been established. Viral clearance is mediated by both innate and adaptive cellular immune response mechanisms.

Innate responses refer to those that neither require previous antigenic exposure nor show enhanced responses during subsequent exposure (memory); examples are cytokines such as the interferons, and effector cells such as natural killer (NK) cells. Innate responses are activated rapidly during viral infection.

Adaptive cellular immune responses refer to those that develop as a result of the first exposure to virus and respond in an antigen-specific manner upon subsequent exposure to the virus; these include B lymphocytes and their associated antibodies, which have been discussed in the previous chapter, and T lymphocytes. In contrast to innate responses, adaptive immune responses develop slowly during the first encounter with antigen. T lymphocytes that recognize any one specific antigen are present at very low frequency in the nonimmune state. Activation of these naïve T lymphocytes first requires that antigen be expressed by specialized antigen-presenting cells — macrophages, B lymphocytes, or dendritic cells, with the latter being most efficient at activating T cell responses. Antigen expression can occur either by direct viral infection of these cells or by uptake of antigen from neighboring virus-infected cells. The T lymphocytes then require several days to achieve full activation and expression of effector functions and in order to expand in number through cell division. After elimination of the target antigen, these T lymphocytes then contract in number and convert to a "memory" phenotype characterized by the potential for more rapid activation and response upon re-exposure to their target antigen.

The innate and adaptive cellular immune responses to dengue virus have been studied predominantly in humans. Limited data exist on the cellular immune responses in animals, although these will be reviewed where relevant. The dearth of data reflects the absence of good animal models of dengue virus infection. Although nonhuman primates develop viremia after subcutaneous injection of dengue virus, as with humans, there is no model for dengue disease in these animals. Laboratory mice, in which immune responses to a variety of viruses have been well characterized, do not typically develop disease or demonstrate quantifiable replication of dengue virus in target cells relevant to human infection. The exceptions to this have been highly immunocompromised mice,[1] which would not model typical immune responses and their influence on infection and disease.

The interactions between dengue virus and the immune system have also been studied in tissue culture. However, these models are limited to the study of a few cell types, and do not reflect well the complex interaction between different components of the immune response. Furthermore, while laboratory strains of dengue virus are able to productively infect many immortalized and primary cell lines, the clinical relevance of these findings is uncertain because dengue viruses appear to have a more limited tropism *in vivo*, involving dendritic cells, monocytes and macrophages, hepatocytes, and possibly fibroblasts and endothelial cells.

Interferons

Type I interferons

The major type I interferons are IFNα and IFNβ. These cytokines are produced by a wide variety of cell types *in vitro* and *in vivo*. Type I interferons have been shown to have broad antiviral activity against DNA and RNA viruses. Studies in the murine model of lymphocytic choriomeningitis virus infection also suggest that type I interferons play a major role in virus-induced cytopenia and bone marrow suppression.[2]

Kurane and Ennis studied the effects of dengue virus on IFNα production by human peripheral blood mononuclear cells (PBMC's) *in vitro*.[3,4] Dengue virus infection directly induced IFNα production by monocytes. Pretreatment of monocytes with these culture supernatants inhibited dengue virus infection. In addition, uninfected PBMC's produced IFNα when exposed to dengue virus-infected monocytes, even after paraformaldehyde fixation. The IFNα-producing cells were heterogeneous, and included both B cells and NK cells. Production of IFNα was readily detected using PBMC's from individuals without evidence of prior flavivirus infection.

In a separate study, Kurane *et al.* investigated the effects of dengue virus infection on primary human dermal fibroblast cultures.[5] These primary human cells were highly permissive to dengue virus infection. Culture supernatants from dengue virus-infected fibroblasts contained high levels of IFNβ. Pretreatment with these culture supernatants inhibited dengue virus infection of fibroblasts *in vitro*, and this inhibitory effect was abolished in the presence of antibody to IFNβ. These results suggested that innate immune responses in the early stages of infection, after subcutaneous injection of dengue virus by the mosquito, could play an important role in controlling dengue virus replication *in vivo*.

Diamond *et al.* have studied the mechanisms by which IFNα and IFNβ inhibit dengue virus infection *in vitro*, using a variety of cell lines, including monocytic cell lines and the HepG2 human hepatoma cell line.[6] Treatment with IFNα or IFNβ resulted in reduced accumulation of negative strand viral RNA. However, viral entry and uncoating were apparently unaffected. These results suggested that there was a block in the initial stages of RNA replication. Similar effects were observed in cells treated with IFNα or IFNβ; this result was not unexpected, since both cytokines signal through the same receptor.

An important recent development has been the discovery of mechanisms used by dengue viruses to block type I interferon activity. Type I interferons activate cells through signal transducer and activator of transcription (STAT) proteins STAT1 and STAT2. Munoz-Jordan *et al.* found that the dengue viral nonstructural proteins NS2A, NS4A, and NS4B had synergistic ability to antagonize IFNα signaling by inhibiting STAT1 activation, thereby blocking IFN-induced gene expression.[7,8] Jacobs *et al.* also reported inhibition of the antiviral effect of IFN and found that dengue virus down-regulated the expression of STAT2.[9] These findings are analogous to the observation that type I interferons play a critical role in the resistance of laboratory mice to dengue infection.[1,10]

There is limited information on the levels of type I interferons *in vivo* during acute dengue virus infection. Kurane *et al.* analyzed serum IFNα levels in 45 Thai children with DHF ($N = 35$) or DF ($N = 10$).[11] Elevated levels of IFNα were found in 80% of children with DHF and 60% of children with DF, but in only 7% of healthy Thai children. Libraty *et al.* also found elevated levels in Thai children with acute dengue virus infection, and noted a close temporal association between peak IFNα levels and peak dengue viral RNA levels in serial plasma samples.[12] The timing of elevations in IFNα levels corresponds to the reported maturation arrest observed in the bone marrow.[13]

Type II interferons

Type II interferon, or IFNγ, is distinct from the other interferons discussed above both biochemically and with regard to its pattern of expression and mechanism of action. The major cellular sources of

IFNγ are T lymphocytes and NK cells. In these cells, IFNγ is produced in response to interaction with antigen, unlike the type I interferons, which are produced by the virus-infected cell. The target cells for IFNγ action are also much more limited than for type I interferons. The IFNγ receptor is expressed at high levels on monocytes and macrophages. Several other cells types, including endothelial cells, are also responsive to IFNγ.

Production of IFNγ by dengue virus-specific T cells in response to dengue viral antigens has been demonstrated *in vitro* using PBMC's from dengue-virus-immune individuals as well as dengue virus-specific T cell clones derived from such individuals.[14,15] CD4+ T lymphocytes appear to be the cells primarily responsible for IFNγ secretion under these conditions, although CD8+ T lymphocytes can also contribute to the production of this cytokine.[16] The specificity of these dengue-virus-specific T lymphocytes is discussed further below.

There are no data from *in vitro* studies to indicate definitively that exposure to dengue virus induces the production of IFNγ in NK cells. However, Kurane *et al.* reported that exposure to dengue-virus-infected cells also induced the production of IFNγ by $CD3^-$ PBMC's from dengue-nonimmune individuals.[17] Furthermore, the observation that mice lacking the receptors for both type I and type II interferons were more susceptible to intraperitoneal challenge with dengue virus than mice lacking only the receptor for type I interferon suggests that some production of IFNγ occurs *in vivo* in mice.[1] Since dengue-virus-naïve mice are unlikely to have significant numbers of dengue-virus-specific T lymphocytes, it is possible that NK cells are a major source of IFNγ in normal mice and type I interferon receptor-deficient mice.

The effects of IFNγ on dengue virus infection appear to be more complex than those of the type I interferons, because both inhibitory and enhancing effects can be demonstrated *in vitro*. Diamond *et al.* found that IFNγ treatment partially inhibited infection of HepG2 cells or human foreskin fibroblasts.[6] Sittisombut *et al.* reported that pretreatment with high concentrations of IFNγ also inhibited infection of monocytes with dengue virus *in vitro*.[18] On the other hand, both Kontny *et al.* and Diamond *et al.* found that IFNγ augmented dengue virus infection of monocytic cell lines in the presence of infection-enhancing antibody.[6,19] This effect was related to an IFNγ-induced increase in the surface expression of Fc_γ receptors, and is most likely to affect viral entry.

Clinical studies on patients with acute dengue virus infection support a role for IFNγ in viral clearance *in vivo*. Studies by Kurane *et al.* and Green *et al.* found elevated plasma levels of IFNγ during acute dengue virus infection.[20,21] Green *et al.* found that plasma IFNγ levels in serial samples from the same individual fluctuated over a wide range; however, mean plasma IFNγ levels throughout the acute infection were within the concentration range that was found to enhance or suppress dengue virus infection *in vitro*.

Natural Killer Cells

NK cells lyse tumor cells and virus-infected cells *in vitro*, and have been shown to play an important role in the control of viral infections *in vivo*, particularly herpesvirus infections.[22] Lysis of target cells by NK cells is not MHC-restricted and does not involve specific recognition of viral gene products on the surface of target cells. Instead, NK cells utilize a variety of receptors, including both inhibitory and stimulatory receptors, to recognize "altered self."[23] They respond in particular to cells that have reduced MHC protein expression on the cell surface. Lysis of target cells by NK cells is enhanced in the presence of type I interferons, and is also enhanced in the presence of antibodies that bind to cell surface proteins. This latter mechanism, antibody-dependent cellular cytotoxicity (ADCC), is mediated through Fc_γ receptors on the NK cell.

Kurane *et al.* demonstrated direct cytotoxicity of dengue-virus-infected Raji cells by PBMC's from nonimmune individuals.[24,25] Target cell lysis was enhanced in the presence of exogenous antidengue antibody. The predominant cell type responsible for both direct cytotoxicity and ADCC, as well as lysis of K562 cells, a cell line highly susceptible to NK cell killing, were nonadherent, $CD3^-$ $Leu11^+$ cells.

The susceptibility of dengue virus-infected cells to NK cell-mediated lysis may be cell-type-specific. Mullbacher and Lobigs observed that infection with flaviviruses, including dengue virus, caused increased surface expression of MHC class I antigens in several murine and hamster cell lines.[26,27] This effect was not related to increased MHC class I gene expression but rather enhanced peptide translocation into the endoplasmic reticulum, resulting in enhanced loading of MHC class I molecules. They reported that these flavivirus-infected cells were less susceptible to lysis by NK cells.

The role of NK cells in clearance of dengue virus *in vivo* is not known. However, several studies have demonstrated activation of circulating NK cells during acute dengue virus infection. Homchampa *et al.* measured the lysis of K562 target cells by PBMC's obtained from Thai children with acute dengue virus infections.[28] The absolute % lysis of K562 cells by PBMC's was not different from normal control subjects. However, when adjusted for the decreased numbers of circulating $HNK1^+$ cells during acute dengue virus infection, the NK cell activity calculated as % cytotoxicity per $HNK1^+$ cell was significantly increased during the febrile period of acute dengue virus infection and during the 1–3 days immediately after defervescence. Green *et al.* measured the expression of CD69, an early activation marker, on PBMC's in Thai children with acute dengue virus infection.[29] They found a marked increase in CD69 expression on $CD16/56^+$ NK cells during the acute infection, with the frequency of activated circulating NK cells approaching 80% in some individuals.

T Lymphocytes

Primary dengue virus infection induces a rich repertoire of virus-specific memory CD4+ and CD8+ T cells in both humans and mice. This dengue-virus-specific T cell repertoire includes a population of T cells specific for the homologous viral serotype (the virus that caused the primary infection) and a population of T cells that are cross-reactive against one or more of the other (heterologous) dengue virus serotypes. Dengue virus-specific T cells are directed at multiple epitopes on the viral proteins. Even among T cell clones recognizing the same epitope, there is often substantial diversity in the T cell receptors on individual clones and in the responsiveness to different dengue virus serotypes. This diversity in the dengue-virus-specific T cell response is likely to have important consequences during secondary dengue virus infection for the response of memory T cells.

CD4+ T lymphocytes

Dengue-virus-specific CD4+ T cell responses to primary dengue virus infection have been characterized *ex vivo* at the population level and also by the study of individual T cell clones. In bulk culture lymphoproliferation assays, primary dengue virus infection induced the highest responses to the homologous viral serotype, but has also induced cross-reactive responses to one or more heterologous serotypes in nearly all patients studied.[14,30,31] Through *ex vivo* assays and analysis of T cell clones, it has been shown that CD4+ T cell epitopes are present on several of the dengue viral proteins (Table 1).

In a series of papers, Kurane *et al.* characterized the CD4+ T cell response to dengue virus in one subject following primary dengue-3 virus infection.[15,32–37] They reported that most dengue virus-specific CD4+ T cell clones recognized the NS3 protein. Within this population, there were at least nine different subpopulations of T cells, based on the specific epitope recognized and the pattern of serotype cross-reactivity (Table 2).

Studying different volunteers, Green *et al.* and Livingston *et al.* characterized CD4+ T cell clones following primary dengue-1 or dengue-4 virus infections, respectively.[38,39] Green *et al.* isolated T cell clones that recognized the NS1–NS2A gene region. This population of cells included clones specific for dengue-1 virus and clones that were also cross-reactive with dengue-3 virus but not dengue-2 or dengue-4 virus. Livingston *et al.*

Table 1. Epitopes recognized by human dengue-virus-specific CD4+ T lymphocytes.

Protein	Amino acids	Sequence	MHC restriction	Reference
C	47–55	VLAFITFLR	DPw4	31
	83–92	GFRKEIGRML	DR1, DPw4	31
NS3	146–154	VIGLYGNGV	DR15	34
	202–211	RKYLPAIVRE	DR15	35
	224–234	TRVVAAEMEEA	DR15	36
	241–249	IRYQTTATK	DR15	35
	255–264	EIVDLMCHAT	DPw2	37
	351–361	WITDFVGKTVW	DR15	35

Amino acid positions are calculated relative to the N terminus of each protein.

Table 2. Complexity of the CD4+ T lymphocyte response to the dengue virus NS3 protein in a single individual.*

			Viruses recognized†					
Epitope‡	**Clone(s)**	**HLA restriction**	**D1**	**D2**	**D3**	**D4**	**WN**	**YF**
146–154	JK4, JK43	DR15	+	+	+	+	+	+
202–211	JK44	DR15	+	+	+			
224–234	JK36, JK46	DR15		+	+	+		
241–249	JK15	DR15			+			
255–266	JK10, JK34	DPw2	+	+	+	+		
	JK28	DPw2	+	+	+	+	+	
	JK26, JK49	DPw2	+	+	+	+	+	+
351–361	JK13	DR15			+			
	JK5	DR15	+		+		+	

* Based on Refs. 34–37. Subject was immunized with live dengue-3 virus.

† D1–D4 — dengue virus serotypes 1–4; WN — West Nile virus; YF — yellow fever virus.

‡ Amino acid position based on dengue-3 NS3 protein sequence.

isolated T cell clones that recognized the E protein; all of the clones studied were specific for dengue-4 virus.

In a different volunteer, following primary dengue-4 virus infection, Gagnon *et al.* found that the major target for CD4+ T cells was the C protein.[31] Two separate epitopes on the C protein were recognized; one epitope recognized by T cells cross-reactive for dengue-2 and dengue-4 viruses, and one epitope recognized by dengue-4 virus-specific T cells. Further studies by Mangada *et al.* identified two epitopes on the NS3 protein recognized by CD4+ T cells from this individual, but the responses to the NS3 epitopes were subdominant.[40] In this volunteer, the dengue-virus-specific CD4+ T cell repertoire was skewed toward use of the Vβ17 gene segment, both in an antigen-stimulated bulk culture and among dengue-virus-specific CD4+ T cell clones.[41]

Limited studies have been done to characterize dengue-virus-specific CD4+ T cell responses in laboratory mice. As in humans, primary dengue virus infection induced the highest proliferative responses to the homologous serotype, but a lower level of proliferation to the other dengue virus serotypes was detected.[42] Bulk culture proliferative responses to recombinant

E and NS1 proteins were also detected.[42,43] Among a small panel of dengue-virus-specific CD4+ T cell clones, most T cell clones were specific for the homologous dengue serotype, but clones cross-reactive with some or all heterologous serotypes were also identified.[44]

These results demonstrate that the immunodominant epitopes for CD4+ T cells vary between subjects, as does the degree of serotype cross-reactivity of CD4+ T cell clones (Table 3). Nevertheless, the general pattern emerges that CD4+ T cell responses directed at the dengue virus non-structural proteins demonstrate a greater degree of serotype cross-reactivity than do CD4+ T cell responses to the dengue virus structural proteins, which reflects the higher amino acid sequence homology among the non-structural proteins.

New immunologic assays have recently permitted the precise quantitation of virus-specific CD4+ T cells in PBMC's. Using one such approach, flow cytometric analysis of IFN-γ production at the single cell level after short-term antigen stimulation,[45] Mangada *et al.* found that dengue-virus-specific CD4+ T cells represented as many as 0.5% of all circulating CD4+ T lymphocytes, even many years after a primary dengue virus infection.[46] Cytokine flow cytometry and staining with HLA class II-peptide tetrameric complexes indicated that these responses are focused

Table 3. Dengue serotype specificity and cross-reactivity among CD4+ T cell clones isolated from different donors.*

			Specificity†		
Donor	**Vaccine‡**	**# Clones**	**Serotype**	**Subcomplex**	**Complex**
1	D3	20	4	6	10
2	D4	12	9	3	0
3	D4	2	1	1	0
4	D4	16	8	8	0
5	D1	17	13	4	0

* Based on Ref. 30, 31, 34–37, 39 and unpublished data.

† Number of T cell clones showing each pattern of specificity: serotype — T cell clones specific for the vaccine (homologous) serotype; subcomplex — T cell clones cross-reactive with 1 or 2 heterologous serotypes; complex — T cell clones cross-reactive with all 3 heterologous serotypes.

‡ The dengue virus serotype of the vaccine received by each donor is shown.

on a few immunodominant (i.e. recognized by 0.02% or more of all CD4+ T cells) epitopes.[40] Consistent with other results described above, the frequency of CD4+ T cells responding to the homologous dengue virus serotype was higher than the frequency of cells responding to heterologous serotypes.

These flow cytometry methods have also demonstrated heterogeneity among virus-specific CD4+ T cells in the response to antigen stimulation.[40] The frequency of CD4+ T cells that produced TNFα in response to *in vitro* antigen stimulation was up to 10 times higher than the corresponding frequency of IFNγ-producing CD4+ T cells. The ratio of TNFα- to IFNγ-producing cells varied among individuals. Furthermore, in any given individual, this ratio varied among the responses to different epitopes, but was consistently higher in the response to heterologous viral serotypes (i.e. serotypes other that the serotype with which the individual was infected or vaccinated). We have proposed that this bias toward a TNFα-predominant response contributes to the increased risk for DHF in secondary infections with heterologous dengue viral serotypes.[47]

Interestingly, most of the human dengue-virus-specific CD4+ T cell clones studied have shown *in vitro* cytolytic activity. Gagnon *et al.* examined the *in vitro* expression of different effector mechanisms by C-protein-specific cytotoxic CD4+ T cell clones.[48] These T cell clones secreted proinflammatory type I cytokines, including IFNγ, IL-2, TNFα, and TNFβ. The T cell clones also expressed both perforin and Fas ligand (FasL/CD95L) mRNA's. While the T cell clones lysed autologous EBV-transformed B lymphoblastoid cells only in the presence of dengue virus or a synthetic peptide corresponding to the target epitope (cognate target cell lysis), activated T cell clones could lyse Jurkat cells in the absence of dengue virus antigens (bystander cell lysis). Using concanamycin A and brefeldin A, which, respectively, inhibit perforin-mediated lysis and Fas–FasL-mediated lysis, it was demonstrated that most of the CD4+ T cell clones lysed cognate target cells through a perforin-dependent mechanism, whereas bystander target cell lysis was usually mediated by Fas–FasL interactions.

Studies in laboratory mice have shown that expression of FasL in the liver can induce hepatocyte apoptosis. The findings of Gagnon *et al.* therefore suggest a model whereby dengue virus infection can cause

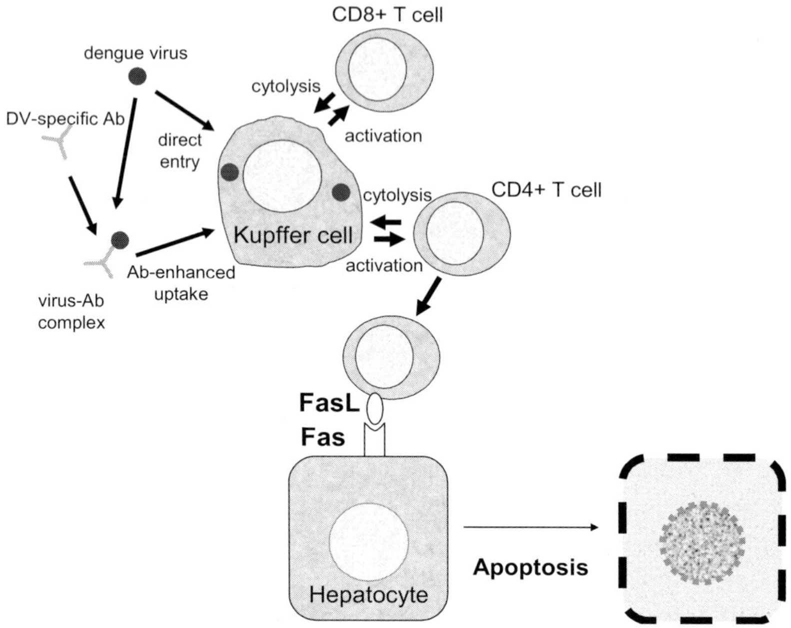

Fig. 1. Model of hepatocyte injury without direct dengue virus infection. Dengue virus infection of monocytic cells (e.g. Kupffer cells), by either direct virus entry or antibody-enhanced virus uptake, leads to presentation of dengue antigens and activation of memory dengue-reactive CD4+ and CD8+ T lymphocytes. Activation of CD4+ T lymphocytes induces expression of Fas ligand (FasL, CD95L), which interacts with Fas (CD95) expressed constitutively on the surface of uninfected hepatocytes, causing hepatocyte apoptosis.

hepatic injury in the absence of direct infection of hepatocytes (Fig. 1). Dengue-virus-specific CD4+ T cells would be activated by interaction with dengue-virus-infected blood monocytes or tissue macrophages, and the surface expression of FasL would be induced. These activated cells may then transit through hepatic sinusoids; interaction with hepatocytes will then induce hepatocyte apoptosis. Infection of Kupffer cells with dengue virus is most likely to facilitate the colocalization of recently activated dengue-virus-specific T cells with hepatocytes, and has been reported in some studies of fatal cases of dengue virus infection.[49]

CD8+ T lymphocytes

Dengue-virus-specific CD8+ T cell responses to primary dengue virus infection have also been characterized *ex vivo* and at the clonal level. Bukowski *et al.* and Mathew *et al.* demonstrated that primary dengue virus infection induces bulk culture CD8+ cytolytic T responses to the homologous viral serotype, as well as cross-reactive recognition of heterologous serotypes.[50,51] Cytolytic responses to the NS3 protein were detected in six of eight subjects studied. Cytolytic responses to the E and NS1 proteins were also common. Serotype cross-reactive responses were primary directed at epitopes on nonstructural viral proteins.

As with CD4+ T cell epitopes, many of the CD8+ T cell epitopes identified to date are derived from the dengue NS3 protein (Table 4). Livingston *et al.* and Zivny *et al.* characterized dengue-virus-specific CD8+ T cell clones in a subject following primary dengue-4 virus infection.[52,53]

Table 4. Epitopes recognized by human and murine dengue-virus-specific CD8+ cytotoxic T lymphocytes.

Protein	Amino acids	Sequence	MHC restriction	Reference
Murine:				
E	331–339	SPCKIPFEI	L^d	44
NS3	298–306	GYISTRVEM	K^d	44
Human:				
E	211–219	FFDLPLPWT	A2	55
	414–422	ILGDTAWDF	B7	66
NS3	71–79	SVKKDLISY	B62	54
	133–143	GTSGSPIIDKK	A11	60
	222–230	APTRVVAAE	B7	65
	235–243	AMKGLPIRY	B62	54
	500–508	TPEGIIPTL	B35	52
	527–535	GESRKTFVE	B7	59
	555–564	INYADRRWCF	A24, A11	61, 66
NS4A	56–64	LLLGLMILL	A2	55
NS4B	111–119	VLLLVTHYA	A2	55
	181–189	LLLMRTSWA	A2	55

Amino acid positions are calculated relative to the N terminus of each protein.

A single CD8+ cytotoxic T cell epitope was mapped on the NS3 protein. However, the T cell response to this epitope included at least three different subpopulations — one recognizing only dengue-4 virus, one recognizing both dengue-2 and dengue-4 viruses but not dengue-1 or dengue-3 virus, and one recognizing all four serotypes. These different specificities could be explained by minor amino acid differences in the target epitope between the four dengue serotypes, as well as variations in the complementarity determining regions of the T cell receptor genes of these clones (unpublished data).

Studying a different subject, Zivny *et al.* characterized CD8+ T cell clones following primary dengue-3 virus infection.[54] Two CD8+ T cell epitopes were identified on the NS3 protein, one of which overlapped with two epitopes recognized by CD4+ T cells from the same individual. Both epitopes were targeted by serotype-specific and serotype-cross-reactive T cell clones. However, there were qualitative differences in the response to heterologous dengue serotypes among the serotype cross-reactive T cells (Table 5). Interaction with antigen-presenting cells expressing the dengue-2 epitope induced cytolytic activity, IFN-γ production, and proliferation of T cells specific for an epitope corresponding to amino acids 234–243 of the NS3 protein. However, T cells specific for an epitope corresponding to amino acids 71–79 of the NS3 protein could only be induced to express cytolytic function upon interaction with cells expressing the dengue-2 sequence. These results predict that secondary dengue-2 virus infection in this individual would induce a more varied set of

Table 5. Qualitative differences in the T cell response to a heterologous dengue serotype at two different epitopes in a dengue-3-immune subject.*

Stimulus			T cell response		
Virus	**NS3 epitope**	**Sequence**	**Cytotoxicity**	**IFN-γ release**	**Proliferation**
D3	71–79	SVKKDLISY	+	+	+
	235–243	AMKGLPIRY	+	+	+
D2	71–79	DVKKDLISY	+	−	−
	235–243	ALRGLPIRY	+	+	+

* Based on Ref. 54.

responses from the dengue-virus-specific T cell repertoire than would be predicted simply on the basis of antigenic cross-reactivity.

Bashyam *et al.* used cytokine flow cytometry to analyze the recognition of variant epitope sequences of the four dengue serotypes by dengue-virus-reactive CD8+ T lymphocytes at the single cell level.[55] The response to four HLA-A2-restricted T cell epitopes was studied in PBMC's from US volunteers who had primary immunization with monovalent candidate live dengue virus vaccines. Seven distinct populations of responding CD8+ T cells could be identified, producing different combinations of the cytokines IFNγ, TNFα, and MIP-1β. Stimulation with variant peptides of the same epitope altered both the overall frequency of responding T cells and the relative proportions of each of the seven subpopulations. Interestingly, in some cases, peptides derived from heterologous viral serotypes stimulated a higher response than the peptide sequence from the homologous serotype. This finding suggests that some epitope variants may be T cell superagonists; *in vivo*, this effect could contribute to the observation that particular sequences of infection with different dengue serotypes may be more likely to induce DHF.[56]

Dengue-virus-specific CD8+ T cell responses in inbred strains of mice were directed at few viral epitopes.[44,57] As in humans, both serotype-specific and serotype-cross-reactive CD8+ cytolytic T cell responses were induced after primary dengue virus infection. In dengue-2-virus–immune BALB/c (H-2^d) mice, dengue-virus-specific CD8+ cytotoxic T lymphocytes recognized epitopes on the E and NS3 proteins. Responses to the epitope on the E protein were specific for the homologous dengue serotype; recognition was even specific at the level of individual viral strains (Fig. 2). Responses to the epitope on the NS3 protein consisted of two subpopulations — T cell clones that recognized dengue-2 and dengue-4 viruses but not dengue-1 and dengue-3 viruses, and other T cell clones that recognized all four serotypes. A single amino acid change was responsible for the different specificities of these NS3-epitope-specific T cells.

As with the CD4+ T cell response to dengue virus, the newer immunologic assays are permitting a more precise and quantitative analysis of the CD8+ T cell response to dengue virus.[58–60] Using flow cytometry assays and HLA-peptide tetramer staining, we and others have found

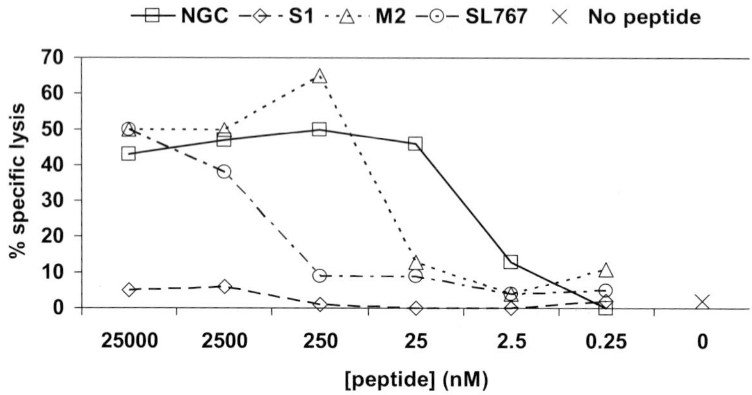

Fig. 2. Some dengue-reactive CD8+ T lymphocytes recognize strain-specific determinants. A murine (H-2^d) cytotoxic T cell clone specific for the E (331–339) epitope of dengue-2 strain New Guinea C (NGC- SPCKIPFEI) was tested for recognition of the corresponding sequences of other dengue-2 strains — S1 (SPCKTPFEI), M2 (SPCKIPLEI), and SL767 (PPCKIPFEI).

that the frequency of T cells responsive to individual epitopes on dengue virus can represent as many as 1% of all circulating CD8+ T lymphocytes during the memory phase.[55,60,61]

These results demonstrate that the dengue-virus-specific CD8+ T cell response is heterogeneous, similar to the CD4+ T cell response. CD8+ T cell responses directed at the nonstructural proteins, particularly the NS3 protein, are frequently serotype-cross-reactive. However, recognition of heterologous viral serotypes can induce qualitatively different responses in different T cell subpopulations.

Factors affecting the dengue-virus-specific T cell repertoire

The relative contributions of host genetic factors (e.g. HLA alleles) and the protein sequences of the virus causing the primary infection in determining the immunodominant epitopes and the serotype cross-reactivity of the response are difficult to distinguish. However, data suggest that both can influence the dengue-virus-specific T cell repertoire. Dharakul *et al.* studied 10 human volunteers who received the same lot of a candidate live dengue-2

Table 6. Influence of the immunizing dengue-virus-serotype on the specificity of the T cell response to the NS3 (298–306) epitope in BALB/c (H-2[d]) mice.*

Mouse immunization		**T cell response to NS3 epitope**		
Virus	**Sequence of epitope**	**D3 (= D1) GYISTRVGM**	**Both (cross-reactive)**	**D2 (= D4) GYISTRVEM**
D2	GYISTRV<u>E</u>M	–	+	+++
D3	GYISTRV<u>G</u>M	–	++++	–

* Based on Ref. 64.

virus vaccine.[62] The subjects differed in their level of serotype cross-reactive proliferation and cytotoxicity responses, likely related to heterogeneity in immune response genes. Effects of HLA genes on the immune response to dengue virus infection are thought to explain the associations between specific HLA alleles and the severity of dengue disease.[58,63]

On the other hand, Spaulding *et al.* examined the pattern of CD8+ T cell responses to a dominant epitope on the NS3 protein in BALB/c mice immunized with different dengue virus serotypes.[64] The predominant epitope-specific T cell response in mice immunized with dengue-2 virus was cross-reactive with dengue-4 virus but not with dengue-1 or dengue-3 virus. Similar responses were detected in mice immunized with dengue-4 virus. However, the epitope-specific T cell response in mice immunized with dengue-3 virus was cross-reactive with all four dengue serotypes. These findings could be explained from the amino acid sequences of the NS3 proteins of the four dengue serotypes. The peptide sequences in dengue-2 and dengue-4 viruses were identical, and differed from the peptide sequences in dengue-1 and dengue-3 viruses at one of the nine residues (Table 6).

Naturally Acquired versus Vaccine-Induced Cellular Immunity

Responses in primary versus secondary dengue virus infections

T lymphocyte responses to dengue virus have been studied in the greatest detail in humans infected with candidate monovalent live attenuated dengue virus vaccines. The major advantages of studies of such experimental infections are that prior flavivirus exposure has been well characterized and the virus causing infection is fully defined. However, these subjects, who have lived in dengue-nonendemic regions, have not been known to experience secondary dengue virus infections. Therefore, there have been no data available correlating immune responses to the primary (experimental) infection with clinical outcomes during subsequent secondary dengue virus infections.

Several recent studies from our group and others have characterized dengue-virus-specific CD8+ T cell responses after natural secondary dengue virus infections.[58–61,65,66] T cell responses in these subjects were directed at epitopes on the nonstructural proteins (principally NS1 and NS3) and were highly cross-reactive with other dengue virus serotypes. These data support the hypothesis that secondary dengue virus infection causes preferential activation and expansion of serotype-cross-reactive memory T lymphocytes (mostly directed at epitopes on the more conserved nonstructural proteins) that were induced during the primary dengue virus infection.

This interpretation is supported by recent data from Mongkolsapaya *et al.*[60,61] They studied the response to HLA-A11- and HLA-A24-restricted T cell epitopes on the NS3 protein in PBMC's of subjects shortly after secondary dengue virus infections by flow cytometry using tetrameric HLA-peptide complexes. Many of the epitope-specific T cells detected showed staining suggestive of higher avidity for the peptide sequences of viral serotypes different from the serotype causing the secondary infection. Although the sequence of infection was not known in these subjects, it is speculated that these T cells had higher avidity for the dengue virus serotype previously encountered. The different patterns of tetramer costaining correlated with differences in functional responses as defined by degranulation, IFNγ production, and TNFα production.

With the identification of immunodominant T cell epitopes on dengue virus proteins, we and others have begun to study the association between the T cell response to dengue virus infection and the severity of dengue disease. We found that the median frequency of T cells specific for an immunodominant HLA-B7-restricted epitope on the NS3 protein was higher in 5 HLA-B7+ subjects who had experienced DHF than in an equal number of subjects who had experienced DF; all but one of these subjects had secondary dengue virus infections.[59] Mongkolsapaya found a similar association between disease severity and the response to an HLA-A11-restricted epitope on the NS3 protein in 19 subjects with secondary dengue virus infections.[60] These data are consistent with the hypothesis that DHF is associated with a greater expansion of dengue-virus-specific T lymphocytes; however, these findings do not yet distinguish whether the T lymphocyte response is a cause of plasma leakage or whether other factors cause both plasma leakage and enhanced immune activation.

T cell responses as potential predictors of disease severity

Prospective population-based cohort studies[67] will provide the best means of distinguishing protective from immunopathological effects of dengue-virus-specific T cells, if these are in fact distinct. In a pilot study, we analyzed *in vitro* proliferation responses and cytokine production to dengue antigens in PBMC's obtained from Thai children prior to secondary dengue-3 virus infections.[68] Preliminary observations from this study suggest that broadly serotype-cross-reactive IFNγ responses may be associated with milder disease, while TNFα production may be associated with more severe disease. Further work will be needed, however, to confirm and extend these associations.

Responses to multivalent dengue vaccines

There is currently no vaccine available for the prevention of dengue virus infection, although a number of candidate vaccines are under investigation. A major factor complicating vaccine development is the need to prevent infection with all four dengue virus serotypes. As a consequence, most groups are pursuing the development of multivalent dengue virus vaccines. As has been noted already, the majority of the data on the T cell response to primary dengue virus infection have been obtained from individuals who were experimentally infected with candidate live virus vaccines, some of which were not considered sufficiently attenuated for further development.

We studied the T cell response to the four dengue virus serotypes in subjects who received a candidate tetravalent live attenuated dengue virus vaccine.[69] *In vitro* proliferation responses, IFN-γ production, and cytolytic responses to dengue virus were detected in all subjects studied. However, responses to the four dengue serotypes were not equal. Proliferation responses and IFN-γ production were highest for dengue-1 and dengue-3 viruses, while responses to dengue-4 virus were detected in only one of six subjects. Proliferation responses were strongly associated with the induction of neutralizing antibodies to the same serotype. Serotype-cross-reactive cytolytic responses directed at the NS3 protein were detected, and were highest after *in vitro* stimulation with dengue-2 and dengue-3 viruses. The imbalance in the T cell response to the four serotypes, together with the observations that neutralizing antibody responses were induced predominantly against dengue-3 virus, have suggested that further work is needed to produce a vaccine that will induce optimal immunity to all four serotypes. Nevertheless, the data on T cell responses to this vaccine indicate that a live attenuated tetravalent vaccine can induce CD4+ and CD8+ T cell responses to multiple dengue serotypes.

Summary

The available data demonstrate that there are multiple cellular mechanisms capable of eliminating cells infected with dengue virus *in vitro*. Studies using immunodeficient mice confirm the importance of these mechanisms but suggest that the various mechanisms are redundant, with each able to effect viral clearance. While clinical studies confirm that the interferon, NK cell, and T lymphocyte responses to dengue virus are all active, their roles in viral clearance, protection from reinfection, or immunopathological response to infection have not been defined fully. Several live attenuated vaccines have succeeded in inducing dengue-virus-specific T lymphocyte responses. Further work is needed to define protective immune responses against dengue virus infection, and to develop a vaccine that induces optimal T cell responses to all four dengue serotypes.

References

1. Johnson AJ, Roehrig JT. New mouse model for dengue virus vaccine testing. *J Virol* 1999;73:783–786.
2. Wilson CC, Kalams SA, Wilkes BM, *et al.* Overlapping epitopes in human immunodeficiency virus type 1 gp120 presented by HLA A, B, and C molecules: effects of viral variation on cytotoxic T-lymphocyte recognition. *J Virol* 1997;71:1256–1264.
3. Kurane I, Ennis FA. Induction of interferon α from human lymphocytes by autologous, dengue virus-infected monocytes. *J Exp Med* 1987;166:999–1010.
4. Kurane I, Ennis FA. Production of interferon alpha by dengue virus-infected human monocytes. *J Gen Virol* 1988;69:445–449.
5. Kurane I, Janus J, Ennis FA. Dengue virus infection of human skin fibroblasts *in vitro*, production of IFN-β, IL-6 and GM-CSF. *Arch Virol* 1992;124:21–30.
6. Diamond MS, Roberts TG, Edgil D, Lu B, Ernst J, Harris E. Modulation of dengue virus infection in human cells by alpha, beta, and gamma interferons. *J Virol* 2000;74:4957–4966.
7. Munoz-Jordan JL, Sanchez-Burgos GG, Laurent-Rolle M, Garcia-Sastre A. Inhibition of interferon signaling by dengue virus. *Proc Natl Acad Sci USA* 2003;100:14333–14338.
8. Munoz-Jordan JL, Laurent-Rolle M, Ashour J, *et al.* Inhibition of alpha/beta interferon signaling by the NS4B protein of flaviviruses. *J Virol* 2005;79:8004–8013.
9. Jones M, Davidson A, Hibbert L *et al.* Dengue virus inhibits alpha interferon signaling by reducing STAT2 expression. *J Virol* 2005;79:5414–5420.
10. Shresta S, Kyle JL, Snider HM, Basavapatna M, Beatty PR, Harris E. Interferon-dependent immunity is essential for resistance to primary dengue virus infection in mice, whereas T- and B-cell-dependent immunity are less critical. *J Virol* 2004;78:2701–2710.
11. Kurane I, Innis BL, Nimmannitya S, Nisalak A, Meager A, Ennis FA. High levels of interferon alpha in the sera of children with dengue virus infection. *Am J Trop Med Hyg* 1993;48:222–229.
12. Libraty DH, Endy TP, Houng HH *et al.* (2002) Differing influences of viral burden and immune activation on disease severity in secondary dengue 3 virus infections. *J Infect Dis* 2002;185:1213–1221.

13. Bierman HR, Nelson ER. Hematodepressive virus diseases of Thailand. *Ann Intern Med* 1965;62:867–884.
14. Kurane I, Innis BL, Nisalak A *et al.* Human T cell responses to dengue virus antigens. Proliferative responses and interferon gamma production. *J Clin Investig* 1989;83:506–513.
15. Kurane I, Meager A, Ennis FA. Dengue virus-specific human T cell clones serotype crossreactive proliferation, interferon gamma production, and cytotoxic activity. *J Exp Med* 1989;170:763–775.
16. Mori M, Kurane I, Janus J, Ennis FA. Cytokine production by dengue virus antigen-responsive human T lymphocytes *in vitro* examined using a double immunocytochemical technique. *J Leukoc Biol* 1997;61: 338–345.
17. Kurane I, Meager A, Ennis FA. Induction of interferon alpha and gamma from human lymphocytes by dengue virus-infected cells. *J Gen Virol* 1986;67:1653–1661.
18. Sittisombut N, Maneekarn N, Kanjanahaluethai A, Kasinrerk W, Viputtikul K, Supawadee J. Lack of augmenting effect of interferon-gamma on dengue virus multiplication in human peripheral blood monocytes. *J Med Virol* 1995;45:43–49.
19. Kontny U, Kurane I, Ennis FA. Gamma interferon augments Fc-gamma receptor-mediated dengue virus infection of human monocytic cells. *J Virol* 1988;62:3928–3933.
20. Kurane I, Innis BL, Nimmannitya S *et al.* Activation of T lymphocytes in dengue virus infections: high levels of soluble interleukin 2 receptor, soluble CD4, soluble CD8, interleukin 2, and interferon-gamma in sera of children with dengue. *J Clin Invest* 1991;88:1473–1480.
21. Green S, Vaughn DW, Kalayanarooj S *et al.* Early immune activation in acute dengue is related to development of plasma leakage and disease severity. *J Infect Dis* 1999;179:755–762.
22. Orange JS, Salazar-Mather TP, Opal SM *et al.* Mechanism of interleukin 12-mediated toxicities during experimental viral infections: role of tumor necrosis factor and glucocorticoids. *J Exp Med* 1995;181:901–914.
23. Moretta A, Biassoni R, Bottino C, Moretta L. Surface receptors delivering opposite signals regulate the function of human NK cells. *Semin Immunol* 2000;12:129–138.

24. Kurane I, Hebblewaite D, Brandt WE, Ennis FA. Lysis of dengue virus-infected cells by natural cell-mediated cytotoxicity and antibody-dependent cell-mediated cytotoxicity. *J Virol* 1984;52:223–230.
25. Kurane I, Hebblewaite D, Ennis FA. Characterization with monoclonal antibodies of human lymphocytes active in natural killing and antibody-dependent cell-mediated cytotoxicity of dengue virus-infected cells. *Immunology* 1986;58:429–436.
26. Mullbacher A, Lobigs M. Up-regulation of MHC class I by flavivirus-induced peptide translocation into the endoplasmic reticulum. *Immunity* 1995;3:207–214.
27. Lobigs M, Blanden RV, Mullbacher A. Flavivirus-induced up-regulation of MHC class I antigens; implications for the induction of CD8+ T-cell-mediated autoimmunity. *Immunol Rev* 1996;152:5–19.
28. Homchampa P, Sarasombath S, Suvatte V, Vongskul M. Natural killer cells in dengue hemorrhagic fever/dengue shock syndrome. *Asian Pac J Allergy Immunol* 1988;6:95–102.
29. Green S, Pichyangkul S, Vaughn DW *et al.* Early CD69 expression on peripheral blood lymphocytes from children with dengue hemorrhagic fever. *J Infect Dis* 1999;180:1429–1435.
30. Green S, Kurane I, Edelman R *et al.* Dengue virus-specific human CD4+ T-lymphocyte responses in a recipient of an experimental live-attenuated dengue virus type 1 vaccine: bulk culture proliferation, clonal analysis, and precursor frequency determination. *J Virol* 1993;67:5962–5967.
31. Gagnon SJ, Zeng W, Kurane I, Ennis FA. Identification of two epitopes on the dengue 4 virus capsid protein recognized by a serotype-specific and a panel of serotype-cross-reactive human CD4+ cytotoxic T-lymphocyte clones. *J Virol* 1996;70:141–147.
32. Kurane I, Brinton MA, Samson AL, Ennis FA. Dengue virus-specific, human CD4+ CD8− cytotoxic T-cell clones: multiple patterns of virus cross-reactivity recognized by NS3-specific T-cell clones. *J Virol* 1991;65:1823–1828.
33. Kurane I, Dai LC, Livingston PG, Reed E, Ennis FA. Definition of an HLA-DPw2-restricted epitope on NS3, recognized by a dengue virus serotype-cross-reactive human CD4+ CD8− cytotoxic T-cell clone. *J Virol* 1993;67:6285–6288.

34. Kurane I, Okamoto Y, Dai LC, Zeng LL, Brinton MA, Ennis FA. Flavivirus-cross-reactive, HLA-DR15-restricted epitope on NS3 recognized by human CD4+ CD8− cytotoxic T lymphocyte clones. *J Gen Virol* 1995;76: 2243–2249.
35. Zeng L, Kurane I, Okamoto Y, Ennis FA, Brinton MA. Identification of amino acids involved in recognition by dengue virus NS3-specific, HLA-DR15-restricted cytotoxic CD4+ T-cell clones. *J Virol* 1996;70: 3108–3117.
36. Kurane I, Zeng L, Brinton MA, Ennis FA. Definition of an epitope on NS3 recognized by human CD4+ cytotoxic T lymphocyte clones cross-reactive for dengue virus types 2, 3, and 4. *Virology* 1998;240:169–174.
37. Okamoto Y, Kurane I, Leporati AM, Ennis FA. Definition of the region on NS3 which contains multiple epitopes recognized by dengue virus serotype-cross-reactive and flavivirus-cross-reactive, HLA-DPw2-restricted CD4+ T cell clones. *J Gen Virol* 1998;79:697–704.
38. Green S, Kurane I, Pincus S, Paoletti E, Ennis FA. Recognition of dengue virus NS1–NS2a proteins by human CD4+ cytotoxic T lymphocyte clones. *Virology* 1997;234:383–386.
39. Livingston PG, Kurane I, Lai CJ, Bray M, Ennis FA. Recognition of envelope protein by dengue virus serotype-specific human CD4+ CD8− cytotoxic T-cell clones. *J Virol* 1994;68:3283–3288.
40. Mangada MM, Rothman AL. Altered cytokine responses of dengue-specific CD4+ T cells to heterologous serotypes. *J Immunol* 2005;175:2676–2683.
41. Okamoto Y, Gagnon SJ, Kurane I, Leporati AM, Ennis FA. Preferential usage of T-cell receptor Vβ 17 by dengue virus-specific human T lymphocytes in a donor with immunity to dengue virus type 4. *J Virol* 1994;68:7614–7619.
42. Rothman AL, Kurane I, Zhang YM, Lai CJ, Ennis FA. Dengue virus-specific murine T-lymphocyte proliferation: serotype specificity and response to recombinant viral proteins. *J Virol* 1989;63:2486–2491.
43. Rothman AL, Kurane I, Lai CJ *et al.* Lymphocyte proliferative responses to dengue virus E and NS1 proteins in BALB/c mice. In: Brown F, Chanock RM, Ginsberg H, Lerner RA (eds.) *Vaccines 90: Modern Approaches to New Vaccines Including Prevention of AIDS*. Cold Spring Harbor Laboratory Press, New York, 1990, pp. 135–138.

44. Rothman AL, Kurane I, Ennis FA. Multiple specificities in the murine CD4+ and CD8+ T-cell response to dengue virus. *J Virol* 1996;70:6540–6546.
45. Murali-Krishna K, Altman JD, Suresh M *et al.* Counting antigen-specific CD8 T cells: a re-evaluation of bystander activation during viral infection. *Immunity* 1998;8:177–187.
46. Mangada MM, Ennis FA, Rothman AL. Quantitation of dengue virus specific CD4+ T cells by intracellular cytokine staining. *J Immunol Methods* 2004;284:89–97.
47. Rothman AL. Dengue: defining protective versus pathologic immunity. *J Clin Invest* 2004;113:946–951.
48. Gagnon SJ, Ennis FA, Rothman AL. Bystander target cell lysis and cytokine production by dengue virus-specific human CD4+ cytotoxic T lymphocyte clones. *J Virol* 1999;73:3623–3629.
49. Hall WC, Crowell TP, Watts DM *et al.* Demonstration of yellow fever and dengue antigens in formalin-fixed paraffin-embedded human liver by immunohistochemical analysis. *Am J Trop Med Hyg* 1991;45:408–417.
50. Bukowski JF, Kurane I, Lai CJ, Bray M, Falgout B, Ennis FA. Dengue virus-specific cross-reactive CD8+ human cytotoxic T lymphocytes. *J Virol* 1989;63:5086–5091.
51. Mathew A, Kurane I, Rothman AL, Zeng LL, Brinton MA, Ennis FA. Dominant recognition by human CD8+ cytotoxic T lymphocytes of dengue virus nonstructural proteins NS3 and NS1.2a. *J Clin Investig* 1996;98:1684–1694.
52. Livingston PG, Kurane I, Dai LC *et al.* Dengue virus-specific, HLA-B35-restricted, human CD8+ cytotoxic T lymphocyte (CTL) clones: recognition of NS3 amino acids 500 to 508 by CTL clones of two different serotype specificities. *J Immunol* 1995;154:1287–1295.
53. Zivny J, Kurane I, Leporati AM *et al.* A single nine-amino acid peptide induces virus-specific, CD8+ human cytotoxic T lymphocyte clones of heterogenous serotype specificities. *J Exp Med* 1995;182:853–863.
54. Zivny J, DeFronzo M, Jarry W *et al.* Partial agonist effect influences the CTL response to a heterologous dengue virus serotype. *J Immunol* 1999;163:2754–2760.
55. Bashyam HS, Green S, Rothman AL. Dengue virus-reactive CD8+ T cells display quantitative and qualitative differences in their response to variant epitopes of heterologous viral serotypes. *J Immunol* 2006;176:2817–2824.

56. Sangkawibha N, Rojanasuphot S, Ahandrik S *et al.* Risk factors for dengue shock syndrome: a prospective epidemiologic study in Rayong,Thailand. I. The 1980 outbreak. *Am J Epidemiol* 1984;120:653–669.
57. Rothman AL, Kurane I, Lai CJ *et al.* Dengue virus protein recognition by virus-specific murine CD8+ cytotoxic T lymphocytes. *J Virol* 1993;67:801–806.
58. Loke H, Bethell DB, Phuong CX *et al.* (2001) Strong HLA class I-restricted T cell responses in dengue hemorrhagic fever: a double-edged sword? *J Infect Dis* 2001;184:1369–1373.
59. Zivna I, Green S, Vaughn DW *et al.* T cell responses to an HLA B*07-restricted epitope on the dengue NS3 protein correlate with disease severity. *J Immunol* 2002;168:5959–5965.
60. Mongkolsapaya J, Dejnirattisai W, Xu X *et al.* Original antigenic sin and apoptosis in the pathogenesis of dengue hemorrhagic fever. *Nat Med* 2003;9: 921–927.
61. Mongkolsapaya J, Duangchinda T, Dejnirattisai W *et al.* T cell responses in dengue hemorrhagic fever: are cross-reactive T cells suboptimal? *J Immunol* 2006;176:3821–3829.
62. Dharakul T, Kurane I, Bhamarapravati N *et al.* Dengue virus-specific memory T cell responses in human volunteers receiving a live attenuated dengue virus type 2 candidate vaccine. *J Infect Dis* 1994;170:27–33.
63. Stephens HA, Klaythong R, Sirikong M *et al.* HLA-A and -B allele associations with secondary dengue virus infections correlate with disease severity and the infecting viral serotype in ethnic Thais. *Tissue Antigens* 2002;60:309–318.
64. Spaulding AC, Kurane I, Ennis FA, Rothman AL. Analysis of murine CD8+ T-cell clones specific for the dengue virus NS3 protein: flavivirus cross-reactivity and influence of infecting serotype. *J Virol* 1999;73:398–403.
65. Mathew A, Kurane I, Green S *et al.* Predominance of HLA-restricted CTL responses to serotype crossreactive epitopes on nonstructural proteins after natural dengue virus infections. *J Virol* 1998;72:3999–4004.
66. Simmons CP, Dong T, Chau NV *et al.* Early T-cell responses to dengue virus epitopes in Vietnamese adults with secondary dengue virus infections. *J Virol* 2005;79:5665–5675.
67. Endy TP, Chunsuttiwat S, Nisalak A *et al.* Epidemiology of inapparent and symptomatic acute dengue virus infection: a prospective study of primary

school children in Kamphaeng Phet, Thailand. *Am J Epidemiol* 2002;156:40–51.

68. Mangada MM, Endy TP, Nisalak A *et al*. Dengue-specific T cell responses in peripheral blood mononuclear cells obtained prior to secondary dengue virus infections in Thai schoolchildren. *J Infect Dis* 2002;185:1697–1703.
69. Rothman AL, Kanesa-thasan N, West K, Janus J, Saluzzo J, Ennis FA. Induction of T lymphocyte responses to dengue virus by a candidate tetravalent live attenuated dengue virus vaccine. *Vaccine* 2001;19:4694–4699.

9

Pathophysiology

Scott B. Halstead

Introduction

The following provide background to this chapter:

(1) *Clinical features of dengue* (Chap. 5). Acute dengue virus infection may be accompanied by physiologic abnormalities affecting multiple body systems, including the vascular system, hematopoesis, blood coagulation and hepatic function. While these various disorders may not evolve from a single direct cause, they must be explained by any comprehensive model of dengue pathophysiology. It should be recalled that only a small percentage of individuals develop DHF during secondary dengue virus infections.
(2) *Host responses.* The crucial role of the complex of innate and acquired immune responses accompanying dengue infection and specifically in controlling and eliminating dengue-infected cells is described in Chap. 8. The by-products of this process generate the phenomena producing dengue disease.
(3) *Viral "virulence."* Although the contribution of natural variation in dengue viruses to the severity of dengue illnesses (virulence) remains to be demonstrated (see below and chaps. 7 and 14), it is nonetheless possible in the laboratory to select viruses that lose the

ability to cause disease (attenuated). Attenuation and avirulence may be two different phenomena. Attenuated viruses have been artificially created through laboratory adaptation or by direct mutagenesis from infectious cDNA clones.[1] Viral attributes, still poorly understood, may contribute importantly to the expression of dengue disease.

(4) *Afferent role of antibodies*. The role played by antibodies in modifying the course of dengue infections and disease (Chap. 7) is evidenced by the association between secondary dengue virus infection with a new serotype and risk of developing DHF as established in multiple prospective studies.[2–5] More tellingly, classical DHF occurs reproducibly in infants born to dengue-immune mothers during initial dengue infections usually in the latter half of the first year of life.[6,7] The relevant risk factor in this patient group is the dengue-hyperimmune status of the infant's mother.[8] Infants born to dengue-immune mothers are at a greater risk of developing DHF during a dengue infection than are children at any age experiencing sequential dengue infections.[9] Only antibody-dependent enhancement of infection (ADE) provides a unified explanation of severe dengue disease during these disparate immunological circumstances.

Dengue Infection Model

Insight into the kinetics of human dengue infections and immune responses may be gained by studying the kinetics of dengue cellular infections in rhesus monkeys. In a major study, 31 animals were infected and then sacrificed at 12–24-hour intervals, and sites of viral infection identified (Fig. 1).[9,10]

Dengue virus was recovered from the site of inoculation in the right forearm during the entire period of infection. Twelve hours after subcutaneous inoculation, virus appeared in the regional lymph node. During the previremic period, virus was distributed to lymph nodes throughout the body and then to the spleen and liver. Toward the end of viremia circulating infected mononuclear leukocytes were observed and shortly later

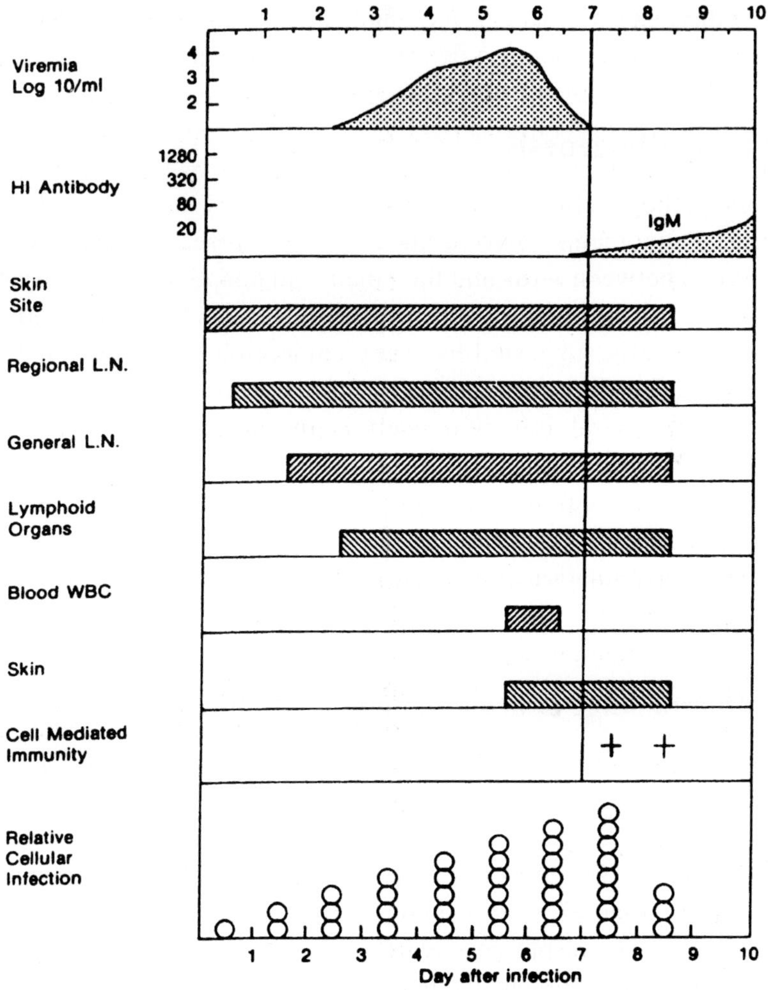

Fig. 1. Schematic distribution of virus recovered from blood and tissues of 31 rhesus monkeys during the course of dengue-2 infections.[9,10]

virus was distributed to skin. Viremia peaked 5–6 days after infection, but tissue infection peaked 3–4 days later. In humans, viremia ceases with defervescence. As evidenced by monkey data, in humans *peak cellular infection occurs at or after defervescence*, i.e., coincident with vascular permeability. Consistent with this model, peak RNAemia in DHF cases has been measured following defervescence.[11]

Immunopathogenesis

Afferent pathogenetic mechanisms are discussed in greater detail in Chap. 7. Phenomena contributing to dengue disease result from the interactions between virus and the innate and adaptive immune systems modulated by individuals by genetic makeup. Genetic control of dengue disease is suggested by many epidemiologic studies, although the specific mechanisms and genes involved have not been fully defined. Since viral infection itself is the stimulus for the cellular immune response, a schema is proposed in which viral replication and direct effects of cellular infection constitute the afferent stage of dengue pathogenesis and the cellular immune response represents the infected cell elimination or efferent stage (Chap. 7). In the middle, these two stages overlap.

The afferent stage of acute dengue virus infection

The absence or the presence of dengue-enhancing antibodies has profound implications for the consequences of dengue infection. Dengue infection in the susceptible host triggers normal innate immune responses, contributing to the generally mild outcome of these infections. As described in Chap. 7, many studies have demonstrated that enhanced infections occur in the presence of enhancing antibodies (ADE). Increased viral production has typically been interpreted to be the result of an increased number of infected Fc-R-bearing cells and possibly the result of an accelerated rate of internalization and cell infection by immune complexes. Now it is known that ADE infections suppress early response by the interferon system resulting in enhanced replication of viruses in each infected cell together with the release of IL-10, which suppresses bystander cell innate immune responses. The ADE phenomenon not only controls the afferent propagation of dengue viruses but rapidly suppresses efferent innate responses of the host.

Cellular targets for dengue viral infection

In vitro studies

Human cells or cell lines which support dengue virus replication *in vitro* include monocytes,[12,13] dendritic cells,[14,15] bone marrow cells,[16] fibroblasts,[17] endothelial cells,[18] epithelial cells,[19] and hepatocytic cell lines.[20,21] However, the proportion of infected cells is often quite low unless highly adapted laboratory strains of dengue viruses are used. It must be recognized that the ability of dengue viruses to replicate in many cell lines *in vitro* does not necessarily reflect *in vivo* tissue tropism in humans.

In vivo studies

In humans, little data has been obtained during the 4–6-day period following inoculation of virus in subcutaneous tissue by the mosquito vector and prior to the onset of fever and viremia. In the rhesus monkey model, following subcutaneous inoculation, dengue virus was detected at the skin inoculation site and in 12 hours in the regional lymph node and at 24 hours disseminated to other lymph nodes, the spleen and the liver, and finally back to skin.[10] There is evidence that the duration of the incubation period is determined by the size of the viral inoculum.[22] Recent data suggest that initial infection in humans occurs in tissue dendritic cells. Dendritic cell infection by dengue viruses was first detected in foot pads of mice inoculated subcutaneously.[23] Studies on US Army volunteers and humans in Indonesia and Thailand have shown that viremia can be detected in virtually all cases early during the febrile phase.[24–27] However, when signs of DHF/DSS become manifest, the majority of patients are no longer viremic.[28,29] High levels of antidengue antibodies usually seen during the late acute phase of secondary dengue infections undoubtedly are responsible for these low isolation rates.[27,30]

Peripheral blood mononuclear cells (PBMC's) are infected with dengue viruses *in vivo,* as has been demonstrated by detection of viral RNA, viral antigen and infectious virus. However, it has been difficult to ascertain which cell type is infected in peripheral blood mononuclear cell fractions as the number of circulating infected cells is low, and viruses in plasma can adhere to cell surfaces. Immature monocytes best support growth of dengue viruses *in vitro*.[12] Virus has been isolated from an adherent PBMC population, but the nonadherent population (most likely lymphocytes) was apparently infected.[31] Others have suggested that peripheral blood B cells are targets of dengue infection *in vivo*.[32]

It has been difficult to determine which tissues are the major sites of dengue virus infection as limited autopsies on DHF/DSS have been performed. In older studies virus isolation rates were low.[29] Infected Langerhans cells have been found in biopsies of the rash observed in an individual receiving a candidate live-attenuated tetravalent dengue vaccine.[14] In skin biopsies from rashes observed in DHF patients, dengue antigen was found predominantly in macrophages.[33] Rosen *et al.* studied 18 fatal DHF cases and detected viral RNA in the majority of liver and spleen specimens and almost half of mesenteric lymph nodes but were able to isolate virus only from a few cases, supporting the conclusion that the tissues contained primarily inactivated virus.[34,35] Virus has been isolated in humans from blood, lymph node, bone marrow, liver and spleen, but the liver appears to have the highest yield of virus.[29,34–36]

Studies on organs obtained by biopsy or necropsy from patients with dengue infections have shown evidence of dengue replication in lymph node, splenic and pulmonary macrophages.[37] Dengue antigens have been detected in Kupffer cells as well as hepatocytes from livers obtained at autopsy.[38–40] No evidence could be found of dengue virus replication in Kupffer cells.[37] Dengue antigens commonly detected in Kupffer cells may derive from phagocytosed dengue-infected human hepatocytes partly destroyed by apoptosis.[40]

In *ex vivo* studies, human splenic macrophages and not T or B cells supported dengue infection and ADE.[41]

Viral factors

As noted above, some laboratory passaged viruses lose the ability to cause disease in humans. Epidemiologic studies show differential association of dengue virus strains with different dengue disease syndromes. For example, DHF is more frequently associated with dengue 2 than other dengue serotypes.[42] In a cohort of children who were seen with febrile disease in hospital outpatient clinics, none were observed with primary infections with dengue-2 or dengue-4 viruses, whereas primary infections constituted approximately one-third of all patients who had dengue-1 or dengue-3 virus infections. It was concluded that primary infections in children with dengue-2 and -4 viruses were clinically inapparent. This hypothesis was confirmed in an outbreak in Cuba in 1997, where nearly all primary dengue-2 infections were silent, while secondary dengue-2 infections (following dengue 1) were nearly all over.[43,44]

An even more striking example of virus strain-related "avirulence" has been reported.[45] In 1995, when the American genotype of dengue 2 was introduced into a population highly immune to dengue 1, disease was mild and no DHF cases occurred. This dengue-2 virus was shown to be genetically similar to viruses circulating decades earlier in the Western Hemisphere, prior to the recognition of DHF and distinct from dengue-2 viruses associated with DHF in Asia.[46] In subsequent studies it was determined that SE Asian dengue-2 viruses could not be neutralized by dengue-1 antibodies *in vitro,* whereas strains of the American dengue-2 genotype viruses were readily neutralized by such sera, often to a high titer.[47] It was suggested that dengue-1 antibodies might down-regulate American dengue-2 virus infections.

American and SE Asian dengue-2 viruses differ genetically in non-translated regions and by three amino acids in structural regions.[46] Recently, it was demonstrated that introduction of the single amino acid substitution in the E protein sequence of the "avirulent" dengue-2 viruses into an infectious dengue-2 cDNA clone reduced the ability of the virus to replicate in monocyte-derived macrophages.[48] Other studies have demonstrated reduced ability of American dengue-2 genotype strains to orally infect and be transmitted by several strains of *Aedes aegypti* mosquitoes.[49]

Consequences of viral infection

Cellular activation as a direct consequence of dengue virus infection has been shown *in vitro* in many different human cell lines, including epithelial cells,[50] endothelial cells,[50,51] myeloid cells,[52] and mast cells,[53] as well as in primary human monocytes[50] and monocyte-derived dendritic cells.[15] In those cell types studied in detail, cellular activation was associated with nuclear translocation of NFκB.[50,51] It has been shown that dengue-virus-induced up-regulation of IL-8 transcription was associated with altered histone acetylation and chromatin rearrangement at the IL-8 promoter.[50] The *in vivo* correlates of cellular activation induced directly by dengue virus infection are currently unknown. Early in infection there would appear to be no major consequences for dengue infection as the incubation period is silent. The earliest evidences of disease — rash, fever, myalgia and headache — occur five or more days after infection and are plausibly related to late innate immunity or early specific immune responses.

Viral burden

A central tenet of ADE-controlled pathophysiology is that disease severity will be correlated with the burden of dengue-infected cells (peak viremia is a surrogate measure of infection burden). Indeed, the best anticipatory correlates of degree of hemoconcentration and thrombocytopenia in dengue infections have been peak levels of virus, viral RNA or NS1.[42,54–56] In animal models DENV infections occurring in the presence of heterotypic dengue antibodies produced significantly higher viremia titers than in immunologically naive subhuman primates.[57–59]

In early studies, researchers were unable to detect enhanced viremia, and in fact reported low virus isolation rates from DHF patients.[28,30] It is now well documented that the low success rate of virus isolation in previous studies was related both to the insensitivity of many early virus isolation methods and more importantly, to the failure to collect specimens early in the course of disease. In the first successful measurement of peak viremia,[42] patients were enrolled as outpatients early in the course of infection and blood was inoculated into *Toxorhynchites splendens*.[60] In 87 Thai children infected with dengue-1 or dengue-2 viruses, initial plasma samples were obtained within 72 hours of onset of fever and children bled for five successive days. Peak viremia titers could be clearly defined in about half of subjects with secondary antibody responses. In others, viremia titers declined by > 0.5 log between the first and the second sample. Maximum plasma viremia titers were significantly higher in subjects with DHF than in those with DF. In addition, maximum plasma viremia titers showed a significant correlation with several measures of disease severity, including the extent of pleural effusion and the degree of thrombocytopenia. In subgroup analyses, these associations were still significant among subjects infected with dengue-2 virus. The same trends were present in subjects infected with dengue-1 virus, although the associations were no longer statistically significant. The phenomenon was observed again in patients infected with dengue-3 viruses measuring viral RNA levels and in concentrations of the nonstructural 1 protein in acute phase blood.[55,56] In Tahiti, using quantitative RT-PCR, patients with DHF had significantly different kinetics of plasma dengue-2 viral RNA levels than subjects with DF.[54]

Liver infection and apoptosis

Dengue viral antigens have been observed in hepatocytes surrounding necrotic foci in a fatal case of dengue fever.[40] Apoptotic hepatocytes were found to colocalize with viral antigen-positive hepatocytes. In fatal human cases from Vietnam, dengue virus infection was detected in the liver by reverse transcription polymerase chain reaction and also demonstrated apoptotic hepatocytes in five fatal dengue-3 cases.[36] In *in vitro* studies dengue-virus-induced apoptosis in a human hepatoma cell line was mediated by NF-κB, and although Kupffer cells were able to ingest dengue viruses *in vitro*, productive infection did not occur and apoptosis ensued.[61,62] These studies suggest that hepatocyte infection and subsequent apoptosis result in liver injury contributing to the liver enlargement and liver enzyme elevations seen in severe dengue illnesses. Soluble dengue NS1 inoculated into mice appears to home to the liver and may potentiate virus replication.[63] In addition, immunocompetent mice infected with dengue 2 developed liver enzyme elevations correlated with the kinetics of T cell responses.[64]

Endothelial cells

It has been speculated that dengue virus infection of endothelial cells plays a role in the plasma leakage syndrome seen in dengue. However, autopsy data show that capillary endothelium are not damaged or infected by dengue viruses.[33,65] And, although endothelial cells in fixed tissues stained for dengue antigens using *in situ* hybridization, no viral replication has been detected from autopsy tissues.[37] Endothelial cells are readily infected by dengue viruses *in vitro* along with many other tissues readily infected in the laboratory. IL-8 and RANTES were induced following dengue virus infection of the human endothelial cell line ECV-304, and infected endothelial cells were found to undergo apoptosis in a TNF-independent fashion.[51] Importantly, endothelial cells can be activated (up-regulation of VCAM-1 and ICAM-1) by TNF-alpha in culture supernatants from dengue-infected monocytes.[66] The mechanism by which endothelial cells are damaged in the pathogenesis of DHF awaits further clarification.

The efferent stage of acute dengue virus infection

Mechanisms for eliminating dengue-infected cells have been reviewed in Chap. 8. ADE infection of monocytes and macrophages alters their efferent responses, down-regulating innate immunity and generating recruitment signals via the release of IL-10. The process of eliminating dengue-virus-infected cells generates a cascade of chemokines and cytokines that constitute the pathophysiology of dengue disease syndromes. This has been aptly termed "a perfect cytokine storm."[67]

Clearance of extracellular virus

Given the probable long duration of dengue-infected cells, the quantity of virus in blood during the course of infection only describes the kinetics of extracellular virus clearance. Using the sensitive technique of intrathoracic inoculation of *Toxorhynchites splendens* mosquitoes,[42,60] it was observed that dengue virus could be detected in nearly all plasma samples collected within 72 hours of the onset of fever. The frequency of virus isolation declined rapidly and virus was usually not present at defervescence.[27] Viremia was of longer duration in individuals experiencing primary dengue virus infections than in those experiencing secondary dengue virus infections, and the slope of the decline in virus titer was greater in individuals with DHF than in those with DF. The most important mechanism clearing extracellular virus should be IgM and IgG dengue antibodies, although T lymphocytes and antibody-dependent cell killing may contribute by attacking and destroying virus-infected cells. As discussed above, it is likely that elimination of dengue-virus-infected cells is delayed and is not complete until well after the end of viremia.

Activation of the immune system

Activation is induced directly in the virus-infected cell (Chap. 8), as well as indirectly, as a result of activation of the innate immune response (see Chap. 4) and by immune effector cells that interact with virus-infected cells and bystander cells. Analysis of serum or plasmas from patients infected with dengue indicates that concentrations of IL-10, TNF-alpha, IL-8, IL-12, IFN-gamma, IFN-alpha, elastase and soluble TNF and IL-2 receptors are increased during DF and/or DHF. These immune markers may be directly associated with specific disease signs, as suggested by the correlations between IL-10 levels and soluble TNF receptor II concentrations and the degree of thromobocytopenia[55] as well as plasma leakage.[68] In BALB/c mice lethally infected with dengue, concentrations of IL-1-beta, TNF-alpha and IL-6 peaked with mortality.[69] Interestingly, when mice were treated with anti-TNF-alpha serum, mortality was reduced by 60%. A role for TNF-alpha is supported by the finding that the TNF-alpha 308A allele, associated with the production of greater concentrations of TNF-alpha, is more common in DHF patients.[70]

Natural killer cells, CD4+ T lymphocytes and CD8+ T lymphocytes are activated by interaction with dengue-virus-infected cells *in vitro*. The molecular basis for the recognition of dengue-virus-infected cells by NK and T cells is described in detail in Chaps. 4 and 8. One by-product of activation of particular subpopulations of lymphocytes is release of soluble forms of various cell surface molecules, including soluble forms of CD4 (sCD4), CD8 (sCD8), and the receptors for IL-2 (sIL-2R), IL-4 (sIL-4R), and TNF (sTNFR-I and sTNFR-II). Elevated circulating levels have been demonstrated of sCD4,[71] sCD8,[63,68,71] sIL-2R,[68,71] sIL-4R (Green, personal communication), sTNFR-I[68] and sTNFR-II[68,71,72] during acute dengue virus infection, suggesting that CD4+ and CD8+ T lymphocytes, as well as NK cells, are activated during the acute illness.

While elevated levels of these activation markers are strongly suggestive of immune activation, they are at best an indirect measure of NK and T cell activation. To directly measure NK and T cell activation *in vivo* during acute dengue virus infection, PBMC's were analyzed by flow cytometry in 51 Thai children with acute dengue illness and detectable plasma dengue viral RNA;[73] 55% were experiencing a secondary dengue virus infection. PBMC's were directly stained *ex vivo* with monoclonal antibodies to phenotypic markers of NK cells (CD16/CD56), CD4 T cells (CD3/CD4) and CD8 T cells (CD3/CD8), as well as cellular activation markers (CD69, CD25 and HLA–DR). Expression of CD69, which is induced early after activation, was detected on a high percentage of circulating NK and CD8+ T cells, in some cases on as many as 80% and 40% of NK and CD8+ T cells, respectively. The percentage of CD4+ T cells expressing CD69, and the percentages of NK and T cells expressing CD25, which is induced later after activation, were low in comparison (< 5% of each subpopulation). The percentages of cells expressing CD69 were highest at the earliest time point studied and had declined when next analyzed one day after defervescence, at which time point plasma dengue virus titers were markedly lower or undetectable. These data on the kinetics of CD69 expression suggest that activation of NK and CD8+ T cells is directly related to interaction with dengue-virus-infected cells *in vivo*.

The marked T cell activation noted above could reflect nonspecific activation due to the high cytokine levels *in vivo* rather than antigen-specific activation of T cells. To distinguish these possibilities, molecular and functional studies were performed to characterize the circulating T cell population. The molecular analyses exploit the vast heterogeneity in T cell receptor (TCR) genes that occurs during T cell development in the thymus. Each individual T cell progenitor expresses a TCR with a unique combination of rearranged TCR variable (V), diversity (D), and joining (J) gene segments as well as random insertions and deletions of nucleotides at the antigen recognition site, or complementarity-determining region 3 (CDR3). Thus, an expansion of a specific T cell clone can be detected based on its unique molecular signal. In an initial analysis, semiquantitative RT-PCR was used to examine the distribution of TCR β chain V gene segment usage in PBMC's obtained during and after acute dengue virus infection in 26 Thai children with acute dengue virus infection.[74] In approximately half of the subjects, increases in the relative expression of one TCR V gene segment were detected which were indicative of an oligoclonal T cell expansion *in vivo*. The specific TCR V gene family that demonstrated increased expression differed in individuals, consistent with the finding that different subjects recognize different viral epitopes, as described in Chap. 8. Interestingly, most of the T cell expansions were apparent in the PBMC's obtained one day after defervescence rather than during the viremic period. Together with the data on CD69 expression described above, these data suggest that T cell activation occurs early in infection and that T cell expansion increases during viremia.

Although oligoclonal T cell expansion is suggestive of antigen-driven rather than nonspecific activation, the molecular analyses do not directly show that the expanded T cells are specific for dengue viral epitopes. Proof of the antigen specificity of T cell expansion is complicated by differences between subjects in the immunodominant target epitopes on dengue virus, which relates in large part to the polymorphism of HLA class I and class II alleles. To address this issue, an IFN-gamma ELISPOT assay was used[75] to conduct a focused analysis of the response to a single, defined CD8+ T cell epitope on the dengue virus NS3 protein[76] in a cohort of HLA-B*07+ Thai children. The peptide-specific T cell frequency was measured in PBMC's obtained during and after acute dengue virus infection in five subjects.[77] Peptide-specific T cells were detectable in PBMC's obtained at least six months after the infection in four of the subjects, at frequencies ranging from 13 to 125 per million PBMC's. In two of these four subjects, peptide-specific T cells were also detectable at similar frequencies in PBMC's obtained during the acute infection or in the early (<1 week postinfection) convalescent stage. Because these studies measured the response to a single T cell epitope, it would be expected that the overall frequency of T cells responding to dengue virus epitopes would be considerably higher. These data therefore support the hypothesis that T cell expansion during acute dengue virus infection represents antigen-driven expansion of dengue-virus-reactive T cells.

CD8+ T cells have been implicated both in viral clearance and in immunopathogenesis in a HepG2-engrafted, severe combined immunodeficient mouse model of dengue infection.[78] Dengue-specific CD4+ T cells were shown to mediate bystander lysis of non-antigen-presenting cells.[79] T cells responses have been shown to recognize a range of epitopes, predominantly on NS3, NS1, NS2a and E proteins. Some studies have indicated that these T cell clones are broadly cross-reactive and match epitopes on all dengue serotypes and even on related flaviviruses However, other studies indicate that, while clones may cross-react, they have significantly weaker binding affinities for other dengue serotypes.[75,80]

Genetic modulation

Several studies have supported genetic variations that are associated with mild or severe dengue disease using HLA class I alleles, with disparate results. Mild disease was associated with HLA A33 in Vietnamese and A*0203 in Thai populations, and severe disease with HLA A24 in Vietnamese and A*0207 in Thai populations.[81,82] Other protective associations include variants of vitamin D receptor (t allele at position 352, which may be associated with a relatively stronger T helper type 1 cellular immune response), FcgammaRIIA (R at position 131, associated with reduced opsonization by IgG2 antibodies and hypothesized to be associated with less antibody-dependent enhancement of infection) and the CD209 promoter, DCSIGN1-336, associated with the decreased expression of CD209 and possibly lower susceptibility of dendritic cells to dengue virus infection.[83,84] A TNF-308 variant, associated with high levels of TNF-alpha production, has been associated with increased dengue disease severity.[70]

Complement and immune complex formation

During secondary dengue infections, high circulating levels of dengue-virus-specific antibodies are achieved during a period of significant viral replication.[27] This creates the potential for the formation of virus–antibody complexes. Immune complexes may contribute to virus clearance, but also to the pathogenesis of DHF through activation of the complement cascade.[85–87] Circulating immune complexes in sera of children with DHF correlate with the consumption of complement proteins C3 and C4 and increased levels of activated complement fragments C3a, C5a and C5-9 complexes.[85,87–91] These observations have been extended by demonstrating the presence of dengue viral RNA by RT-PCR in immune complexes isolated from the plasma of children with DHF.[92] The majority of plasma dengue viral RNA was found to be bound to antibody *in vivo* in subjects with secondary dengue infections, whereas only a small fraction of viral RNA was bound to antibody in subjects with primary dengue infections.

Immune complex formation and complement activation may not involve circulating dengue virions alone. The nonstructural protein NS1 has been known to be released from dengue-virus-infected cells and to be an important target for the antibody response to dengue infection. Two papers have reported the detection of soluble NS1 protein in the plasma of subjects with acute dengue infections, especially those with secondary infections.[93,94]

Immune inactivation/suppression

The widespread immune activation in the response to acute dengue virus infection also stimulates homeostatic counterbalancing mechanisms. Two mechanisms are known to result in immunologic deactivation: anergy and apoptosis. Anergic T cells are viable but unresponsive to further stimulation. PBMC's obtained from Thai children during an acute dengue virus infection or in the early convalescent stage (<1 week postinfection) had markedly reduced *in vitro* proliferation responses.[95] The suppression of *in vitro* proliferation was a global effect, as shown by the reduced response to mitogens, both phytohemagglutinin

(PHA) and antibody to CD3, dengue virus antigens, and other T cell antigens such as tetanus toxoid. The ability of the T cells in the acute phase PBMC's to respond to these stimuli could be restored by the addition of IL-2 or IL-7, but not by IL-4 or IL-12. Interestingly, the *in vitro* proliferation was also restored by the addition of autologous T-cell-depleted PBMC's obtained six months or more after the acute infection, suggesting that the defect in proliferation was caused by altered function of the accessory cell population, perhaps as a result of the production of immunosuppressive cytokines (discussed further below).

A similar suppression of *in vitro* proliferation responses has been described in association with other acute viral infections, particularly measles, EBV and CMV.[96–98] Several cellular defects have been described in these other viral infections, affecting both the T cells as well as the accessory (antigen-presenting) cells. In the case of measles virus, it has been suggested that the suppression of *in vitro* proliferation is associated with the observed increased susceptibility to other infections and the increased mortality that persists for weeks to months after measles.[99] In contrast, an increase in susceptibility to other infections has not been reported after acute dengue virus infection; therefore, the clinical significance of the observed suppression of *in vitro* proliferation responses remains uncertain.

In addition to its potential involvement in tissue injury, described above, apoptosis is a part of the normal immune response to viral infection. Studies in some animal models have shown that the majority of activated T cells undergo apoptosis during the stage of resolution of the acute viral infection, and that only a small minority of T cells enter the memory lymphocyte pool;[100] this phenomenon is essential for maintenance of normal total numbers of T cells in the body. Lymphocytes can be induced to undergo apoptosis by at least two signals: absence of growth factors (e.g. IL-2) as antigen is eliminated and the stimulus to activation disappears, and activation-induced cell death in which T cells are hyperstimulated by high antigen expression. Apoptosis was studied in cytospin preparations of PBMC's obtained at several time points during acute dengue virus infection and in the early convalescent stage.[101] Apoptotic PBMC's were detected using the

TUNEL (Terminal deoxynucleotidyltransferase-mediated dUTP Nicked End Labeling) method; cells were also stained with anti-CD8 for identification of apoptotic CD8+ T cells. Elevated numbers of apoptotic cells were detected in PBMC's of children with acute dengue virus infection, and the number of apoptotic cells was highest prior to the onset of plasma leakage. Approximately half of the apoptotic PBMC's stained positively with anti-CD8. These data confirm that T cell activation in acute dengue-virus-infection stimulates normal homeostatic mechanisms.

Apoptosis of a wide array of dengue virus infected host cells has been reported.[21,51,102–106]

Effector mechanisms in dengue disease

Generally, protein levels have been measured in plasma or serum without corrections being made for losses into interstitial spaces.[107,108] There are several areas in which the data on cytokine levels and complement activation in DHF are lacking and studies performed to date may represent a fraction of the biologically active factors released during acute infection. Both plasma and pleural fluid levels of IL-8 and MCP-1 have been studied in a small number of DHF cases. Levels were elevated in both fluids, but were much higher in pleural fluid samples, suggesting significant local expression of these factors.[51] A further difficulty in interpreting the data from studies of circulating cytokine levels arises from the fact that many studies have included single serum or plasma samples from patients hospitalized with established DHF. Such data provide little information about the kinetics of cytokine production during the acute infection, particularly prior to the onset of plasma leakage. A number of studies fail to include mildly ill subjects with DF for comparison. Occasionally, subjects were obtained from different facilities — patients from an outpatient clinic being compared with hospitalized DHF cases. The best data are derived from prospective studies in which children are enrolled at an early stage of illness, within the first three days after the onset of fever.[109] It is crucial to begin to measure cytokine levels prior to the onset of plasma leakage so as to identify the chain of antecedent events. Numerous studies have indicated that concentrations of cytokines, mediators and soluble

Table 1. Cytokines and mediators present in the serum of dengue patients or released from cells circulating in patients with dengue fever (DF) and dengue hemorrhagic fever (DHF).

Source	Cytokine/mediator	References
Serum	IL-1β, IL-2, **IL-6**, **IL-8**, **IL-10**, *IL-12*, *IL-13*, **TNFα**, IFNα, IFNγ, MIF, RANTES Elastase, soluble TNF receptors, soluble IL-2 receptor, soluble endothelial growth receptor-2, complement: *C3*, **C3a**, **C5a**	55, 68, 126, 161–163; 85, 91, 164–170, 159
PBMC	**Properdin**, **RANTES**, **MIP-1β**, **IL-8**, IL-2, IL-4, IL-5, **IL-10**, TNFα, IFNγ, IFNα	160, 171, 159

Bold: Values higher in DHF than in DF.

Italics: Values higher in DF than in DHF.

In patients with DF there was an up-regulation of IFN genes and the complement inhibitor, C-59, while in DHF there was strong T and B cell activation, cytokine production, complement activation and T cell apoptosis.

receptors may be significantly increased during the acute stage of dengue illness (Table 1).

Strikingly, the kinetics of individual cytokines in the circulation differ and change considerably over the course of infection. Type 1 and type 2 cytokines predominate at different times during acute infection. Plasma IL-12 levels were highest at the time of study enrollment and declined steadily over the course of illness.[110] IFN-gamma levels increased and then decreased, but in all cases levels were highest during the febrile period.[68] In contrast, IL-10 levels rose steadily over the course of illness and were at their highest at the end of the febrile period.[110]

Vascular permeability

Not only is vascular permeability the hallmark of DHF, but there is some evidence to suggest that damage in dengue infections may begin at endothelial surfaces. The current concept of normal microvascular ultrafiltration suggests that intrinsic permeability is regulated by the endothelial surface glycocalyx as much as by endothelial cells themselves.[111] This highly anionic proteoglycan matrix is located on the luminal surface of all vascular endothelial beds and anchored in the plasma membrane of the endothelial cells. The glycocalyx has proved difficult to visualize using conventional histopathological techniques. It forms an electrostatic barrier repelling negatively charged plasma proteins away from the endothelial cell surface and effectively restricts access to underlying cellular transport mechanisms. Preliminary evidence from children with dengue shock syndrome indicates that molecular size and charge characteristics determine which molecules are preferentially lost from the circulation, and suggests that a transient disturbance in the function of the endothelial glycocalyx layer may occur during dengue infections.[107] Endothelial cells themselves are dislodged into circulation.[112] How is the glycocalyx layer damaged? Possibilities include dengue virus, one of the dengue nonstructural proteins, or one of the components of the immune response may act directly with the glycocalyx layer to alter temporarily the characteristics of the fiber matrix. Heparan sulfate, an important constituent of the structure to which dengue virus can adhere, may be involved in this process.[113] In the few studies of dengue-infected patients in which endothelial architecture has been examined, the absence of detectable abnormalities at the cellular level or evidence of viral infection, despite profound disturbances in permeability, supports the idea that the pathological process results from infections that occur elsewhere.[37,114] New techniques to visualize the surface glycocalyx layer are becoming available and may prove interesting in this context.[111]

Extensive cell death or damage does not appear to be responsible for the increase in permeability.[37,114] Indeed, recovering DHF patients regain normal endothelial function relatively quickly, implying a reversible effect.[108,115] It appears that plasma leakage is a result of transient endothelial permeability caused by one or more soluble mediators released by the endothelium or by immune cells. TNF-alpha and products of complement activation showed transient elevations around the time of defervescence, when plasma leakage is most evident.[68] Treatment of endothelial cells with TNF-alpha or culture supernatants from dengue-infected macrophages resulted in an increase in cell permeability in the absence of infection.[66,115,116]

Numerous cytokines and mediators have been shown, or suggested, to be able to induce endothelial permeability, including IL-1-alpha and IL-1-beta,[117] IL-2,[118] IL-6,[119] IL-8,[120] TNF-alpha,[66,115] IFN-gamma,[121] histamine, platelet-activating factor[122] and vascular endothelial growth factor (VEGF).[123,124] Many of these factors have been shown to be elevated during dengue infection or released by cells infected by dengue viruses.[19,55,125–127] In addition, levels of histamine have been shown to be elevated in DHF patients[128,129] while dengue-infected PBMC's or *ex vivo* PBMC's from dengue patients produced PAF.[130] VEGF in the serum of DHF patients were significantly higher than those in the serum of DF patients or health controls.[131]

Thrombocytopenia

Thrombocytopenia is almost invariable in patients with dengue, with several mechanisms thought to be involved. Early bone marrow suppression, combined with increased peripheral destruction of platelets during the febrile and early convalescent phase of the disease, can lead to profound thrombocytopenia, with platelet nadirs as low as 5000/mm^3 recorded in some cases.[132,133] A pronounced pancytopenic suppression of bone marrow has been observed during the early febrile phase of dengue infection.[134,135] Toward the end of the febrile period the bone marrow recovers, leaving a residual megakaryocyte arrest observed at autopsy.[136] It has been suggested that dengue produces transient suppression of hematopoesis via direct infection,[137] or that infection of cells in the mononuclear phagocyte lineage produces macrophage inflammatory protein 1-alpha.[138] In the lymphocytic choriomeningitis mouse model, IFN-alpha produces a profound depression of all elements of the bone marrow;[139] IFN-alpha is produced early in dengue infection.[140] It has not yet been shown that dengue virus grows in megakaryocytes, *in vivo*.[37] Alternatively, it has been demonstrated that dengue virus binds to platelets in the presence of virus-specific antibody, suggesting immune-mediated clearance.[141] A considerable literature exists describing the antiplatelet activity of antibodies raised to dengue-2 NS1 protein.[142] Pros and cons in relation to a role in dengue pathogenesis for heterophilic anti-NS1 antibodies are discussed in greater detail in Chap. 14.

There is new evidence that dengue viruses replicate in human platelets, *in vivo* (Perng, personal communication). However, during the recovery period platelet numbers rise promptly as production increases in the now-hypercellular marrow. In the absence of significant bleeding (i.e. bleeding sufficient to warrant consideration of blood transfusion), *prophylactic platelet transfusions do not improve outcome* but offer a very definite risk of both acute and long-term complications[143] (S. Kalayanarooj, personal communication).

Coagulopathy

The coagulopathy associated with dengue infections is well described, but unfortunately the underlying mechanisms still remain unclear, although liver damage is likely a major factor. Severe bleeding occurs only rarely in children (almost invariably in association with profound shock), and thrombotic complications are not seen. An increase in APTT and a reduction in fibrinogen levels are fairly consistent findings, together with thrombocytopenia, and these abnormalities tend to correlate with overall severity.[144,145] However, the evidence for classical disseminated intravascular coagulation (DIC) in most cases is not convincing.[145,146] Concentrations of procoagulant markers are elevated to some degree (usually mild), with significant reductions in anticoagulant protein concentrations, but the findings with respect to the fibrinolytic pathway are conflicting. In general the evidence points toward enhanced fibrinolytic activity, and this could reflect a direct interaction between the virus and plasminogen, one of the key proteins in this pathway; several groups have noted the presence of plasminogen cross-reactive antibodies during and after dengue infection (see discussion of "autoimmune" pathogenesis mechanisms accompanying dengue infections in Chap. 14). Release of heparan sulfate or chondroitin sulfate (molecules similar in structure to heparin which can mimic its function as an anticoagulant) from the glycocalyx may also contribute to the overall picture.[107] In most patients the coagulopathy is relatively minor and resolves within a few days as the virus is cleared. However, in some children, usually those with severe shock, these minor derangements are compounded by the effects of prolonged hypotension and tissue hypoxia, and major hemorrhage occurs, often from the gastrointestinal tract. It is probable that in some of these patients true DIC develops. As yet little information is available regarding the coagulopathy in adults but bleeding tends to be more prominent in this group, sometimes in the absence of shock. Systematic investigation of hemostatic mechanisms of dengue disease in adults may prove valuable.

Shock

Important changes in blood pressure occur as dengue shock syndrome worsens, leading to increased peripheral vascular resistance with decreased cardiac output and normal or low central venous pressure.[147] Shock is not due to congestive heart failure, but to venous pooling. With increasing cardiovascular compromise, the diastolic pressure rises toward the systolic and the pulse pressure narrows. Finally, there is decompensation and both pressures disappear abruptly.

Models of Dengue Pathophysiology

Altered secondary antibody responses

Serotype-specific as well as -cross-reactive T cells have been detected in PBMC's of individuals with acute dengue virus infection. Studies of CD4 and CD8 T cell responses after primary dengue virus infection showed that CD4 T cells made greater amounts of IFN-gamma following stimulations with homotypic virus while the production of TNF-alpha was higher than that of IFN-gamma stimulation by heterotypic virus or CD4 epitopes.[148] CD8 T cells have been found to exhibit partial agonist responses. A heterologous dengue virus sensitized target cells to lysis but did not result in the production of cytokines.[75] It has been hypothesized that serotype-cross-reactive T cells raised to the first infection dengue serotype are of low avidity and predominate during a secondary dengue infection — a phenomenon termed "original antigenic sin." Arguments for and against a special role for heterologous T cell responses as a cause of expanded infected cell populations and exaggerated host responses are found in Chap. 14.

Accelerated T cell responses

While acknowledging a role for increased antigenic mass early in dengue infections, many workers emphasize as being critical to disease causation the acceleration and enhancement of the secondary T cell response. Because nearly all dengue immunological studies have been performed on children or adults, secondary dengue infections vastly predominate. In T cell responses a high percentage of CD8+ T cells expressing the early activation marker CD 69 was noted. The percentage of CD 69+ cells among CD8+ T cells was significantly higher in individuals with DHF than in those with dengue fever. The percentage of CD69+ cells was highest at the earliest time point studied during the febrile period, indicating that T cell activation preceded plasma leakage.[73] Complementary B cell pathogenic roles have been proposed:

(1) Secondary infection of an individual who has sub- or nonneutralizing antibody titers against DENV leads to a booster antibody response and a steep rise in antibody levels.
(2) Antibodies against DENV bind to and direct a selective attack of the complement system onto cells expressing viral antigens on their surface. DHF and DSS are the direct and indirect consequences of complement activation on these cells."[149–153]

Enhanced infections

The initial concept of ADE was that both the number and the rate of infection of FcR-bearing cells by dengue-virus-enhancing antibody complexes were increased, resulting in an expanded cell mass and a correspondingly severe infection.[154] The enhanced mass of dengue-infected cells would be expected to elicit a comparable immune response of both T and B cell arms. This enhanced immune response and correspondingly high cytokine levels are predicted to be observed during primary as well as secondary enhanced infections.[155] Many of these predictions have been satisfied. New data reveal that enhancing dengue antibodies play a much more significant role in directing disease responses. This role was first illustrated when mouse macrophages were infected by Ross River virus complexed with enhancing antibodies (ADE infection), and a 25-fold increase in infectivity in 12 hours was observed.[156] This infection suppressed the activation of IFN transcription factors, STAT-1 and NF-κB complexes. This observation was confirmed for dengue using a human macrophage cell line, THP-1. Dengue 2 ADE infection increased virion production 10-fold in 24 hours compared with cells infected with virus only.[157] Suppression of STAT-1 phosphorylation and IRF-1 gene expression and increased production of IL-10 were observed. Mature human dendritic cells infected by dengue 2 ADE showed increased protein synthesis, increased viral RNA production and increased release of virions compared with cells infected with virus only.[158] These cells released increased quantities of TNF-alpha and IL-6 compared with cells infected with DENV2 only.

These studies mirror findings in DHF cases. In 14 adult Vietnamese with DHF, the abundance of type I interferon gene transcripts in PBMC's from 6 DSS patients was lower than in PBMC's from 8 nonshock cases.[159] In DF and DHF cases in children, differences in the gene expression profile in PBMC's were noted during the acute phase of dengue infection.[160] cDNA array screening demonstrated that from DF patients the genes most strongly up-regulated were interferon-inducible or interferon-induced genes. These accounted for 47% of the altered genes, suggesting a significant role for IFN during dengue fever, confirmed by the robust production of IFN in DF PBMC's by ELISA. Type I interferon levels in the DF plasma were significantly higher than in the DHF plasma. The abundance of interferon and interferon-inducible genes in DF patients correlated with lower levels of peak viremia found in DF compared with DHF patients. In patients with DF there was an up-regulation of IFN genes and the complement inhibitor, C-59, while in DHF there was reduced IFN production together with strong T and B cell activation, cytokine production, complement activation and T cell apoptosis. As reviewed in Chap. 8, type II interferons are produced during dengue infections. In addition to a role in reducing and containing cellular dengue infection, IFN-gamma treatment of Fc-R-bearing cells increases Fc-R's and, as a result, measurably increases ADE infection.

A working hypothesis of dengue pathogenesis consistent with evidence is that severe disease in infants with primary infections and in older individuals with secondary infections is the result of antibody-enhanced infection of mononuclear phagocytes (Fig. 2). Immune complex infection suppresses cellular innate immune responses, increasing intracellular infection and generating inflammatory cytokines and chemokines that collectively result in enhanced disease. Cellular infection might also be augmented during the course of infection by the ability of T-lymphocyte-released

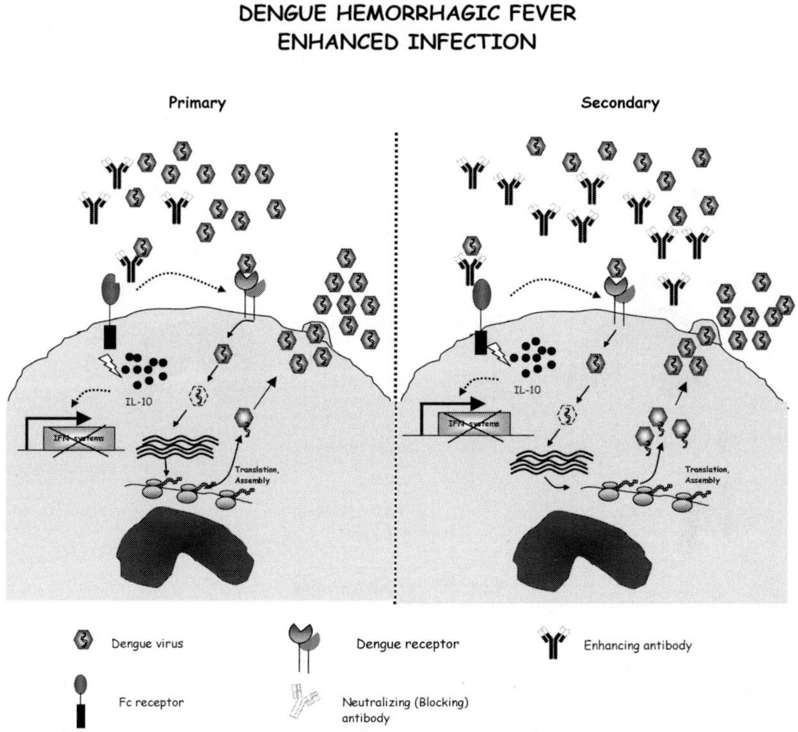

Fig. 2. Schema describing the fundamental cellular mechanism underlying antibody-dependent enhancement of dengue infection and disease. Antibodies — whether passively or actively acquired — at the appropriate concentration to combine with dengue viruses and trigger the suppression of intracellular innate immunity result in a cascade of effectors that lead to the production and release of a supernormal population of dengue viruses. Those antibodies capable of neutralization when circulating at appropriate concentrations either block the formation of enhancing immune complexes or block dengue infection. The result may be no infection or infection of moderate severity. *Figure design courtesy of Dr. Mary Marovich, Chief, Department of Vaccine Research and Development, Walter Reed Army Institute of Research, Rockville, MD.*

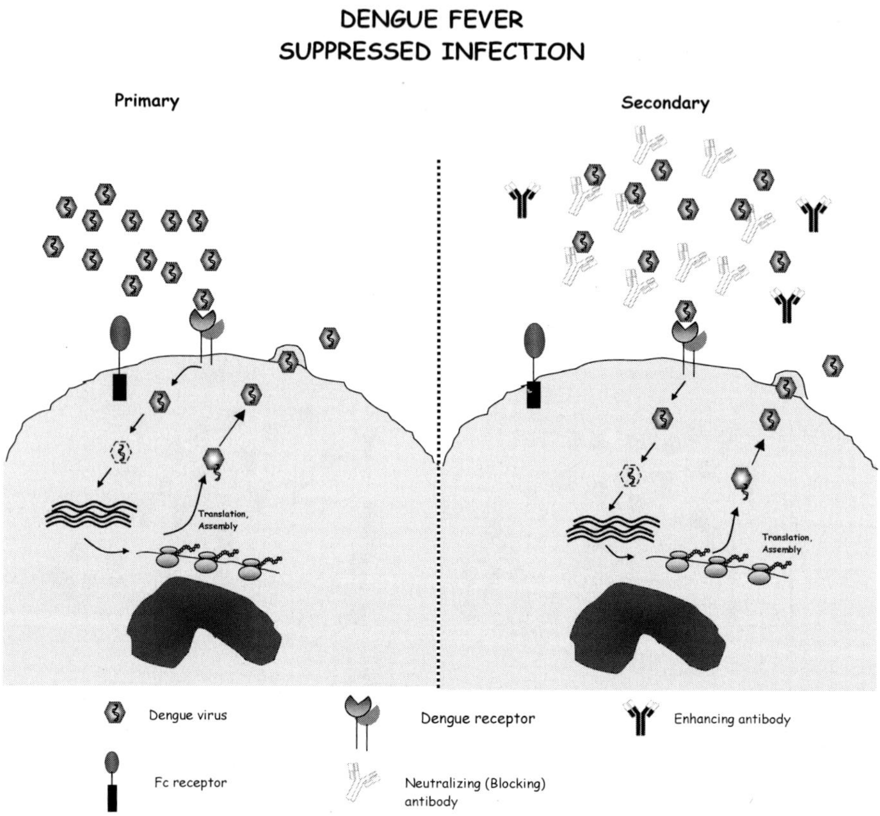

Fig. 2. (*Continued*)

IFN-gamma to increase numbers of Fc receptors on monocytes and macrophages. Liver infection and a pathogenic role for NS1 add to the complexity. In patients with dengue fever, type I interferon production and activated NK cells contribute to limiting disease severity.

References

1. Durbin AP, Karron RA, Sun W, Vaughn DW, Reynolds MJ, Perreault JR *et al.* Attenuation and immunogenicity in humans of a live dengue virus type-4 vaccine candidate with a 30 nucleotide deletion in its 3′-untranslated region. *Am J Trop Med Hyg* 2001;65:405–413.

2. Sangkawibha N, Rojanasuphot S, Ahandrik S, Viriyapongse S, Jatanasen S, Salitul V *et al.* Risk factors in dengue shock syndrome: a prospective epidemiologic study in Rayong, Thailand. I. The 1980 outbreak. *Am J Epidemiol* 1984;120:653–669.
3. Burke DS, Nisalak A, Johnson DE, Scott RM. A prospective study of dengue infections in Bangkok. *Am J Trop Med Hyg* 1988;38(1):172–180.
4. Thein S, Aung MM, Shwe TN, Aye M, Zaw A, Aye K *et al.* Risk factors in dengue shock syndrome. *Am J Trop Med Hyg* 1997;56(5):566–572.
5. Graham RR, Juffrie M, Tan R, Hayes CG, Laksono I, Ma'roef C *et al.* A prospective seroepidemiologic study on dengue in children four to nine years of age in Yogyakarta, Indonesia I. Studies in 1995–1996. *Am J Trop Med Hyg* 1999;61(3):412–419.
6. Halstead SB, Lan NT, Myint TT, Shwe TN, Nisalak A, Soegijanto S *et al.* Infant dengue hemorrhagic fever: research opportunities ignored. *Emerg Infect Dis* 2002;12:1474–1479.
7. Hung NT, Lei HY, Lan NT, Lin YS, Huang KJ, Lien LB *et al.* Dengue hemorrhagic fever in infants: a study of clinical and cytokine profiles. *J Infect Dis* 2004;189:221–232.
8. Kliks SC, Nimmannitya S, Nisalak A, Burke DS. Evidence that maternal dengue antibodies are important in the development of dengue hemorrhagic fever in infants. *Am J Trop Med Hyg* 1988;38(2):411–419.
9. Halstead SB. Immunological parameters of togavirus disease syndromes. In: Schlesinger RW (ed.) *The Togaviruses: Biology, Structure, Replication.* Academic, New York, 1980, pp. 107–173.
10. Marchette NJ, Halstead SB, Falkler WA Jr, Stenhouse A, Nash D. Studies on the pathogenesis of dengue infection in monkeys. III. Sequential distribution of virus in primary and heterologous infections. *J Infect Dis* 1973;128(1):23–30.
11. Wang WK, Chao DY, Kao CL, Wu HC, Lin YC, Li CM *et al.* High levels of plasma dengue viral load during defervescence in patients with dengue hemorrhagic fever: implications for pathogenesis. *Virology* 2003;305:330–338.
12. Halstead SB, O'Rourke EJ. Dengue viruses and mononuclear phagocytes. I. Infection enhancement by non-neutralizing antibody. *J Exp Med* 1977;146: 201–217.
13. Kou Z, Quinn M, Chen H, Rodrigo WW, Rose RC, Schlesinger JJ *et al.* Monocytes, but not T or B cells, are the principal target cells for dengue

virus (DV) infection among human peripheral blood mononuclear cells. *J Med Virol* 2008;80(1):134–146.

14. Wu SJ, Grouard-Vogel G, Sun W, Mascola JR, Brachtel E, Putvatana R *et al.* Human skin Langerhans cells are targets of dengue virus infection [see comments]. *Nat Med* 2000;6(7):816–820.
15. Libraty DH, Pichyangkul S, Ajariyakhajorn C, Endy TP, Ennis FA. Human dendritic cells are activated by dengue virus infection: enhancement by gamma interferon and implications for disease pathogenesis. *J Virol* 2001;75(8):3501–3508.
16. Nakao S, Lai CJ, Young NS. Dengue virus, a flavivirus, propagates in human bone marrow progenitors and hematopoietic cell lines [see comments]. *Blood* 1989;74(4):1235–1240.
17. Kurane I, Janus J, Ennis FA. Dengue virus infection of human skin fibroblasts *in vitro* production of IFN-beta, IL-6 and GM-CSF. *Arch Virol* 1992;124:21–30.
18. Andrews BS, Theofilopolous AN, Peters CJ, Loskutoff DJ, Brandt WE, Dixon FJ. Replication of dengue and Junin virues in cultured rabbit and human endothelial cells. *Infect Immun* 1978;20:776–781.
19. Huang YH, Lei HY, Liu HS, Lin YS, Liu CC, Yeh TM. Dengue virus infects human endothelial cells and induces IL-6 and IL-8 production. *Am J Trop Med Hyg* 2000;63:71–75.
20. Lin YL, Liu CC, Lei HY, Yeh TM, Lin YS, Chen RM *et al.* Infection of five human liver cell lines by dengue-2 virus. *J Med Virol* 2000;60(4):425–431.
21. Marianneau P, Cardona A, Edelman L, Deubel V, Despres P. Dengue virus replication in human hepatoma cells activates NF-kappaB which in turn induces apoptotic cell death. *J Virol* 1997;71(4):3244–3249.
22. Nishiura H, Halstead SB. Natural history of dengue virus (DEN)-1 and DEN-4 infections: reanalysis of classical studies. *J Infect Dis* 2007;195: 1007–1013.
23. Taweechaisupapong S, Sriurairatana S, Angsubhakorn S, Yoksan S, Khin MM, Sahaphong S *et al.* Langerhans cell density and serological changes following intradermal immunisation of mice with dengue 2 virus. *J Med Microbiol* 1996;45:138–145.
24. Siler JF, Hall MW, Hitchens AP. Dengue: Its history, epidemiology, mechanism of transmission, etiology, clinical manifestations, immunity, and prevention. *Philippine J Sci* 1926;29:1–304.

25. Simmons JS, St. John JH, Reynolds FHK. Experimental studies of dengue. *Philippine J Sci* 1931;44(1–2):1–252.
26. Gubler DJ, Suharyono W, Lubis I, Eram S, Suliantisaroso J. Epidemic dengue hemorrhagic fever in rural Indonesia. I. Virological and epidemiological studies. *Am J Trop Med Hyg* 1979;28(4):701–710.
27. Vaughn DW, Green S, Kalayanarooj S, Innis BL, Nimmannitya S, Suntayakorn S *et al*. Dengue in the early febrile phase: viremia and antibody responses. *J Infect Dis* 1997;176:322–330.
28. Nimmannitya S, Halstead SB, Cohen S, Margiotta MR. Dengue and chikungunya virus infection in man in Thailand, 1962–1964. I. Observations on hospitalized patients with hemorrhagic fever. *Am J Trop Med Hyg* 1969;18(6):954–971.
29. Nisalak A, Halstead SB, Singharaj P, Udomsakdi S, Nye SW, Vinijchaikul K. Observations related to pathogenesis of dengue hemorrhagic fever. III. Virologic studies of fatal disease. *Yale J Biol Med* 1970;42:293–310.
30. Gubler DJ, Suharyono W, Tan R, Abidin M, Sie A. Viraemia in patients with naturally acquired dengue infection. *Bull World Health Organ* 1981;59(4):623–630.
31. Scott RM, Nisalak A, Cheamudon U, Seridhoranakul S, Nimmannitya S. Isolation of dengue viruses peripheral blood leukocytes of patients with hemorrhagic fever. *J Infect Dis* 1980;141(1):1–6.
32. King AD, Nisalak A, Kalayanrooj S, Myint KSA, Pattanapanyasat K, Nimmannitya S *et al*. B cells are the principal circulating mononuclear cells infected by dengue virus. *Southeast Asian J Trop Med Public Health* 1999;30(4):718–728.
33. Boonpucknavig S, Boonpucknavig V, Bhamarapravati N, Nimmannitya S. Immunofluorescence study of skin rash in patients with dengue hemorrhagic fever. *Arch Pathol Lab Med* 1979;103:463–466.
34. Rosen L, Drouet MT, Deubel V. Detection of dengue virus RNA by reverse transcription-polymerase chain reaction in the liver and lymphoid organs but not in the brain in fatal human infection. *Am J Trop Med Hyg* 1999;61(5):720–724.
35. Rosen L, Khin MM, U T. Recovery of virus from the liver of children with fatal dengue: reflections on the pathogenesis of the disease and its possible analogy with that of yellow fever. *Res Virol* 1989;140:351–360.

36. Huerre MR, Lan NT, Marianneau P, Hue NB, Khun H, Hung NT *et al.* Liver histopathology and biological correlates in five cases of fatal dengue fever in Vietnamese children. *Virchows Arch* 2001;438(2):107–115.
37. Jessie K, Fong MY, Devi S, Lam SK, Wong KT. Localization of dengue virus in naturally infected human tissues, by immunohistochemistry and *in situ* hybridization. *J Infect Dis* 2004;189:1411–1418.
38. Hall WC, Crowell TP, Watts DM, Barros VL, Kruger H, Pinheiro F *et al.* Demonstration of yellow fever and dengue antigens in formalin- fixed paraffin-embedded human liver by immunohistochemical analysis. *Am J Trop Med Hyg* 1991;45(4):408–417.
39. Innis BL, Myint KSA, Nisalak A, Ishak KG, Nimmannitya S, Laohapand T *et al.* Acute liver failure is one important cause of fatal dengue infection. *Southeast Asian J Trop Med Public Health* 1990;21:695–696.
40. Couvelard A, Marianneau P, Bedel C, Drouet MT, Vachon F, Henin D *et al.* Report of a fatal case of dengue infection with hepatitis: demonstration of dengue antigens in hepatocytes and liver apoptosis. *Hum Pathol* 1999;30(9):1106–1110.
41. Blackley S, Kou Z, Chen H, Quinn M, Rose RC, Schlesinger JJ *et al.* Primary human splenic macrophages, but not T or B cells, are the principal target cells for dengue virus infection *in vitro*. *J Virol* 2007;81(24): 13325–13334.
42. Vaughn DW, Green S, Kalayanarooj S, Innis BL, Nimmannitya S, Suntayakorn S *et al.* Dengue viremia titer, antibody response pattern, and virus serotype correlate with disease severity. *J Infect Dis* 2000;181(1):2–9.
43. Guzman MG, Kouri G, Valdes L, Bravo J, Alvarez M, Vazquez S *et al.* Epidemiologic studies on dengue in Santiago de Cuba, 1997. *Am J Epidemiol* 2000;152(9):793–799.
44. Vaughn DW. Invited commentary: dengue lessons from Cuba. *Am J Epidemiol* 2000;152(9):800–803.
45. Watts DM, Porter KR, Putvatana P, Vasquez B, Calampa C, Hayes CG *et al.* Failure of secondary infection with American genotype dengue 2 to cause dengue haemorrhagic fever [see comments]. *Lancet* 1999;354(9188): 1431–1434.
46. Leitmeyer KC, Vaughn DW, Watts DM, Salas R, Villalobos I, de Chacon *et al.* Dengue virus structural differences that correlate with pathogenesis. *J Virol* 1999;73(6):4738–4747.

47. Kochel TJ, Watts DM, Halstead SB, Hayes CG, Espinosa A, Felices V, Caceda R, Bautista T, Montoya Y, Douglas S, Russell KL. Effect of dengue-1 antibodies on American dengue-2 viral infection and dengue haemorrhagic fever. *Lancet* 2002;360:310–312.
48. Pryor MJ, Carr JM, Hocking H, Davidson AD, Li PX, Wright PJ. Replication of dengue virus type 2 in human monocyte-derived macrophages: comparisons of isolates and recombinant viruses with substitutions at amino acid 390 in the envelope glycoprotein. *Am J Trop Med Hyg* 2001;65:427–434.
49. Anderson JR, Rico-Hesse R. *Aedes aegypti* vectorial capacity is determined by the infecting genotype of dengue virus. *Am J Trop Med Hyg* 2006; 75:886–892.
50. Bosch I, Xhaja K, Estevez LR, G., Melichar H, Warke RV, Fornier MV *et al.* Increased production of IL-8 in primary human monocytes and in human epithelial and endothelial cell lines after dengue virus challenge. *J Virol* 2002;76:5588–5597.
51. Avirutnan P, Malasit P, Seliger B, Bhakdi S, Husmann M. Dengue virus infection of human endothelial cells leads to chemokine production, complement activation, and apoptosis. *J Immunol* 1998;161(11):6338–6346.
52. Spain-Santana TA, Marglin S, Ennis FA, Rothman AL. MIP-1 alpha and MIP-1 beta induction by dengue virus. *J Med Virol* 2001;65(2):324–330.
53. King CA, Marshall JS, Alshurafa H, Anderson R. Release of vasoactive cytokines by antibody-enhanced dengue virus infection of a human mast cell/basophil line. *J Virol* 2000;74(15):7146–7150.
54. Murgue B, Roche C, Chungue E, Deparis X. Prospective study of the duration and magnitude of viraemia in children hospitalised during the 1996–1997 dengue-2 outbreak in French Polynesia. *J Med Virol* 2000; 60(4):432–438.
55. Libraty DH, Endy TP, Houng HS, Green S, Kalayanarooj S, Suntayakorn S *et al.* Differing influences of virus burden and immune activation on disease severity in secondary dengue-3 virus infections. *J Infect Dis* 2002;185(9):1213–1221.
56. Libraty DH, Young PR, Pickering D, Endy TP, Kalayanarooj S, Green S *et al.* High circulating levels of the dengue virus nonstructural protein NS1 early in dengue illness correlate with the development of dengue hemorrhagic fever. *J Infect Dis* 2002;186:1165–1168.

57. Halstead SB, Shotwell H, Casals J. Studies on the pathogenesis of dengue infection in monkeys. II. Clinical laboratory responses to heterologous infection. *J Infect Dis* 1973;128(1):15–22.
58. Halstead SB. *In vivo* enhancement of dengue virus infection in rhesus monkeys by passively transferred antibody. *J Infect Dis* 1979;140(4):527–533.
59. Goncalvez AP, Engle RE, St Claire M, Purcell RH, Lai CJ. Monoclonal antibody-mediated enhancement of dengue virus infection *in vitro* and *in vivo* and strategies for prevention. *Proc Natl Acad Sci USA* 2007;104:9422–9427.
60. Rosen L. The use of toxorhynchites mosquitoes to detect and propagate dengue and other arboviruses. *Am J Trop Med Hyg* 1981;30(10):177–183.
61. Marianneau P, Flamand M, Deubel V, Despres P. Apoptotic cell death in response to dengue virus infection: the pathogenesis of dengue haemorrhagic fever revisited. *Clin Diagn Virol* 1998;10(2–3):113–119.
62. Marianneau P, Steffan AM, Royer C, Drouet MT, Jaeck D, Kirn A *et al.* Infection of primary cultures of human Kupffer cells by Dengue virus: no viral progeny synthesis, but cytokine production is evident. *J Virol* 1999;73(6):5201–5206.
63. Alcon-LePoder S, Drouet MT, Roux P, Frenkiel MP, Arborio M, Durand-Schneider AM *et al.* The secreted form of dengue virus nonstructural protein NS1 is endocytosed by hepatocytes and accumulates in late endosomes: implications for viral infectivity. *J Virol* 2005;79(17):11403–11411.
64. Chen HC, Lai SY, Sung JM, Lee SH, Lin YC, Wang WK *et al.* Lymphocyte activation and hepatic cellular infiltration in immunocompetent mice infected by dengue virus. *J Med Virol* 2004;73:419–431.
65. Boonpucknavig V, Bhamarapravati N, Boonpucknavig S, Futrakul P, Tanpaichtr S. Glomerular changes in dengue hemorrhagic fever. *Arch Pathol Lab Med* 1976;100:206–212.
66. Anderson R, Wang S, Osiowy C, Issekutz AC. Activation of endothelial cells via antibody-enhanced dengue virus infection of peripheral blood monocytes. *J Virol* 1997;71(6):4226–4232.
67. Pang T, Cardosa MJ, Guzman MG. Of cascades and perfect storms: the immunopathogenesis of dengue haemorrhagic fever-dengue shock syndrome (DHF/DSS). *Immunol Cell Biol* 2007;85(1):43–45.
68. Green S, Vaughn DW, Kalayanarooj S, Nimmannitya S, Suntayakorn S, Nisalak A *et al.* Early immune activation in acute dengue illness is related

to development of plasma leakage and disease severity. *J Infect Dis* 1999;179(4):755–762.

69. Atrasheuskaya A, Petzelbauer P, Fredeking TM, Ignatyev G. Anti-TNF antibody treatment reduces mortality in experimental dengue virus infection. *FEMS Immunol Med Microbiol* 2003;35:33–42.
70. Fernandez-Mestre MT, Gendzekhadze K, Rivas-Vetencourt P, Layrisse Z. TNF-alpha 308A allele, a possible severity risk factor of hemorrhagic manifestatoin in dengue fever patients. *Tissue Antigens* 2004;64:468–472.
71. Kurane I, Innis BL, Nimmannitya S, Nisalak A, Meager A, Janus J *et al.* Activation of T lymphocytes in dengue virus infections: high levels of soluble interleukin 2 receptor, soluble CD4, soluble CD8, interleukin 2, and interferon-gamma in sera of children with dengue. *J Clin Invest* 1991;88:1473–1480.
72. Bethell DB, Flobbe K, Cao XT, Day NP, Pham TP, Buurman WA *et al.* Pathophysiologic and prognostic role of cytokines in dengue hemorrhagic fever. *J Infect Dis* 1998;177(3):778–782.
73. Green S, Pichyangkul S, Vaughn DW, Kalayanarooj S, Nimmannitya S, Nisalak A *et al.* Early CD69 expression on peripheral blood lymphocytes from children with dengue hemorrhagic fever. *J Infect Dis* 1999;180(5): 1429–1435.
74. Gagnon SJ, Leporati A, Green S, Kalayanarooj S, Vaughn DW, Stephens HA *et al.* T cell receptor Vbeta gene usage in Thai children with dengue virus infection. *Am J Trop Med Hyg* 2001;64(1–2):41–48.
75. Zivny J, DeFronzo M, Jarry W, Jameson J, Cruz J, Ennis FA *et al.* Partial agonist effect influences the CTL response to a heterologous dengue virus serotype. *J Immunol* 1999;163(5):2754–2760.
76. Mathew A, Kurane I, Green S, Stephens HA, Vaughn DW, Kalayanarooj S *et al.* Predominance of HLA-restricted cytotoxic T-lymphocyte responses to serotype-cross-reactive epitopes on nonstructural proteins following natural secondary dengue virus infection. *J Virol* 1998;72(5):3999–4004.
77. Zivna I, Green S, Vaughn DW, Kalayanarooj K, Stephens HAF, Chandanayingyong D *et al.* T cell responses to an HLA-B*07-restricted epitope on the dengue NS3 protein correlate with disease severity. *J Immunol* 2002;168:5959–5965.
78. An J, Zhou DS, Zhang JL, Morida H, Wang JL, Yasui K. Dengue-specific CD8+ T cells have both protective and pathogenic roles in dengue virus infection. *Immunol Lett* 2004;95:167–174.

79. Gagnon SJ, Ennis FA, Rothman AL. Bystander target cell lysis and cytokine production by dengue virus- specific human CD4(+) cytotoxic T-lymphocyte clones. *J Virol* 1999;73(5):3623–3629.
80. Mongkolsapaya J, Dejnirattisai W, Xu XN. Original antigenic sin and apoptosis in the pathogenesis of dengue hemorrhagic fever. *Nat Med* 2003;9:921–927.
81. Loke H, Bethell DB, Phuong CX, Dung M, Schneider J, White NJ *et al.* Strong HLA class I-restricted T cell responses in dengue hemorrhagic fever: a double-edged sword? *J Infect Dis* 2001;184(11): 1369–1373.
82. Stephens HA, Klaythong R, Sirikong M, Vaughn DW, Green S, Kalayanarooj K *et al.* HLA-A and -B allele associations with secondary dengue virus infections correlate with disease severity and infecting viral serotype in ethnic Thais. *Tissue Antigens* 2002;60:309–318.
83. Loke H, Bethell D, Phuong CX, Day NP, White NJ, Farrar J *et al.* Susceptibility to dengue hemorrhagic fever in Vietnam: evidence of an association with variation in the vitamin D receptor and Fc gamma receptor IIa genes. *Am J Trop Med Hyg* 2002;67:102–106.
84. Sakuntabhai A, Turbpaiboon C, Casademont I, Chuansumrit A, Lowhnoo T, Kajaste-Rudnitski A *et al.* A variant in the CD209 promoter is associated with severity of dengue disease. *Nat Genet* 2005;37(5):507–513.
85. Bokisch VA, Top FH Jr, Russell PK, Dixon FJ, Muller-Eberhard HJ. The potential pathogenic role of complement in dengue hemorrhagic shock syndrome. *N Engl J Med* 1973;289:996–1000.
86. Halstead SB. Antibody, macrophages, dengue virus infection, shock, and hemorrhage: a pathogenetic cascade. *Rev Infect Dis* 1989;11(Suppl 4): S830–S839.
87. Halstead SB. Pathogenesis of dengue: challenges to molecular biology. *Science* 1988;239:476–481.
88. Sobel AT, Bokisch VA, Muller-Eberhard HJ. C1q deviation tests for the detection of immune complexes, aggregates of IgG, and bacterial products in human serum. *J Exp Med* 1975;142:139–150.
89. Theofilopoulos AN, Wilson CB, Dixon FJ. The Raji cell radioimmune assay for detecting immune complexes in human sera. *J Clin Investig* 1976;67:169–182.

90. Ruangjirachuporn W, Boonpucknavig S, Nimmannitya S. Circulating immune complexes in serum from patients with dengue haemorrhagic fever. *Clin Exp Immunol* 1979;36:46–53.
91. Malasit P. Complement and dengue haemorrhagic fever/shock syndrome. *Southeast Asian J Trop Med Public Health* 1987;18(3):316–320.
92. Warke RV, Xhaja K, Martin KJ, Fournier MF, Shaw SK, Brizuela N *et al.* Dengue virus induces novel changes in gene expression of human umbilical vein endothelial cells. *J Virol* 2003;77:11822–11832.
93. Young PR, Hilditch PA, Bletchly C, Halloran W. An antigen capture enzyme-linked immunosorbent assay reveals high levels of the dengue virus protein NS1 in the sera of infected patients. *J Clin Microbiol* 2000;38(3):1053–1057.
94. Alcon S, Talarmin A, Debruyne M, Falconar A, Deubel V, Flamand M. Enzyme-linked immunosorbent assay reveals high levels of the dengue virus protein NS1 in the sera of infected patients. *J Clin Microbiol* 2002;40:376–381.
95. Mathew A, Kurane I, Green S, Vaughn DW, Kalayanarooj S, Suntayakorn S *et al.* Impaired T cell proliferation in acute dengue infection. *J Immunol* 1999;162(9):5607–5615.
96. Rouse BT, Horohov DW. Immunosuppression in viral infections. *Rev Infect Dis* 1986;8:850–873.
97. Hirsch RL, Griffen DE, Johnson RT. Cellular immune responses during complicated and uncomplicated measles virus infections of man. *Clin Immunol Immunopathol* 1984;31:1–12.
98. Carney WP, Hirsch MS. Mechanisms of immunosuppression in cytomegalovirus mononucleosis. II. Virus–monocyte interactions. *J Infect Dis* 1981;144:47–54.
99. Borrow P, Oldstone MBA. Measles virus–mononuclear cell interactions. *Curr Top Microbiol Immunol* 1995;191:85–100.
100. Welsh RM, McNally JM. Immune deficiency, immune silencing and clonal exhaustion of T cell reponses during viral infections. *Curr Opin Microbiol* 1999;2:382–387.
101. Myint KSA, Mongkolsirichaikul D, Manomuth C, Kalayanarooj S, Vaughn DW, Nisalak A *et al.* Apoptosis of peripheral blood mononuclear cells (PBMCS) in children with acute dengue. In: *Annual Meeting of the*

American Society of Tropical Medicine and Hygiene, 28 November–2 December 1999, Washington, D.C, p. 211.

102. Despres P, Flamand M, Ceccaldi PE, Deubel V. Human isolates of dengue type 1 virus induce apoptosis in mouse neuroblastoma cells. *J Virol* 1996; 70:4090–4096.
103. Despres P, Frenkiel MP, Ceccaldi PE, Duarte Dos Santos C, Deubel V. Apoptosis in the mouse central nervous system in response to infection with mouse-neurovirulent dengue viruses. *J Virol* 1998;72(1):823–829.
104. Marianneau P, Flamand M, Courageot MP, Deubel V, Despres P. [Apoptotic cell death in response to dengue virus infection: what are the consequences of viral pathogenesis?]. *Ann Biol Clin (Paris)* 1998;56(4):395–405.
105. Marianneau P, Flamand M, Deubel V, Despres P. Apoptotic cell death in response to dengue virus infection: the pathogenesis of dengue haemorrhagic fever revisited [In Process Citation]. *Clin Diagn Virol* 1998;10(2–3):113–119.
106. Thongtan T, Panyim S, Smith DR. Apoptosis in dengue virus infected liver cell lines HepG2 and Hep3B. *J Med Virol* 2004;72(3):436–444.
107. Wills BA, Oragui EE, Dung NM, Loan HT, Chau NV, Farrar JJ *et al.* Size and charge characteristics of the protein leak in dengue shock syndrome. *J Infect Dis* 2004;190:810–818.
108. Bethell DB, Gamble J, Pham PL, Nguyen MD, Tran TH, Ha TH *et al.* Noninvasive measurement of microvascular leakage in patients with dengue hemorrhagic fever. *Clin Infect Dis* 2001;32(2):243–253.
109. Kalayanarooj S, Vaughn DW, Nimmannitya S, Green S, Suntayakorn S, Kunentrasai N *et al.* Early clinical and laboratory indicators of acute dengue illness. *J Infect Dis* 1997;176(2):313–321.
110. Green S, Vaughn DW, Kalayanarooj S, Nimmannitya S, Suntayakorn S, Nisalak A *et al.* Elevated plasma interleukin-10 levels in acute dengue correlate with disease severity. *J Med Virol* 1999;59(3):329–334.
111. Pries AR, Kuebler WM. Normal endothelium. *Handb Exp Pharmacol* 2006;176(Pt 1):1–40.
112. Cardier JE, Rivas B, Romano E, Rothman AL, Perez-Perez C, Ochoa M *et al.* Evidence of vascular damage in dengue disease: demonstration of high levels of soluble cell adhesion molecules and circulating endothelial cells. *Endothelium* 2006;13(5):335–340.

113. Rehm M, Zahler S, Lotsch M, Welsch U, Conzen P, Jacob M *et al.* Endothelial glycocalyx as an additional barrier determining extravasation of 6% hydroyethyl starch or 5% albumin solutions in the coronary vascular bed. *Anesthesiology* 2004;100:1211–1223.
114. Sahaphong S, Riengrojpitak S, Bhamarapravati N, Chirachariyavej T. Electron microscopic study of the vascular endothelial cell in dengue hemorrhagic fever. *Southeast Asian J Trop Med Public Health* 1980;11: 194–204.
115. Jacobs M, Levin M. An improved endothelial barrier model to investigate dengue hemorrhagic fever. *J Virol Methods* 2002;104:173–185.
116. Carr JM, Hocking H, Bunting K, Wright PJ, Davidson AD, Gamble J *et al.* Supernatants from dengue virus type-2 infected macrophages induce permeability changes in endothelial cell monolayers. *J Med Virol* 2003;69:521–528.
117. Burke-Gaffney A, Keenan AK. Modulation of human endothelial cell permeability by combinations of the cytokines interleukin-1, tumor necrosis factor-alpha and interferon-gamma. *Immunopharmacology* 1993;25:1–9.
118. Rosenstein N, Ettinghausen SE, Rosenberg SA. Extravasation of intravascular fluid mediated by the systemic administration of recombinant interleukin 2. *J Immunol* 1986;137:1735–1742.
119. Maruo N, Morita I, Shirao M, Murota S. IL-6 increased endothelial permeability *in vitro*. *Endocrinology* 1992;131:710–714.
120. Talavera D, Castillo AM, Cominguez MC, Gutierrez AE, Meza I. IL8 release, tight junction and cytoskeleton dynamic reorganization conducive to permeability increase are induced by dengue virus infection of microvascular endothelial monolayers. *J Gen Virol* 2004;85:1801–1813.
121. Beynon HL, Haskard DO, Davies KA, Haroutunian R, Walport MJ. Combinations of low concentrations of cytokines and acute agonists synergize in increasing the permeability of endothelial monolayers. *Clin Exp Immunol* 1993;91:314–319.
122. Sirois MG, de Lima WT, de Brum Fernandes AI, Johnson RJ, Plante GE, Sirois P. Effect of PAF on rat lung vascular permeability, role of platelets and polymorphonuclear leucocytes. *Br J Pharmacol* 1994;111:1111–1116.
123. Murohara K, Horowitz JR, Silver M, Tsurumi Y, Chen D, Sullivan A *et al.* Vascular endothelial growth factor/vascular permeability factor enhances

vascular permeability via nitric oxide and prostcyclin. *Circulation* 1998;97:99–107.

124. Rollin S, Lernieux C, Maliba R, Favier J, Villeneuve LR, Allen BG *et al.* VEGF-mediated endothelial P-selection translocation role of BEGF receptors and endogenous PAF synthesis. *Blood* 2004;103: 3789–3797.
125. Hober D, Nguyen TL, Shen L, Ha DQ, Huong VT, Benyoucef S *et al.* Tumor necrosis factor alpha levels in plasma and whole-blood culture in dengue-infected patients: relationship between virus detection and pre-existing specific antibodies. *J Med Virol* 1998;54(3):210–218.
126. Juffrie M, van Der Meer GM, Hack CE. Inflammatory mediators in dengue virus infection in children: interleukin-8 and its relationship to neutraophil degranulation. *Infect Immun* 2000;68:702–707.
127. Chaturvedi UC, Elbishbishi EA, Agarwal R, Raghupathy R, Nagar R, Tandon R *et al.* Sequential production of cytokines by dengue virus-infected human peripheral blood leukocyte cultures. *J Med Virol* 1999; 59(3):335–340.
128. Tuchinda M, Dhorranintra B, Tuchinda P. Histamine content in 24-hour urine in patients with dengue haemorrhagic fever. *Southeast Asian J Trop Med Public Health* 1977;8(1):80–83.
129. Phan DT, Ha NT, Thuc LT, Diet NH, Phu LV, Ninh LY *et al.* Some changes in immunity and blood in relation to clinical states of dengue hemorrhagic fever patients in Vietnam. *Haematologia (Budap)* 1991;24:13–21.
130. Yang KD, Wang CL, Shaio MF. Production of cytokines and platelet activating factor in secondary dengue virus infections. *J Infect Dis* 1995;172:604–605.
131. Tseng CS, Lo HW, Teng HC, Lo WC, Ker CG. Elevated levels of plasma VEGF in patients with dengue hemorrhagic fever. *FEMS Immunol Med Microbiol* 2005;43(1):99–102.
132. Mitrakul C, Poshyachinda M, Futrakul P, Sangkawibha N, Ahandrik S. Hemostatic and platelet kinetic studies in dengue hemorrhagic fever. *Am J Trop Med Hyg* 1977;26(5):975–984.
133. Mitrakul C. Bleeding problem in dengue haemorrhagic fever: platelets and coagulation changes. *Southeast Asian J Trop Med Public Health* 1987;18: 407–412.

134. Nelson ER, Bierman HR, Chulajata R. Hematologic findings in the 1960 hemorrhagic fever epidemic (dengue) in Thailand. *Am J Trop Med Hyg* 1964;13:642–649.
135. Bierman HR, Nelson ER. Hematodepressive virus diseases of Thailand. *Ann Intern Med* 1965;62(5):867–883.
136. Bhamarapravati N, Tuchinda P, Boonyapaknavik V. Pathology of Thailand haemorrhagic fever: a study of 100 autopsy cases. *Ann Trop Med Parasitol* 1967;61:500–510.
137. La Russa VF, Innis BL. Mechanisms of dengue virus-induced bone marrow suppression. *Baillieres Clin Haematol* 1995;8:249–270.
138. Murgue B, Cassar O, Deparis X, Guigon M, Chungue E. Implication of macrophage inflammatory protein-1alpha in the inhibition of human haematopoietic progenitor growth by dengue virus. *J Gen Virol* 1998;79(Pt 8):1889–1893.
139. Binder D, Fehr J, Hengartner H, Zinkernagel RM. Virus-induced transient bone marrow aplasia: major role of interferon-alpha/beta during acute infection with the noncytopathic lymphocyteic choriomeningitis virus. *J Exp Med* 1997;185:517–530.
140. Kurane I, Innis BL, Nimmannitya S, Nisalak A, Meager A, Ennis FA. High levels of interferon-alpha in the sera of children with dengue virus infection. *Am J Trop Med Hyg* 1993;48:222–229.
141. Wang S, He RT, Patarapotikul J, Innis BL, Anderson R. Antibody-enhanced binding of dengue-2 virus to human platelets. *Virology* 1995;213:254–257.
142. Lin CF, Lei HY, Liu CC, Liu HS, Yeh TM, Wang ST *et al.* Generation of IgM anti-platelet autoantibody in dengue patients. *J Med Virol* 2001;63(2):143–149.
143. Lum CS, Goh YT, Chan WK, EL-Amin A-L, Lam SK. Risk factors for hemorrhage in severe dengue infections. *J Pediatr* 2002;140:629–631.
144. Krishnamurti C, Kalayanarooj S, Cutting MA, Peat RA, Rothwell SW, Reid TJ *et al.* Mechanisms of hemorrhage in dengue without circulatory collapse. *Am J Trop Med Hyg* 2001;65(6):840–847.
145. Wills BA, Oragui EE, Stephens AC, Daramola OA, Dung NM, Loan HT *et al.* Coagulation abnormalities in dengue hemorrhagic fever: serial investigations in 167 Vietnames children with dengue shock syndrome. *Clin Infect Dis* 2002;35:277–285.

146. Pongpanich B. Hemodynamic changes in shock associated with dengue haemorrhagic fever. *Southeast Asian J Trop Med Public Health* 1987;18:326–330.
147. Pongpanich B, Kumponpant S. Studies of dengue hemorrhagic fever. V. Hemodynamic studies of clinical shock associated with dengue hemorrhagic fever. *J Pediatr* 1973;83:1073–1077.
148. Mangada MM, Rothman AL. Altered cytokine responses of dengue-specific CD4+ T cells to heterologous serotypes. *J Immunol* 2005;175:2676–2683.
149. Rothman AL, Ennis FA. Immunopathogenesis of dengue hemorrhagic fever. *Virology* 1999;257(1):1–6.
150. Rothman AL. Immunology and immunopathogenesis of dengue disease. *Adv Virus Res* 2003;60:397–419.
151. Rothman A. Dengue:defining protective versus pathologic immunity. *J Clin Investig* 2004;113:946–951.
152. Green S, Rothman A. Immunopathological mechanisms in dengue and dengue hemorrhagic fever. *Curr Opin Infect Dis* 2006;19:429–436.
153. Fink J, Gu F, Vasudevan SG. Role of T cells, cytokines and antibody in dengue fever and dengue haemorrhagic fever. *Rev Med Virol* 2006;16:263–275.
154. Halstead SB, O'Rourke EJ. Antibody-enhanced dengue virus infection in primate leukocytes. *Nature* 1977;265(5596):739–741.
155. Halstead SB. The Alexander D. Langmuir Lecture: the pathogenesis of dengue — molecular epidemiology in infectious disease. *Am J Epidemiol* 1981;114(5):632–648.
156. Mahalingam S, Lidbury BA. Suppression of lipopolysaccharide-induced antiviral transcription factor (STAT-1 and NF-kB) complexes by antibody-dependent enhancement of macrophage infection by Ross River virus. *Proc Natl Acad Sci USA* 2002;99:13819–13824.
157. Chareonsirisuthigul T, Kalayanarooj S, Ubol S. Dengue virus (DENV) antibody-dependent enhancement of infection upregulates the production of anti-inflammatory cytokines, but suppresses anti-DENV free radical and pro-inflammatory cytokine production, in THP-1 cells. *J Gen Virol* 2007;88(Pt 2):365–375.
158. Boonnak K, Slike BM, Burgess TH, Mason RM, Wu SJ, Sun P *et al*. Role of dendritic cells in antibody-dependent enhancement of dengue virus infection. *J Virol* 2008;82(8):3939–3951.

159. Simmons CP, Popper S, Dolocek C, Chau TN, Griffiths M, Dung NT *et al.* Patterns of host genome-wide gene transcript abundance in the peripheral blood of patients with acute dengue hemorrhagic fever. *J Infect Dis* 2007;195(8):1097–1107.
160. Ubol S, Masrinoul P, Jaijarnwanich J, Kalayanarooj SM, Chareonsirisuthikul T, Kasisith J. Differences in PBMC-global gene expression indicate a significant role in the innate responses in DF but not DHF progression. *J Infect Dis* 2008;197(10):1459–1467.
161. Azeredo EL, Zagne SM, Santiago MA, Gouvea AS, Santana AA, Neves-Souza PC *et al.* Characterisation of lymphocyte response and cytokine patterns in patients with dengue fever. *Immunobiology* 2001;204: 494–507.
162. Raghupathy R, Chaturvedi UC, Al-Sayer H, Elbishbishi EA, Agarwal R, Nagar R *et al.* Elevated levels of IL-8 in dengue hemorrhagic fever. *J Med Virol* 1998;56(3):280–285.
163. Pasca WS, Agarwal R, Elbishbishi EA, Chaturvedi UC, Nagar R, Mustafa AS. Role of interleukin-12 in patients with dengue hemorrhagic fever. *FEMS Immunol Med Microbiol* 2000;28:151–155.
164. Srikiatkhachorn A, Ajanyakhajorn C, Endy TP. Virus indived decline in soluble endothelial growth receptor 2 is associated with plasma leakage in dengue hemorrhagic fever. *J Virol* 2007;81:1592–1600.
165. Chen LC, Lei HY, Liu CC, Shiesh SC, Chen SH, Liu HS *et al.* Correlation of serum levels of macrophage migration inhibitory factor with disease severity and clinical outcome in dengue patients. *Am J Trop Med Hyg* 2006;74(1):142–147.
166. Avirutnan P, Punyadee N, Noisakran S, Komoltri C, Thiemmeca S, Auethavornanan K *et al.* Vascular leakage in severe dengue virus infections: a potential role for the nonstructural viral protein NS1 and complement. *J Infect Dis* 2006;193:1078–1088.
167. Hober D, Poli L, Roblin B, Gestas P, Chungue E, Granic G *et al.* Serum levels of tumor necrosis factor-alpha (TNF-alpha), interleukin-6 (IL-6), and interleukin-1-beta (IL-1-beta) in dengue-infected patients. *Am J Trop Med Hyg* 1993;48:324–331.
168. Kuno G, Bailey RE. Cytokine responses to dengue infection among Puerto Rican patients. *Mem Inst Oswaldo Cruz (Rio de Janeiro)* 1994;89:179–182.

169. Mustafa AS, Elbishbishi EA, Agarwal R, Chaturvedi UC. Elevated levels of interleukin-13 and IL-18 in patients with dengue hemorrhagic fever. *FEMS Immunol Med Microbiol* 2001;30(3):229–233.
170. Lin YL, Liu CC, Chuang JL, Lei HY, Yeh TM, Lin YS *et al.* Involvement of oxidative stress, NF-IL-6 and RANTES expression in dengue-2-virus-infected human liver cells. *Virology* 2000;276:114–126.
171. Gagnon SJ, Mori M, Kurane I, Green S, Vaughn DW, Kalayanarooj S *et al.* Cytokine gene expression and protein production in peripheral blood mononuclear cells of children with acute dengue virus infections. *J Med Virol* 2002;67:41–46.

10

Diagnosis of Dengue Virus Infections

Timothy P. Endy, Ananda Nisalak and David W. Vaughn

Introduction

Dengue virus (DENV) is a group of four antigenically distinct serotypes (DENV-1, DENV-2, DENV-3, and DENV-4) sharing major antigens within the group and with other mosquito and tick-borne flaviviruses.[1,2] In countries where dengue is now or is becoming endemic, the cocirculation of two or more flaviviruses such as DENV and Japanese encephalitis (JE) virus in Asia; dengue, JE and West Nile in India; and dengue and yellow fever in the Americas complicates efforts to make a specific serologic diagnosis of acute DENV infection. From public health, clinical care and research perspectives, the accurate diagnosis of DENV infection is important today and likely to increase in importance in the future.

By the time of the first international meeting on dengue fever (DF) and dengue hemorrhagic fever (DHF) in 1964, these diseases were recognized as a serious public health problem in Southeast Asia. In 1956, when two new dengue viruses were identified, Hammon and Sather found it difficult to make serotype-specific diagnoses in patients with dengue illnesses because of the problem of antibody cross reactivity.[3] At the time, three antibody assays were in use: suckling mouse virus neutralization test,[4] complement fixation (CF)[5] and hemagglutination inhibition (HI).[6,7] In individuals who were experiencing their first DENV infection, HI antibodies could be detected using dengue viral antigens different from the virus-causing infection. Using the CF test, antibody responses were delayed (CF detects IgG antibodies only) but were quite specific to the infecting virus type. The neutralization test was and is the most specific of the three tests. But neutralization tests performed in mice were cumbersome and expensive. Over the ensuing 40 years and despite the introduction of newer serological tests, some of these same problems have remained.

This chapter reviews laboratory tests currently available for the diagnosis of DENV infection with appraisal of their strengths and weaknesses.

Antibody and Virus Patterns in Dengue Virus Infection

During the process of probing for capillaries, dengue-infected mosquitoes inject saliva containing viruses which attach to and replicate in skin dendritic cells.[8] From the time of inoculation, DENV replicates and disseminates to various lymphoid organs, producing measurable viremia beginning approximately 3–5 days later and lasting another 4–5 days (Fig. 1). The onset of fever and symptoms occurs approximately 24 hours after the onset of viremia and lasts a mean of 4.5 days, followed by defervescence.[9–14] The day of defervescence is an immunologically important

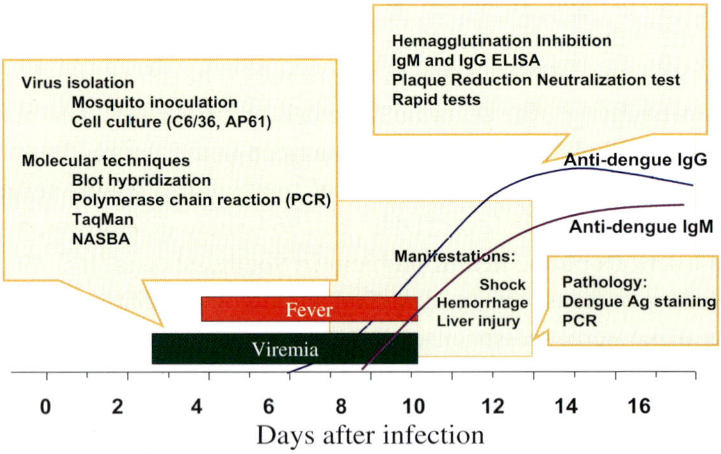

Fig. 1. Schema for selection of dengue diagnostic tests based upon the day after the mosquito bite during a secondary dengue infection.

landmark in the course of DENV disease. In DHF, defervescence coincides with clinically detectable plasma leakage.[15,16]

Primary dengue

Initial or primary DENV infections occur in individuals without previous exposure to a flavivirus. In patients experiencing a primary infection, anti-dengue antibodies, initially of the IgM class, evolve relatively slowly.[14] For at least three weeks, the molar ratio of IgM to IgG is high ($> 1.8{:}1.0$).[17] Dengue antibodies during a primary infection, whether detected by HI, CF, or neutralizing antibody assays, develop slowly to relatively low titers.[15,18]

Secondary dengue

A secondary dengue antibody response accompanies a DENV infection in an individual with prior exposure to one or more flaviviruses, whether by natural infection or by immunization. Homologous immunity (resistance to challenge with the same dengue virus causing an earlier infection) is lifelong. After a short period of cross-protection, infections with different dengue viruses are possible. Secondary responses are IgG antibodies which appear early, often during the febrile period, rise rapidly and are referred to as anamnestic (memory) responses.[14,17,19] There is also an IgM component to a secondary antibody response. The quantity of IgM is variable with an IgM to IgG ratio ≤ 1.8 as measured by IgM capture ELISA.[17] It is possible that the IgM dengue antibody response reflects the number of new epitopes presented by the infecting virus. For example, individuals immune to JE who then experience a DENV infection develop more IgM antibodies than do individuals experiencing a second DENV infection. The rate and amount of antibody produced during a second DENV infection are used to define serological responses to DENV infection.[15]

Viremia pattern

Definitive diagnosis of DENV infection requires isolation of the virus or detection of virus genome. Viremia correlates closely with fever, with peak levels (titers up to 10 logs/ml) occurring 2–3 days following the onset of fever, typically 2–3 days prior to defervescence.[20] During secondary dengue infections with DENV-1, DENV-2, and DENV-3, peak viremia titers correlate with ensuing disease severity.[20,21] The importance of the peak viremia titer beyond its potential as a prognostic sign is its importance as an estimate of total maximum achieved cellular infection. If standardized, the quantity of virus/antigen in acute phase blood might be used to predict immediate risk of severe disease (DHF). Viremia level as originally determined by Vaughn and colleagues was measured by making serial dilutions of viremic plasma from dengue patients and inoculating these into *Toxorhynchites splendens* mosquitoes, scored as negative or positive using fluorescence microscopy or RT-PCR, a time-consuming and tedious process. New molecular tools are available for measuring dengue viral load.[21,22] The use of NS-1 or RNA detection systems, which do not depend on identifying living virions, may extend the sensitivity of the technique.[23]

Diagnostic Pathway in Patients with Suspected Acute Dengue

Figure 1 illustrates strategies for diagnosing an acute DENV infection in patients with suspected dengue. Diagnostic options include assays that detect the presence of the virus (including virus antigen and virus genome) versus assays that detect the host's response to the virus (antibody). Virus detection includes recovery of live virus in animals, mosquitoes, or cell culture or the detection of viral antigens or nucleic acids. The former are slow and require facilities to maintain living hosts; the latter are rapid but require expensive equipment and reagents. Antibodies can be detected in a few minutes or after a week or more. With specimens collected at the appropriate time of illness, and attempting to detect virus/antigen and antibody, it should be possible to diagnose virtually every case of suspected dengue. Use of rigorous methods is required for dengue research, but such methods may not be appropriate at all levels in the health care chain. Ideally, every country with endemic dengue should maintain a center of diagnostic excellence. Ability to undertake the differential diagnosis of denguelike outbreaks, e.g. leptospirosis, is also important.

For suspected dengue patients, serum specimens for virus isolation should be collected early in the febrile phase and stored at –70 degrees Celsius. Specimens for serologic testing can be stored at –20 degrees Celsius. An acute phase serum specimen for serology should be collected prior to 7 days after onset of fever and paired with a convalescent serum drawn at least 7 days after acute serum, optimally 14–21 days after onset of fever.[15] A common practice in many dengue-hyperendemic countries is to draw blood upon hospital admission and then at discharge, usually 1–2 days after defervescence. Most of these patients will be experiencing secondary infections, and serum samples obtained at short intervals will nonetheless detect rapidly rising IgG antibody titers. For patients with primary infections, a post-illness (convalescent) specimen is needed to confirm or rule out DENV infection if virus isolation is not possible or results are negative.

For travelers returning to nonendemic countries with suspected dengue, conventional paired acute and convalescent samples, spaced at a 14-day interval, should be collected and sent to the appropriate level health department with a request to forward them to the national authorities for testing.

Serological Assays

Hemagglutination inhibition (HI) assay

Simplified methods of measuring antibodies were introduced by Sabin and colleagues who discovered the ability of DENV antigens to agglutinate certain types of erythrocytes.[4] Normal erythrocytes in a suspension settle to the bottom of the test tube or well and form a compact, dense button of red blood cells after 30 minutes to one hour. Agglutination of red cells using virus antigen prevents normal settling, producing instead a uniform lattice of cross-linked cells covering the lower part of the tube or well (hemagglutination). Casals and Brown in 1954 modified the technique using acetone and ether to extract (purify) the hemagglutination antigens (HA's) and developed the classical hemagglutination inhibition (HI) assay.[24] In this assay, reactive antibodies bind to the HA, preventing lattice formation and permitting red blood cells to form a button at the

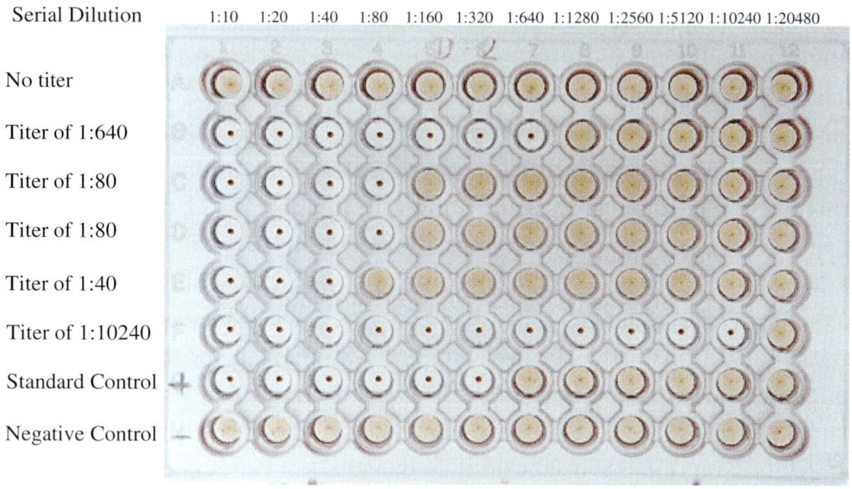

Fig. 2. Result of a hemagglutination inhibition test (dengue 2 virus antigen) on acute and convalescent sera from six patients with suspected dengue. Normal erythrocytes settle to the bottom of the well and form a compact, dense button of red blood cells after 30 minutes to 1 hour. Addition of virus antigen results in the agglutination of red cells (hemagglutination — a lattice-like pattern is seen in negative control wells). Dengue antibodies inhibit agglutination, permitting erythrocytes to settle into a tight red blood cell button. (Courtesy of Virology Department, Armed Forces Research Institute of Medical Sciences, Bangkok, Thailand.)

bottom of the well (Fig. 2). Antibody titers are determined by the highest dilution producing inhibition of HA. Porterfield and Clarke introduced goose red blood cells (RBC's) instead of human O cells, reducing the problem of nonspecific HA inhibitors.[25–27] The HI test was adapted to microtiter plates in 1980 and is still used in some laboratories to measure dengue antibodies in clinical or epidemiological studies.[15]

Agglutination of red cells is dependent upon pH and antigen concentration. Dengue antigens are commonly produced from a variety of sources, including suckling mouse brain and cell culture. Test methods in use at the Armed Forces Research Institute of Medical Sciences (AFRIMS) in Bangkok, Thailand, are as follows: 1–2-day-old suckling mice (*Mus musculus*) are inoculated intracranially, with 0.02 ml of a dengue serotype-specific virus suspension. Mice are observed twice a day for the first signs of encephalitis (failure to eat, as evidenced by lack of milk in the stomach, color change, wasting, or tremors), which occurs 3–10 days after inoculation. For DENV-1 this usually occurs at day 5, for DENV-2, day 4; for DENV-3, day 7; for DENV-4, day 4; and for Japanese encephalitis virus, day 3. Harvested brain is made into a 20% (weight/volume) suspension in an 8% sucrose solution and homogenized. The homogenate is then acetone-extracted, condensed, washed, and resuspended in sterile normal saline. All antigen is assayed for the number of HA units. Goose RBC's are obtained from adult female geese (*Anser cinereus*). Blood is diluted in Alsever's solution (glutaraldehyde-fixed goose RBC's may also be used[28]). RBC's are then washed with dextrose-gelatin-veronal solution and are brought up to a final 8% solution just prior to assay use. pH optima for each dengue serotype are as follows: pH of 6.2 for DENV-1; 6.4 for DENV-2; 6.6 for DENV-3; and 6.8 for DENV-4.

Nonspecific inhibitors must be removed as most sera contain inhibitors of RBC agglutination in the absence of antibody. These are removed using acetone, ether, kaolin, or diethylaminoethyl-sephadex (DEAE-Sephadex).[29,30] At AFRIMS, acetone extraction is the usual method and is performed by adding heat-inactivated serum to cold acetone and decanting the mixture. A drop of goose red cells is added to the sera, mixed, and removed to eliminate other nonspecific agglutinators.

The HI assay is performed in microtiter plates. Twofold serial dilutions of the test sera and standard positive and negative controls are made using microtiter equipment. All wells receive 0.025 ml containing 8–16 HA units of each dengue antigen and the plates are covered and incubated at 4°C overnight. To reduce the impact of test-to-test variability, all sera from a single patient should be tested in the same assay. The following morning the test plates are allowed to reach room temperature and 0.05 ml of an 8% goose red cell stock solution diluted 1:24 in the proper pH buffer is added to each well. The plate is allowed to sit undisturbed for one hour at room temperature and agglutination scored (Fig. 2). Criteria for negatives, positives, primary and secondary HI responses with respect to the time during illness that each serum specimen was obtained from patients follow criteria developed by the WHO (Table 1[15]).

The response to a primary DENV infection is characterized by the slow evolution of HI antibody. Because the HI assay does not identify the

Table 1. World Health Organization criteria for interpreting dengue hemagglutination inhibition assay results.[15]

Change in antibody titer	Sample interval	Antibody titer at time of convalescence	Interpretation
≥ to a 4-fold rise	Paired sera with ≥ 7 days of separation	≤ 1:1280	Acute primary flavivirus infection
≥ to a 4-fold rise	With any specimen	≥ 1:2560	Acute secondary flavivirus infection
≥ to a 4-fold rise	Paired sera with < 7 days of separation	≤ 1:1280	Acute flavivirus infection, indeterminate primary/ secondary
No change in titer	With any specimen	> 1:2560	Recent secondary flavivirus infection
No change in titer	Paired sera with ≥ 7 days of separation	≤ 1:1280	Not dengue
No change in titer	Paired sera with < 7 days of separation	≤ 1:1280	None
Uncertain	One seraum specimen	≤ 1:1280	None

immunoglobulin isotype, a primary antibody response is inferred from the absence of antibodies in sera obtained before five days after the onset of fever and a fourfold or greater antibody rise during convalescence. Conventionally, acute phase sera are obtained during the febrile period and convalescent sera 14 days later. By WHO criteria, if antibody titers rise at least fourfold in acute and convalescent sera are spaced at least 7 days apart, a dengue illness is documented. Secondary antibody responses are characterized by the rapid evolution of HI antibodies. All antibodies are broadly flavivirus-reactive, so that a specific diagnosis of DENV-1, -2, -3 or -4 (or JE) infection is not possible using this test alone. Tests suggesting a secondary DENV infection are acute and convalescent sera (or convalescent sera only) exhibiting a fourfold or greater rise in antibody titer to a titer exceeding 1:1280 or a high fixed titer exceeding 1:1280.[31] If the end titer is 1:1280 or less, the interpretation is an acute primary DENV infection. For many patients with a secondary DENV infection, a fourfold increase will be seen well before 7 days. If a single specimen or paired specimens show a titer of 1:2560 or higher without a fourfold increase, the interpretation is a recent DENV infection, possibly acute or having occurred during the previous couple of months.

The HI test, especially during secondary infections, fails to discriminate between infections with flaviviruses such as dengue, JE, and West Nile. More specific tests must be applied to diagnostic or epidemiological specimens in countries where dengue and other flaviviruses circulate. Despite these limitations, the HI assay is a powerful technique that is useful for the diagnosis of acute primary and secondary DENV infections. It is robust and can be performed with minimal laboratory equipment, reagents, and expense. Reagents can be produced locally for use in dengue diagnostic laboratories.

Plaque reduction neutralization test (PRNT)

In 1952, Dulbecco and colleagues described a chick embryo fibroblast monolayer plaque assay for several viruses, including Western equine encephalitis and Newcastle disease.[32] This technique allowed quantification of viruses *in vitro* as plaque-forming units (PFU's). In 1959, Henderson and Taylor described a method for detecting arboviral plaques in an agar overlay stained with neutral red and demonstrated its utility in measuring serum antibody neutralization titers.[33] The standard neutralization test prior to the availability of tissue culture was performed by the intracerebral inoculation of suckling or weanling mice with a constant dilution of serum mixed with log dilutions of virus. Results were expressed as the log neutralization index. A DENV neutralization assay using challenge virus resistance was developed in BS-C-1, PS, and LLC-MK2 cells by Halstead, Nisalak and Sukhavachana.[34,35] Finally, in 1967, a direct plaque assay was developed to measure DENV neutralizing antibody.[18] This assay, the dengue plaque reduction neutralization test (PRNT), used probit analysis to measure plaque reduction at a 50% endpoint ($PRNT_{50}$). This technique introduced an efficient and reproducible assay for measuring dengue antibodies that were often serotype-specific neutralizing antibodies and became the standard assay for measuring dengue antibodies. Variations of the technique were introduced, such as a micrometabolic inhibition test using BHK-21 cells and a microculture plaque reduction test utilizing the LLC-MK2 cell line[36]; microplate cultures using BHK-21 cells[37]; a focus reduction method using peroxidase–antiperoxidase staining of BHK-21 cells[38]; a screening test using a single dilution and a 70% plaque reduction endpoint[39]; and a simplified PRNT assay using BHK-21 cells.[40] Many of these assays are being used to determine serological responses to dengue vaccines,[41] in seroepidemiologic studies[42] and pathogenesis studies.[16]

In the PRNT, it is necessary to select a cell line capable of supporting plaque formation with each of the four dengue viruses, preferably with recent isolates. Experience has shown that LLC-MK2, Vero, and BHK21 cells are useful. The PRNT involves incubating a constant number of PFU's with serial dilutions of serum. The standard PRNT protocol at AFRIMS utilizes LLC-MK2 cells in six-well culture plates. Serial four-fold test serum dilutions and positive and negative serum controls are added to an equal volume of virus suspension diluted to contain 50 PFU's in the inoculum volume, placed on a shaker, and then incubated at 37°C for 1 hour. The mixtures are placed in an ice-bath and inoculated onto cells and allowed to absorb onto cells for 1 hour at room temperature. The first overlay medium is added, containing an agar–nutrient mixture, and incubated at 37°C for 7 days, after which a second overlay is added, containing neutral red, and incubated at 35°C overnight. Plaques are counted per well on a fluorescent light box. Figure 3 illustrates a plaque reduction test including an example of plaque overlap. Because of plaque overlap, if the input dose is too high, plaques that can be counted may be significantly below input PFU's, resulting in a reduction in the antibody titer. Titers of neutralizing antibody are most accurately measured at $PRNT_{50}$. Titers are determined using the log probit method. Other endpoints are in use for screening purposes. Use of higher percent plaque reduction endpoints will increase specificity but decrease sensitivity of the assay. No standards have been adopted for interpreting the $PRNT_{50}$ assay. Within one to two weeks following a primary DENV infection, an individual may raise IgM neutralizing antibodies fairly specific to the infecting type such that the highest antibody titer is to the homologous DENV. Within three months serum neutralizing antibodies are predominantly IgG. There is some evidence that titers of homologous neutralizing antibodies increase while heterotypic antibody titers decrease over long periods of time.[43]

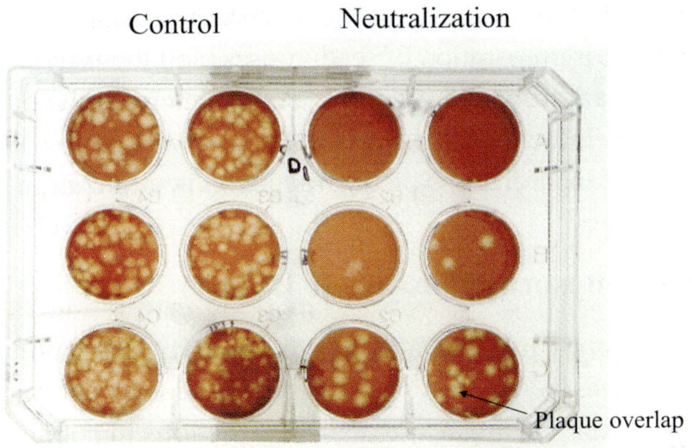

Fig. 3. Dengue type 1 plaque reduction neutralization test. Antibody reduces the number of plaques compared with those in control wells. The arrow illustrates a common source of error in interpreting data — two closely adjacent plaques are read as one plaque, a phenomenon called "plaque overlap." (Courtesy of Virology Department, Armed Forces Research Institute of Medical Sciences, Bangkok, Thailand.)

Whether this phenomenon can be simply explained by changes in antibody avidity is not known.

Two successive infections with different dengue viruses results in the rapid production of IgG neutralizing antibodies detected early, often several days before defervescence. Antibodies are heteroreactive, often to all four dengue viruses and to nondengue flaviviruses such as Japanese encephalitis. Immediately postinfection, the phenomenon of "original antigenic sin", is observed, in which antibodies with the highest titers are directed at the first infecting dengue virus. Gradually, with the passage of time, titers of cross-reactive antibodies wane, the highest titers being directed at viruses causing past infection. Neutralizing antibody response patterns vary unpredictably, possibly as the result of individual genetically determined responsiveness and in part due to differences in viral antigenicity. Lower and more specific antibody titers follow inapparent infections. In vaccine studies, a titer of 1:10 or higher is regarded as significant, suggesting protective immunity.

The PRNT itself is a complex *in vitro* bioassay that is also subject to a high degree of inter- and intra- assay variation. Common sources of variation are the cell line used for plaquing, seed virus preparation and passage history, cell culture used for preparation of seed viruses, and the use of complement. Results varied markedly, making it difficult to interpret the data and identify prior DENV infections.[44] Efforts are under way at the World Health Organization to standardize the plaque reduction neutralization assay.

Enzyme immunosorbent assays (EIA's)

Innis and colleagues in 1989 adapted the anti-JE IgM antibody capture EIA of Burke and colleagues[45] to dengue. They described a serological assay capable of diagnosing acute dengue infections in countries where JE and dengue viruses cocirculate.[17] Using IgM capture and IgG EIA ratios it is possible to distinguish primary from secondary dengue infections and from dengue and JE by including JE antigen in the EIA. This assay is now the standard for testing new dengue antibody assays.[46]

EIA's are in wide use for the serodiagnosis of viral infections owing to their relative simplicity, high degree of reproducibility, and ability to automate plate washers and scanners. In the Innis assay,[17] 96 well plates are sensitized with goat antihuman IgM or IgG antibody (Fig. 4). Control and test sera are diluted 1:100 and placed in the wells overnight. IgM or IgG in human test sera is captured on the respective plates and followed by tetravalent dengue antigen (16 HA units each of DENV-1, DENV-2, DENV-3 and 8 HA units of DENV-4) for the dengue EIA or JE antigen (50 HA units) for the JE EIA. This is followed by an anti-flavivirus horseradish peroxidase conjugate with substrate producing a quantitative colorimetric result read by an EIA plate reader. A binding index (BI) is calculated using the optical density (OD) of the test sample minus the OD of the negative control, all divided by the OD of the weak positive control minus the OD of the negative control. The BI multiplied by 100 generates EIA units, of which ≥ 40 units are considered positive for the IgM capture assay. The sensitivity of diagnosing acute dengue is 78% on hospital admission sera (usually clinical illness day 4 and prior to defervescence) and 97% in paired sera with a specificity of 100%. The value of this assay in addition to diagnosing acute dengue is in distinguishing between acute dengue and acute JE. A ratio of anti-dengue IgM to anti-JE IgM of ≥ 1.0 is consistent with acute dengue,

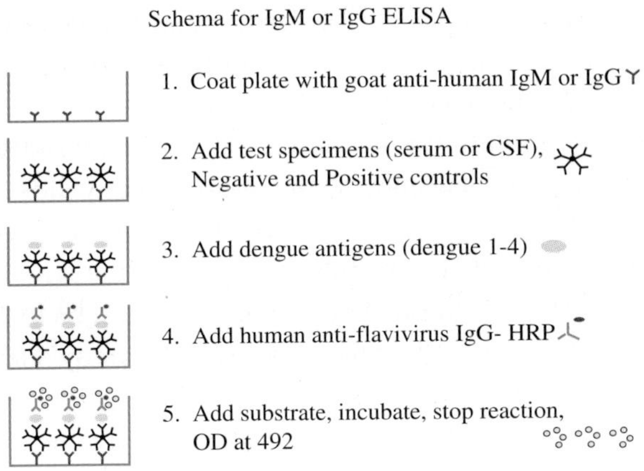

Fig. 4. Illustration of the IgM and IgG capture enzyme immunoassay (ELISA). Antibodies are detected by an antiantibody conjugated with a chemical dye. Intensity of color is read in a colorimeter and translated into antibody units. (Courtesy of Virology Department, Armed Forces Research Institute of Medical Sciences, Bangkok, Thailand.)

whereas a ratio of ≤ 1.0 is consistent with an acute JE virus infection. The added value of this assay is to distinguish primary from secondary DENV infection. Sera defined as consistent with primary or secondary dengue by the HI test were used to standardize ratios of IgM/IgG; a ratio of IgM to IgG units of ≥ 1.8 suggests a primary DENV infection, while a ratio of < 1.8 is diagnostic of a secondary dengue. Dengue EIA and the HI tests, when compared using WHO interpretive criteria, demonstrated a high coefficient of correlation.[15,47]

Quality controls should to be adopted to decrease inter- and intra-assay variation. Sources of interassay variation include: (1) working dilution of the antiflavivirus IgG enzyme conjugate; (2) duration of the chromogen-substrate reaction; and (3) the plate coating sensitization step (amount of anti-isotype antibody bound). The serum samples and antigens are used in quantities that saturate the capture system, with the result that the key components of the assay are the anti-isotype antibody adsorbed to plates and the IgG-enzyme conjugate. The amount of bound anti-isotype antibody is affected by the quantity used, the duration of the sensitization reaction, and the type of plates used. Reduction of variability in the sensitization step can be achieved by using the same plates that were used to optimize the assay. Plates should be presensitized plates in batches and stored at –20°C. They can be kept at –20°C for up to one month without significant loss of bound anti-isotype antibody. It is also important to determine the optimal dilution of the antiflavivirus IgG-HRP conjugate. This is accomplished by determining the dilution that yields an OD at 492 nm of 0.4 (established cutoff for a positive test) with a 1:100 dilution of the weak positive control serum. Intra-assay variation should be 10% or less. Within one test, all four positive standards should be in the OD range of 0.25–0.55. Values above this range decrease the assay's sensitivity and values below this range decrease specificity.

Use of cell-culture-derived dengue antigens and dengue monoclonal antibodies as control sera rather than patient-derived control serum provides a test system as sensitive and specific as the assay described by Innis *et al.*[48–50] Commercial kits are on the market and these compare favorably with the original EIA described by Innis and colleagues.[47,51–57] The dengue IgM and IgG capture EIA assay marketed by PanBio (Dengue Duo) exhibited a sensitivity of 99% and a specificity of 92% compared to the Innis assay.[52]

An EIA has recently been tested on saliva for the diagnosis of acute infection.[58] Saliva contains both dengue-specific IgM and IgG antibodies. A test on saliva specimens using the PanBio Dengue Duo ELISA obtained a sensitivity of 92% and a specificity of 100% compared to the serum-based Innis EIA.[59]

The dengue IgM/IgG EIA allows high throughput. Reagents are commercially available or readily available from reference laboratories. The advantage of the antidengue IgM and IgG isotype capture enzyme immunoassay is its ability to detect elevated levels of IgM in a sample collected late during the acute phase of illness. This is the time during a dengue illness when most DHF patients are in the hospital. This makes it possible to use a single specimen dengue IgM capture test to diagnose an acute dengue infection. It is important for the clinician to understand that it is not until three days after defervescence that all patients with DENV infections will have diagnostic quantities of dengue IgM in their sera. An advantage of the EIA test is that human sera contain no inhibitor. This means that no pretreatment of test sera is needed. Also, the assay can be used to distinguish between acute infections with different flaviviruses.

A further advantage of the EIA is its ability to detect IgM in the CSF in the diagnosis of dengue encephalitis. Some dengue EIA kits on the market demonstrate high levels of sensitivity and specificity for acute DENV infection.

Indirect fluorescent antibody (IFA) test

This assay to detect dengue-specific IgM and IgG using fluorescent antibody is used primarily in research laboratories. It was developed by Vathanophas and colleagues in 1973.[60] It utilizes a solid phase (usually dengue-virus-infected cells that are cold-acetone-fixed onto slides) that binds human dengue antibody, which in turn is detected by fluorescein-conjugated antihuman antibody. Visible fluorescence seen on the slides constitutes a positive antibody test. Serial dilutions of test serum are used to measure the amount of antibody present as an antibody titer. This method is limited due to its time-intensiveness, subjective reading, reliance on infected cells as the antibody capture agent, and lack of automation. The major advantage of this assay is the relatively low cost required to test a few samples.[61]

Western blot

Western blot is based upon the ability of proteins to travel through a polyacrylamide gel powered by an electric current and then be transferred to nitrocellulose sheets.[62,63] Molecular weight and charge determine the speed with which proteins in an electric field result in the migration of structural and nonstructural proteins, predominantly E, NS1, and pre-M. As a final step, proteins are transferred by a horizontal electrical current to a paper strip, usually nitrocellulose. Human sera can be applied onto the paper strip, permitting antibodies to bind to bands of dengue protein which have been differentially migrated according to molecular weight. Antibodies interact with E, NS1, and NS3 proteins. Dengue antibodies studied by Western blot are cross-reactive within the dengue group, which limits the usefulness of this method for identifying specific etiology. Western blot is used in research to study the production of antibodies to DENV-specific proteins following infections with wild-type viruses or vaccines.[64–67] The Western blot method provides the format for rapid diagnostic assays using nitrocellulose strips.

Rapid antibody assays

In 1988, Cardosa and colleagues described a dot enzyme immunoassay (DEIA) for use in clinical settings.[68] The DEIA used a single acute phase serum and was more sensitive for diagnosis of acute dengue than the HI test. This and similar assays are an adaptation of the Western blot test. DENV proteins are transferred to a solid phase, nitrocellulose paper or dipstick and exposed to DENV-specific antibody and then an antihuman antibody detection system. The result is a band indicating the presence of antibodies to DENV. Dot–blot tests are rapid (usually 4–6 hours or less) and capable of detecting IgM and IgG dengue antibodies and therefore discriminating between primary and secondary DENV infections. They require minimal expertise or laboratory equipment (centrifuge and water bath). In the assay developed by Cardosa and colleagues, strips containing DENV antigen are incubated with human sera, permitting the binding of dengue-specific antibody to the test strip.[68] A biotinylated antihuman IgG, or IgM, depending on the assay, detects bound immunoglobulin as a color band. A positive IgM alone signifies an acute primary DENV infection, and positive IgM plus IgG indicates an acute secondary DENV infection. Another test, the dengue rapid test, has been developed by PanBio, LTD.[69,70] This immunochromatographic test detects dengue-specific IgM and IgG antibodies in less than seven minutes. When compared with standard serologic techniques using sera from dengue patients obtained on admission and discharge from the hospital, this test was 100%-sensitive in identifying a DENV infection and reliably distinguished primary from secondary DENV. This assay now uses recombinant antigens and is available in a dipstick format.[57,71,72] Other rapid assays are coming on the market.[52] The limitations of rapid antibody-based assays are decreased specificity in distinguishing antibodies to other flaviviruses such as JE virus. All acute illness phase tests that rely on detection of specific IgM antibodies may read as false negatives if sera are collected too early in the illness. Dengue IgM antibodies evolve differently in different individuals and may not be detected until seven or more days after the onset of illness and may be blunted in secondary DENV infections.

Virus Isolation and Serotype Identification

Intracerebral inoculation of suckling mice

DENV was first isolated in laboratory animals in August 1942, during the dengue epidemic in the Nagasaki–Sasebo region of Japan.[73] Blood taken from patients within 48 hours after the first rise of temperature was inoculated intracerebrally (IC) into suckling white mice. Infected mice developed debility, tremors, and hind limb paralysis. Until the 1960's, IC inoculation of suckling mice was the standard method for isolation of dengue viruses. Inoculation of suckling or weanling mice also was used to prepare dengue antigens (for a description of the technique see the section on HA and HI assays).

Mosquito inoculation

Recovery of dengue viruses by inoculation of mosquitoes was developed into a standard laboratory method in the 1970's.[74] Several mosquito species have been used, including *Toxorhynchites splendens*, *Aedes albopictus*, and *Aedes aegypti*.[75–77] *Toxorhynchites* mosquitoes have the advantage of being larger, easier to inoculate, and with no risk of laboratory infections as these insects do not feed on blood.[75] At AFRIMS, *Toxorhynchites splendens* mosquitoes are injected with 0.02 μl of human sera intrathoracically (Fig. 5). After an incubation period of 10–14 days, DENV is detected in the mosquito head using a polyvalent serum and an immunofluorescent method. If positive, the mosquito body is triturated and inoculated onto C6/36 cells for virus expansion and typing. Isolation rates are higher than direct inoculation onto cell lines and nearly equivalent to detection of viral RNA by RT-PCR.[20] The need for insectaries and class 3 mosquito holding facilities restricts the use of this virus isolation method.

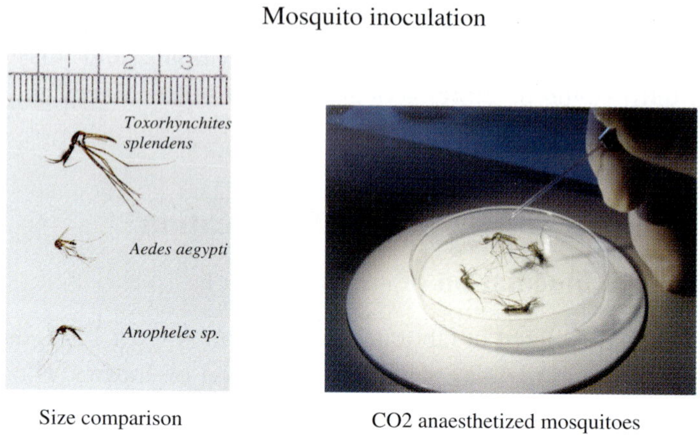

Fig. 5. Inoculation of *Toxorhynchites splendens* mosquitoes is used to isolate dengue viruses. *Aedes aegypti* and *anopheles* species mosquitoes are shown for size comparison. Scale in centimeters. (Courtesy of Virology Department, Armed Forces Research Institute of Medical Sciences, Bangkok, Thailand.)

Cell culture inoculation

A variety of insect and mammalian cell lines are permissive to DENV infection. Widely used insect cells include mosquito-derived cell lines: AP-61 from *Ae. pseudoscutellaris,* C6/36 from *Ae. Albopictus*, and TRA-284 from *Tx. Amboinensis.*[78–80] Of the three, C6/36 is the most widely used. Commonly used mammalian cells are derived from hamsters, BHK-21, or primates, LLC-MK2, and Vero. Some mammalian cells regularly demonstrate cytopathic changes and plaque formation.[81] Mammalian cells are commonly used in the plaque reduction neutralization assay, and mosquito cells for virus isolation.[74]

Dengue virus serotype identification

Previous single DENV infections can be identified in late convalescent sera by measuring neutralizing antibodies, which are typically monotypic or relatively monotypic.[45] The same principle can be used to generate monospecific antisera in monkeys. While molecular approaches using RT-PCR have evolved significantly in recent years, the standard for

DENV serotype identification remains the use of serotype-specific monoclonal antibodies in an immunofluorescence assay or an antigen capture EIA format.[82] Common monoclonal antibodies used for serotype identification include 1F1 for DENV-1, 3H5 for DENV-2, 5D4 for DENV-3, and 1H10 for DENV-4.[83,84]

The antigen capture EIA for the identification of DENV has been demonstrated to be a simple and reliable technique.[85] At AFRIMS, an antigen capture EIA utilizes virus isolated in *Toxorhynchites splendens* mosquitoes after expansion using C6/36 cells. Immulon U plates are sensitized with goat antimouse IgG in each well. Dengue-serotype-specific mouse monoclonal antibodies (4G2 antiflavivirus, 1F1 anti-DENV-1, 3H5 anti-DENV-2, 10C10 anti-DENV-3, 1H10 anti-DENV-4, and 2H2 anti-dengue group) are bound onto the plate, followed by capture of the unknown DENV serotype. A colorimetric reaction is formed after the addition of anti-flavivirus-horseradish-peroxidase and its substrate. DENV-serotype-specific mouse-brain-derived antigens (DENV-1 Hawaii, DENV-2 NGC, DENV-3 H87, DENV-4 914669) are used in parallel as positive controls. Optical density (OD) is read at a wavelength of 492 nm and the results are interpreted by comparing with positive and negative controls, a positive control OD is always ≥ 0.20 and a negative control OD is < 0.20. Matching the highest OD reading to the positive dengue control serotype identifies the DENV serotype.[85]

Detection of dengue viral antigens

The use of surrogate markers of dengue virus replication is a recent innovation in the detection of dengue viremia, exploiting the phenomenon that cellular infection with the dengue viruses results in the production and release of the nonstructural protein NS-1. In humans, NS1 localizes on the surface of infected cells and is released into the blood. An enzyme-linked immunosorbent assay has been developed to detect and quantify the NS-1 in the plasma of infected patients.[86] The use of rabbit polyclonal and monoclonal antibodies as the capture and detection antibodies, immunoaffinity-purified NS1 derived from dengue 2 virus–infected cells were used as a standard to detect NS-1 protein with a sensitivity of 4 ng/ml. In clinical samples, this assay was able to detect and quantify

NS1 during acute infection. Clinical validation of this assay in kit form using a sandwich format microplate EIA demonstrated a sensitivity of this assay in diagnosing acute dengue infection of 88.7% and a specificity of 100%.[87] The half-life of NS1 antigen in the blood of infected patients is longer than that of intact virus and has been detected in sera up to18 days after the onset of symptoms, with peak antigen detected at days 6–10 after the onset of fever. There is an overall detection sensitivity of 82% and specificity of 98.9%.[88] Serotype-specific anti-NS-1 does not cross-react with other dengue or flavivirus NS1's.[88,89] The amount of NS-1 protein in the blood of acutely infected dengue patients correlated with dengue viral load, days of viremia, and disease severity. Patients with dengue hemorrhagic fever had higher NS1 levels in acute phase blood than did patients with dengue fever.[90,91] Elevated free NS1 of greater than or equal to 600 ng/ml within 72 hours of illness onset identified patients at risk for developing DHF.

Commercial NS-1 tests provide an inexpensive, high-throughput assay that is easy to standardize and validate with minimal inter- and intra-assay variability. Furthermore, it has been possible to detect NS-1 protein as long as 18 days after the onset of clinical illness. The detection of NS-1 appears to offer an advantage over viral isolation or detection of viral RNA in establishing a specific diagnosis of dengue infection. Early detection of NS-1 in blood may permit the identification of patients with dengue illness, and standardized quantitation of circulating NS1 could play a role in predicting the severity of ensuing disease.

Immunohistochemistry

The identification of dengue-infected cells in biopsy or autopsy tissues requires the use of immunohistology, direct and indirect fluorescent antibody staining, or the use of enzyme conjugates, peroxidase alkaline phosphatase.[92–94]

Tissue specimens may be fixed in Millonig's formalin for 2 hours, irradiated in a microwave oven, and then embedded in paraffin. Viral antigen is detected by the immunoalkaline phosphatase method described by Hall and others.[94] Tissue must be deparaffinized in absolute alcohol and water, immersed in a 0.05% solution of protease VIII, stained with HistoMark Blue (Kirkegaard and Perry Laboratories, Gaithersburg, MD, USA), and blocked with normal horse serum and bovine serum albumin. The specimen is incubated overnight with polyclonal mouse dengue antibody followed by a secondary biotinylated horse antimouse IgG streptaviradin-alkaline phosphatase, AS-B1 phosphate, hexazotized new fuchsin, and levamisol as the chromogenic substrate. The tissue specimen is then counterstained with Mayer's hematoxylin.

Genome-Based Assays

Hybridization probes and polymerase chain reaction (PCR)

In 1987, Henchal and colleagues described molecular techniques for identifying dengue virus in blood using slot–blot nucleic acid hybridization with a radiolabeled cDNA probe.[95] In 1991, reverse transcription (RT) of viral RNA to DNA followed by amplification of DNA fragments by the polymerase chain reaction (PCR) allowed the rapid (less than 12 hours) detection of DENV in patient sera. A nested technique allowed the serotype-specific diagnosis of DENV infection.[96–98]

Henchal's slot–blot nucleic acid hybridization technique used a radiolabeled cDNA probe to detect as few as 11 plaque-forming units of each of the four DENV serotypes.[95] Nucleic acid hybridization is not affected by antibodies directed against virions. Virus-specific RNA is released into blood at the same time that viral antigen can be detected in infected cells. The Henchal technique was modified by Ruiz and colleagues as a microplate hybridization method.[99] DENV RNA was isolated from serum or tissue samples and immobilized onto wells, followed by hybridization with a biotin-labeled cDNA probe with signal detection by peroxidase conjugation. When tested on a panel of viremic seras, this assay was found to have a sensitivity of 95% and a specificity of 100% for all four DENV serotypes.

Henchal and colleagues developed a universal set of sense and antisense oligomeric DNA primers that matched all known DENV sequences.[96] Their RT-PCR was found to be 80% sensitive and 100% specific for diagnosing acute DENV infections when compared to virus isolation using inoculation of living mosquitoes. Modifications made to this assay by Lanciotti, Morita and others employing nested RT-PCR techniques have increased sensitivity and reduced assay time to less than 6 hours.[97,98,100–106]

The dengue RT-PCR assay provides a rapid, sensitive, diagnostic tool for detecting DENV in specimens from infected individuals.[107] The primary limitation on the application of RT-PCR to the diagnosis of DENV infection in patients is that tests must be performed on acute phase samples. DENV viremia terminates around the time of defervescence, although dengue RNA has been detected in blood for several days thereafter.[20] Other limitations include the need for a laboratory equipped with an ultracentrifuge, thermocycler, and electrophoresis equipment. The use of positive and negative controls is essential, along with strict adherence to specified techniques that are required to eliminate cross-contamination with RNA and DNA to produce false positive results. Despite these limitations, dengue RT-PCR is a powerful tool for diagnosing dengue-serotype-specific viremia. Newer RT-PCR techniques are being developed that may be more practical for the developing dengue diagnostic laboratory, including pocket thermocyclers with gel cartridges containing all the essential reagents that can be used in the field and require minimal technical expertise.

Nucleic acid sequence-based amplification (NASBA)

NASBA is an isothermal RNA amplification method that uses electrochemiluminescence to detect mRNA utilizing the NuclisensTM basic kit and the Nuclisens Reader (Organon Teknika). Unlike RT-PCR, which relies on the conversion of RNA into cDNA and then amplification, NASBA directly amplifies RNA using primers and capture probes at isothermal temperatures. NASBA has been successfully used in other pathogens, such as malaria, cytomegalovirus, and human immunodeficiency virus.[108–111] Recently, it has been applied to the diagnosis of dengue.[112] Using spiked sera, NASBA had a detection threshold of 1–10 PFU/ml. When it was tested against clinical samples, a threshold of 25 PFU/ml was observed, as well as a 100% serotype concordance with viral isolation, and a sensitivity of 98.3% and a specificity of 100%.

NASBA, though still in development, may prove to be a useful tool for diagnosis during the early viremic phase of acute DENV infection.

Fluorogenic-probe-based 5′ exonuclease assay (Taqman)

The fluorogenic-probe-based 5′ exonuclease assay (Taqman), using the Perkin–Elmer Applied Biosystems automated sequence detection system 7700, has been successfully employed to diagnose and quantify a number of human pathogens, including many viruses.[113–118] This technique is based on the use of a fluorescent-tagged probe that hybridizes with the target cDNA sequence (follows the RT step). A fluorescent signal is released through the 5′-3′ exonuclease activity of DNA *Taq* polymerase.[119] This allows real-time monitoring of the targeted PCR product and, with an internal control, a quantitative measurement. Taqman has been successfully used to detect and quantify DENV infection.[22,120–122] It may prove to be a useful technique for rapidly diagnosing DENV infection, and in particular for rapidly quantifying viremia and its correlate of ensuing dengue disease severity.

Future Directions

Understanding the pattern of immune responses to first or subsequent DENV infections in the context of the clinical illness is essential for developing and validating tools to diagnose acute DENV infection. Antibody-based assays will not be positive early in the course of disease; patients with suspected dengue and negative antibody-based test results should not be sent home believing that the cause of their fever is not dengue. The patients must be warned about the signs of plasma leakage and appropriately followed. Serotype-specific diagnosis is difficult during convalescence and the severe limitations imposed by broad antibody responses in patients with previous DENV infection must be remembered. With these concepts in mind, is there an ideal assay for diagnosing DENV infection? First, the assay must be both sensitive and specific, with a high predictive value despite a low incidence of disease. This will be important in countries where dengue is an emerging disease. Such a test must have low cross-reactivity with other cocirculating flaviviruses, such as JE, yellow fever, or West Nile virus. The assay must be reproducible, with low inter- and intra-assay variability, and inexpensive so that developing countries with dengue epidemics may be able to use it. Likewise, the assay must be simple to perform, with minimal training and diagnostic equipment. Such an assay will identify serotype-specific dengue antigen during the viremic period, and IgM and IgG during the late acute or early convalescent period. Given these criteria, one can visualize a rapid diagnostic test that will take minutes to perform using a finger prick of whole blood or saliva placed on a card with one space for a dengue-serotype-specific antigen capture and/or surrogate marker for dengue virus such as NS-1, and another for the detection of dengue-specific IgM and IgG during the late acute or convalescent period. Future assays will go beyond confirming or refuting dengue as the etiology to distinguish multiple etiologies of fever. Etiologic diagnosis based on gene expression in response to infection needs to be evaluated. These challenges for flavivirologists and commercial companies, if met, should expand opportunities to control and manage this global health problem.

References

1. Innis BL. Dengue and dengue hemorrhagic fever. In: Porterfield JS (ed.) *Kass Handbook of Infectious Diseases: Exotic Virus Infections*. Chapman & Hall, London, 1995, pp. 103–146.
2. Calisher CH, Karabatsos N, Dalrymple JM *et al*. Antigenic relationships between flaviviruses as determined by cross-neutralization tests with polyclonal antisera. *J Gen Virol* 1989;70:37–43.
3. Hammon W, Sather G. Problems of typing dengue viruses. *Mil Med* 1964;129:130–135.
4. The dengue group of viruses and its family relationships. *Bacteriol Rev* 1950;14(3):225–232.
5. Casals J. Acetone-ether extracted antigens for complement-fixation with certain neurotropic viruses. *Proc Soc Exp Biol* 1949;70:339–343.
6. Clarke DH, Casals J. Techniques for hemagglutination and hemagglutination inhibition with arthropod-borne viruses. *Am J Trop Med Hyg* 1958;7: 561–573.

7. Shope RE. Antigen and antibody detection and update on the diagnosis of dengue. *Southeast Asian J Trop Med Public Health* 1990;21:642–645.
8. Wu SJ, Grouard-Vogel G, Sun W *et al.* Human skin Langerhans cells are targets of dengue virus infection. *Nat Med* 2000;6:816–820.
9. Chandler A, Rice L. Observations on the etiology of dengue fever. *Am J Trop Med Hyg* 1923;3:233–262.
10. Sabin AB. Recent advances in our knowledge of dengue and sandfly fever. *Am J Trop Med Hyg* 1955;4:198–207.
11. Siler JF, Hall MW, Hitchens AP. Dengue: its history, epidemiology, mechanism of transmission, etiology, clinical manifestations, immunity, and prevention. *Philippine J Sci* 1926;29:1–304.
12. Simmons JS. Dengue fever. *Am J Trop Med Hyg* 1931;11:77–102.
13. Kalayanarooj S, Vaughn DW, Nimmannitya S *et al.* Early clinical and laboratory indicators of acute dengue illness. *J Infect Dis* 1997;176: 313–321.
14. Vaughn DW, Green S, Kalayanarooj S *et al.* Dengue in the early febrile phase: viremia and antibody responses. *J Infect Dis* 1997;176:322–330.
15. Anonymous. *Dengue Haemorrhagic Fever: Diagnosis, Treatment, Prevention and Control*, 2nd ed. World Health Organization, Geneva, 1997.
16. Green S, Vaughn DW, Kalayanarooj S *et al.* Early immune activation in acute dengue illness is related to development of plasma leakage and disease severity. *J Infect* Dis 1999;179:755–762.
17. Innis BL, Nisalak A, Nimmannitya S *et al.* An enzyme-linked immunosorbent assay to characterize dengue infections where dengue and Japanese encephalitis co-circulate. *Am J Trop Med Hyg* 1989;40:418–427.
18. Russell PK, Nisalak A, Sukhavachana P, Vivona S. A plaque reduction test for dengue virus neutralization antibodies. *J Immunol* 1967;99:285–290.
19. Halstead SB, Rojanasuphot S, Sangkawibha N. Original antigenic sin in dengue. *Am J Trop Med Hyg* 1983;32:154–156.
20. Vaughn DW, Green S, Kalayanarooj S *et al.* Dengue viremia titer, antibody response pattern, and virus serotype correlate with disease severity. *J Infect Dis* 2000;181:2–9.
21. Libraty DH, Endy TP, Houng HS *et al.* Differing influences of virus burden and immune activation on disease severity in secondary dengue-3 virus infections. *J Infect Dis* 2002;185:1213–1221.

22. Houng HH, Chung-Ming Chen R, Vaughn DW, Kanesa-thasan N. Development of a fluorogenic RT-PCR system for quantitative identification of dengue virus serotypes 1–4 using conserved and serotype-specific 3′ noncoding sequences. *J Virol Methods* 2001;95:19–32.
23. Wang WK, Chao DY, Kao CL *et al.* High levels of plasma dengue viral load during defervescence in patients with dengue hemorrhagic fever: implications for pathogenesis. *Virology* 2003;305:330–338.
24. Casals J, Brown LV. Hemagglutination with arthropod-borne viruses. *J Exp Med* 1954;99:429–449.
25. Porterfield J. The haemagglutination-inhibition test in the diagnosis of yellow fever in man. *Trans R Soc Trop Med Hyg* 1954;48:261–266.
26. Porterfield J. Use of goose cells in haemagglutination tests with arthropod-borne viruses. *Nature* 1957;180:1201–1202.
27. Clarke D, Casals J. Improved methods for hemagglutination studies with arthropod-borne viruses. *Proc Soc Exp Biol Med* 1955;88:96–99.
28. Wolff KL, Trent DW, Karabatsos N, Hudson BW. Use of glutaraldehyde-fixed goose erythrocytes in arbovirus serology. *J Clin Microbiol* 1977;6:55–57.
29. Monath TP, Lindsey HS, Nuckolls JG *et al.* Comparison of methods for removal of nonspecific inhibitors of arbovirus hemagglutination. *Appl Microbiol* 1970;20:748–753.
30. Altemeier WA, III, Mundon FK, Top FH, Jr, Russell PK. Method of extracting viral hemagglutination-inhibiting antibodies from the nonspecific inhibitors of serum. *Appl Microbiol* 1970;19:785–790.
31. Burke DS, Nisalak A, Johnson DE, Scott RM. A prospective study of dengue infections in Bangkok. *Am J Trop Med Hyg* 1988;38:172–180.
32. Dulbecco R. Production of plaques in monolayer tissue cultures by single particles of an animal virus. *Proc Natl Acad Sci USA* 1952;38:747–752.
33. Henderson J, Taylor R. Arthropod-borne virus plaques in agar overlaid tube cultures. *Proc Soc Exp Biol Med* 1959: 257–259.
34. Halstead SB, Sukhavachana P, Nisalak A. Assay of mouse adapted dengue viruses in mammalian cell cultures by an interference method. *Proc Soc Exp Biol Med* 1964;115:1062–1068.
35. Sukhavachana P, Nisalak A, Halstead SB. Tissue culture techniques for the study of dengue viruses. *Bull World Health Organ* 1966;35:65–66.

36. Sukhavachana P, Yuill TM, Russell PK. Assay of arbovirus neutralizing antibody by micro methods. *Trans R Soc Trop Med Hyg* 1969;63: 446–455.
37. Fujita N, Tamura M, Hotta S. Dengue virus plaque formation on microplate cultures and its application to virus neutralization (38564). *Proc Soc Exp Biol Med* 1975;148:472–475.
38. Okuno Y, Igarashi A, Fukai K. Neutralization tests for dengue and Japanese encephalitis viruses by the focus reduction method using peroxidase–anti-peroxidase staining. *Biken J* 1978;21:137–147.
39. Morens DM, Halstead SB, Larsen LK. Comparison of dengue virus plaque reduction neutralization by macro and "semi-micro" methods in LLC-MK2 cells. *Microbiol Immunol* 1985;29:1197–1205.
40. Morens DM, Halstead SB, Repik PM *et al.* Simplified plaque reduction neutralization assay for dengue viruses by semimicro methods in BHK-21 cells: comparison of the BHK suspension test with standard plaque reduction neutralization. *J Clin Microbiol* 1985;22:250–254.
41. Jacobs M, Young P. Dengue vaccines: preparing to roll back dengue. *Curr Opin Investig Drugs* 2003;4:168–171.
42. Halstead SB, Streit TG, Lafontant JG *et al.* Haiti: absence of dengue hemorrhagic fever despite hyperendemic dengue virus transmission. *Am J Trop Med Hyg* 2001;65:180–183.
43. Guzman MD, Alvarez M, Rodriguez-Roche R, Bernardo L, Montes T, Vazquez S, Morier L, Alvarez A, Gould EA, Kouri G, Halstead SB *et al.* Neutralizing antibodies after infection with dengue 1 virus. *Emerg Infect Dis* 2007;13(2):282–286.
44. Kuno G, Gubler DJ, Oliver A. Use of original antigenic sin theory to determine the serotypes of previous dengue infections. *Trans R Soc Trop Med Hyg* 1993;87:103–105.
45. Burke DS, Nisalak A. Detection of Japanese encephalitis virus immunoglobulin M antibodies in serum by antibody capture radioimmunoassay. *J Clin Microbiol* 1982;15:353–361.
46. Vaughn DW, Nisalak A, Solomon T *et al.* Rapid serologic diagnosis of dengue virus infection using a commercial capture ELISA that distinguishes primary and secondary infections. *Am J Trop Med Hyg* 1999;60: 693–698.

47. Chungue E, Marche G, Plichart R *et al.* Comparison of immunoglobulin G enzyme-linked immunosorbent assay (IgG-ELISA) and haemagglutination inhibition (HI) test for the detection of dengue antibodies: prevalence of dengue IgG-ELISA antibodies in Tahiti. *Trans R Soc Trop Med Hyg* 1989;83:708–711.
48. Cardosa MJ, Tio PH, Nimmannitya S *et al.* IgM capture ELISA for detection of IgM antibodies to dengue virus: comparison of 2 formats using hemagglutinins and cell culture derived antigens. *Southeast Asian J Trop Med Public Health* 1992;23:726–729.
49. Lam SK, Devi S, Pang T. Detection of specific IgM in dengue infection. *Southeast Asian J Trop Med Public Health* 1987;18:532–538.
50. Kuno G, Gomez I, Gubler DJ. An ELISA procedure for the diagnosis of dengue infections. *J Virol Methods* 1991;33:101–113.
51. Groen J, Koraka P, Velzing J *et al.* Evaluation of six immunoassays for detection of dengue virus-specific immunoglobulin M and G antibodies. *Clin Diagn Lab Immunol* 2000;7:867–871.
52. Cuzzubbo A, Vaughn DW, Nisalak A *et al.* Commercial assays for the serological diagnosis of dengue infections. *Arbovirus Res Australia* 1997;7:56–60.
53. Sang CT, Cuzzubbo AJ, Devine PL. Evaluation of a commercial capture enzyme-linked immunosorbent assay for detection of immunoglobulin M and G antibodies produced during dengue infection. *Clin Diagn Lab Immunol* 1998;5:7–10.
54. Cuzzubbo AJ, Vaughn DW, Nisalak A *et al.* Comparison of PanBio Dengue Duo IgM and IgG capture ELISA and Venture Technologies Dengue IgM and IgG dot blot. *J Clin Virol* 2000;16:135–144.
55. Kit Lam S, Lan Ew C, Mitchell JL *et al.* Evaluation of a capture screening enzyme-linked immunosorbent assay for combined determination of immunoglobulin M and G antibodies produced during dengue infection. *Clin Diagn Lab Immunol* 2000;7:850–852.
56. Wu SJ, Kung CG, Chen TB *et al.* Comparison of two rapid diagnostic assays for the detection of immunoglobulin M antibodies to dengue virus. *Clin Diagn Lab Immunol* 1999;7:106–110.
57. Lam SK, Fong MY, Chungue E *et al.* Multicentre evaluation of dengue IgM dot enzyme immunoassay. *Clin Diagn Virol* 1996;7:93–98.
58. Cuzzubbo AJ, Vaughn DW, Nisalak A *et al.* Detection of specific antibodies in saliva during dengue infection. *J Clin Microbiol* 1998;36:3737–3739.

59. Artimos de Oliveira S, Rodrigues CV, Camacho LA *et al.* Diagnosis of dengue infection by detecting specific immunoglobulin M antibodies in saliva samples. *J Virol Methods* 1999;77:81–86.
60. Vathanophas K, Hammon WM, Atchison RW, Sather GE. Attempted type specific diagnosis of dengue virus infection by the indirect fluorescent antibody method directed at differentiating IgM and IgG responses. *Proc Soc Exp Biol Med* 1973;142:697–702.
61. Boonpucknavig S, Vuttivirojana O, Siripont J *et al.* Indirect fluorescent antibody technique for demonstration of serum antibody in dengue hemorrhagic fever cases. *Am J Clin Pathol* 1975;64:365–371.
62. Burnette WN. "Western blotting": electrophoretic transfer of proteins from sodium dodecyl sulfate–polyacrylamide gels to unmodified nitrocellulose and radiographic detection with antibody and radioiodinated protein A. *Anal Biochem* 1981;112:195–203.
63. Towbin H, Staehelin T, Gordon J. Electrophoretic transfer of proteins from polyacrylamide gels to nitrocellulose sheets: procedure and some applications. *Proc Natl Acad Sci USA* 1979;76:4350–4354.
64. Churdboonchart V, Bhamarapravati N, Peampramprecha S, Sirinavin S. Antibodies against dengue viral proteins in primary and secondary dengue hemorrhagic fever. *Am J Trop Med Hyg* 1991;44:481–493.
65. Churdboonchart V, Harisdangkul V, Bhamarapravati N. Letter: Countercurrent immunoelectrophoresis for rapid diagnosis of dengue haemorrhagic fever. *Lancet* 1974;2:841.
66. Kuno G, Vorndam AV, Gubler DJ, Gomez I. Study of anti-dengue NS1 antibody by Western blot. *J Med Virol* 1990;32:102–108.
67. Cardosa M, Wang S, Sum M, Tio P. Antibodies against prM protein distinguish between previous infection with dengue and Japanese encephalitis viruses. *BMC Microbiol* 2002;2:9.
68. Cardosa MJ, Hooi TP, Shaari NS. Development of a dot enzyme immunoassay for dengue 3: a sensitive method for the detection of anti-dengue antibodies. *J Virol Methods* 1988;22:81–88.
69. Vaughn DW, Nisalak A, Kalayanarooj S *et al.* Evaluation of a rapid immunochromatographic test for the diagnosis of dengue infection. *J Clin Microbiol* 1998;36:234–238.
70. Lam SK, Devine PL. Evaluation of capture ELISA and rapid immunochromatographic test for the determination of IgM and IgG antibodies produced during dengue infection. *Clin Diagn Virol* 1998;10:75–81.

71. Parida MM, Upadhyay C, Saxena P *et al.* Evaluation of a dipstick ELISA and a rapid immunochromatographic test for diagnosis of dengue virus infection. *Acta Virol* 2001;45:299–304.
72. Cuzzubbo AJ, Endy TP, Nisalak A *et al.* Use of recombinant envelope proteins for serological diagnosis of dengue virus infection in an immunochromatographic assay. *Clin Diagn Lab Immunol* 2001;8:1150–1155.
73. Hotta S. Experimental studies in dengue I. Isolation, identification and modification of the virus. *J Infect Dis* 1952;90:1–9.
74. Rosen L, Gubler DJ. The use of mosquitoes to detect and propagate dengue viruses. *Am J Trop Med Hyg* 1974;23:1153–1160.
75. Rosen L, Shroyer DA. Comparative susceptibility of five species of *Toxorhynchites* mosquitoes to parenteral infection with dengue and other flaviviruses. *Am J Trop Med Hyg* 1985;34:805–809.
76. Rosen L, Roseboom LE, Gubler DJ *et al.* Comparative susceptibility of mosquito species and strains to oral and parenteral infection with dengue and Japanese encephalitis viruses. *Am J Trop Med Hyg* 1985;34:603–615.
77. Gubler DJ, Nalim S, Tan R *et al.* Variation in susceptibility to oral infection with dengue viruses among geographic strains of *Aedes aegypti*. *Am J Trop Med Hyg* 1979;28:1045–1052.
78. Igarashi A. Isolation of a Singh's *aedes albopictus* cell clone sensitive to dengue and chikungunya viruses. *J Gen Virol* 1978;40:531–544.
79. Kuno G. Persistent infection of a nonvector mosquito cell line (TRA-171) with dengue viruses. *Intervirology* 1982;18:45–55.
80. Varma MGR, Pudney M, Leake CJ. Cell lines from larvae of *Aedes* (*stegomyia*) *malayensis colless* and *Aedes* (s) *pseudoscutellaris* (*theobald*) and their infection with some arboviruses. *Trans R Soc Trop Med Hyg* 1974;68:374–382.
81. Yuill TM, Sukhavachana P, Nisalak A, Russell PK. Dengue-virus recovery by direct and delayed plaques in LLC-MK2 cells. *Am J Trop Med Hyg* 1968;17:441–448.
82. Henchal EA, McCown JM, Seguin MC *et al.* Rapid identification of dengue virus isolates by using monoclonal antibodies in an indirect immunofluorescence assay. *Am J Trop Med Hyg* 1983;32:164–169.
83. Kuno G, Gomez I, Gubler DJ. Detecting artificial anti-dengue IgM immune complexes using an enzyme-linked immunosorbent assay. *Am J Trop Med Hyg* 1987;36:153–159.

84. Malergue F, Chungue E. Rapid and sensitive streptavidin-biotin amplified fluorogenic enzyme-linked immunosorbent-assay for direct detection and identification of dengue viral antigens in serum. *J Med Virol* 1995;47:43–47.
85. Kuno G, Gubler DJ, Santiago de Weil NS. Antigen capture ELISA for the identification of dengue viruses. *J Virol Methods* 1985;12:93–103.
86. Young P, Hilditch PA, Bletchly C, Halloran W. An antigen capture enzyme-linked immunosorbent assay reveals high levels of the dengue virus protein NS1 in the sera of infected patients. *J Clin Microbiol* 2000;38(3): 1053–1057.
87. Dussart P, Labeau B, Lagathu G, Louis P, Nunes MR, Rodrigues SG, Storck-Herrmann C, Cesaire R, Morvan J, Flamand M, Baril L. Evaluation of an enzyme immunoassay for detection of dengue virus NS1 antigen in human serum. *Clin Vaccine Immunol* 2006;13(11):1185–1189.
88. Xu H, Di B, Pan YX, Qiu LW, Wang YD, Hao W, He LJ, Yuen KY, Che XY. Serotype 1-specific monoclonal antibody-based antigen capture immunoassay for detection of circulating nonstructural protein NS1: implications for early diagnosis and serotyping of dengue virus infections. *J Clin Microbiol* 2006;44(8):2872–2878.
89. Shu PY, Chen LK, Chang SF, Su CL, Chien LJ, Chin C, Lin TH, Huang JH. Dengue virus serotyping based on envelope and membrane and nonstructural protein NS1 serotype-specific capture immunoglobulin M enzyme-linked immunosorbent assays. *J Clin Microbiol* 2004;42(6):2489–2494.
90. Libraty DH, Young PR, Pickering D, Endy TP, Kalayanarooj S, Green, S, Vaughn DW, Nisalak A, Ennis FA, Rothman AL. High circulating levels of the dengue virus nonstructural protein NS1 early in dengue illness correlate with the development of dengue hemorrhagic fever. *J Infect Dis* 2002;186(8):1165–1168.
91. Wang WK. High levels of plasma dengue viral load during defervescence in patients with dengue hemorrhagic fever: implications for pathogenesis. *Virology* 2003;305:330–338.
92. Boonpucknavig S, Bhamarapravati N, Nimmannitya S *et al.* Immunofluorescent staining of the surfaces of lymphocytes in suspension from patients with dengue hemorrhagic fever. *Am J Pathol* 1976;85:37–47.
93. Boonpucknavig S, Vuttiviroj O, Boonpucknavig V. Infection of young adult mice with dengue virus type 2. *Trans R Soc Trop Med Hyg* 1981;75: 647–653.

94. Hall WC, Crowell TP, Watts DM *et al.* Demonstration of yellow fever and dengue antigens in formalin-fixed paraffin-embedded human liver by immunohistochemical analysis. *Am J Trop Med Hyg* 1991;45:408–417.
95. Henchal EA, Narupiti S, Feighny R *et al.* Detection of dengue virus RNA using nucleic acid hybridization. *J Virol Methods* 1987;15:187–200.
96. Henchal EA, Polo SL, Vorndam V *et al.* Sensitivity and specificity of a universal primer set for the rapid diagnosis of dengue virus infections by polymerase chain reaction and nucleic acid hybridization. *Am J Trop Med Hyg* 1991;45:418–428.
97. Lanciotti RS, Calisher CH, Gubler DJ *et al.* Rapid detection and typing of dengue viruses from clinical samples by using reverse transcriptase–polymerase chain reaction. *J Clin Microbiol* 1992;30:545–551.
98. Morita K, Tanaka M, Igarashi A. Rapid identification of dengue virus serotypes by using polymerase chain reaction. *J Clin Microbiol* 1991;29: 2107–2110.
99. Ruiz BH, Zamora MP, Liu SG. Detection of dengue viral RNA by microplate hybridization. *J Virol Methods* 1995;54:97–108.
100. Morita K. *The Identification of Dengue Virus Using PCR Methods in Molecular Biology*. J. P. Clapp Humana, Totowa NJ, 1996, pp. 127–132.
101. Deubel V. The contribution of molecular techniques to the diagnosis of dengue infection. In: Gubler DJ and Kuno G (eds.) *Dengue and Dengue Hemorrhagic Fever*. CAB, New York, 1997, pp. 335–366.
102. Sudiro TM, Ishiko H, Green S *et al.* Rapid diagnosis of dengue viremia by reverse transcriptase–polymerase chain reaction using 3′-noncoding region universal primers. *Am J Trop Med Hyg* 1997;56:424–429.
103. Chungue E, Roche C, Lefevre MF *et al.* Ultra-rapid, simple, sensitive, and economical silica method for extraction of dengue viral RNA from clinical specimens and mosquitoes by reverse transcriptase–polymerase chain reaction. *J Med Virol* 1993;40:142–145.
104. Brown JL, Wilkinson R, Davidson RN *et al.* Rapid diagnosis and determination of duration of viraemia in dengue fever using a reverse transcriptase polymerase chain reaction. *Trans R Soc Trop Med Hyg* 1996; 90:140–143.
105. Yenchitsomanus PT, Sricharoen P, Jaruthasana I *et al.* Rapid detection and identification of dengue viruses by polymerase chain reaction (PCR). *Southeast Asian J Trop Med Public Health* 1996;27:228–236.

106. De Paula SO, Lopes da Fonseca BA. Optimizing dengue diagnosis by RT-PCR in IgM-positive samples: comparison of whole blood, buffy-coat and serum as clinical samples. *J Virol Methods* 2002;102:113–117.
107. Chan SY, Kautner I, Lam SK. Detection and serotyping of dengue viruses by PCR: a simple, rapid method for the isolation of viral RNA from infected mosquito larvae. *Southeast Asian J Trop Med Public Health* 1994;25:258–261.
108. Witt DJ, Kemper M, Stead A *et al.* Analytical performance and clinical utility of a nucleic acid sequence-based amplification assay for detection of cytomegalovirus infection. *J Clin Microbiol* 2000;38:3994–3999.
109. Blok MJ, Lautenschlager I, Goossens VJ *et al.* Diagnostic implications of human cytomegalovirus immediate early-1 and pp67 mRNA detection in whole-blood samples from liver transplant patients using nucleic acid sequence-based amplification. *J Clin Microbiol* 2000;38:4485–4491.
110. Berndt C, Muller U, Bergmann F *et al.* Comparison between a nucleic acid sequence-based amplification and branched DNA test for quantifying HIV RNA load in blood plasma. *J Virol Methods* 2000;89:177–181.
111. Schoone GJ, Oskam L, Kroon NC *et al.* Detection and quantification of *Plasmodium falciparum* in blood samples using quantitative nucleic acid sequence-based amplification. *J Clin Microbiol* 2000;38:4072–4075.
112. Wu SJ, Lee EM, Putvatana R *et al.* Detection of dengue viral RNA using a nucleic acid sequence-based amplification assay. *J Clin Microbiol* 2001;39:2794–2798.
113. Morris T, Robertson B, Gallagher M. Rapid reverse transcription-PCR detection of hepatitis C virus RNA in serum by using the TaqMan fluorogenic detection system. *J Clin Microbiol* 1996;34:2933–2936.
114. Hawrami K, Breuer J. Development of a fluorogenic polymerase chain reaction assay (TaqMan) for the detection and quantitation of varicella zoster virus. *J Virol Methods* 1999;79:33–40.
115. Jordens JZ, Lanham S, Pickett MA *et al.* Amplification with molecular beacon primers and reverse line blotting for the detection and typing of human papillomaviruses. *J Virol Methods* 2000;89:29–37.
116. Schutten M, van den Hoogen B, van der Ende ME *et al.* Development of a real-time quantitative RT-PCR for the detection of HIV-2 RNA in plasma. *J Virol Methods* 2000;88:81–87.

117. Loeb KR, Jerome KR, Goddard J *et al.* High-throughput quantitative analysis of hepatitis B virus DNA in serum using the TaqMan fluorogenic detection system. *Hepatology* 2000;32:626–629.
118. Lanciotti RS, Kerst AJ, Nasci RS *et al.* Rapid detection of West Nile virus from human clinical specimens, field-collected mosquitoes, and avian samples by a TaqMan reverse transcriptase-PCR assay. *J Clin Microbiol* 2000;38:4066–4071.
119. Holland PM, Abramson RD, Watson R, Gelfand DH. Detection of specific polymerase chain reaction product by utilizing the 5′–3′ exonuclease activity of *Thermus aquaticus* DNA polymerase. *Proc Natl Acad Sci USA* 1991;88: 7276–7280.
120. Laue T, Emmerich P, Schmitz H. Detection of dengue virus RNA in patients after primary or secondary dengue infection by using the TaqMan automated amplification system. *J Clin Microbiol* 1999;37:2543–2547.
121. Warrilow D, Northill JA, Pyke A, Smith GA. Single rapid TaqMan fluorogenic probe based PCR assay that detects all four dengue serotypes. *J Med Virol* 2002;66:524–528.
122. Callahan JD, Wu SJ, Dion-Schultz A *et al.* Development and evaluation of serotype- and group-specific fluorogenic reverse transcriptase PCR (TaqMan) assays for dengue virus. *J Clin Microbiol* 2001;39:4119–4124.

11

The Control of Dengue Vectors

Norman G. Gratz and Scott B. Halstead

Introduction

Dengue and the severe syndromes dengue hemorrhagic fever and dengue shock syndrome (DHF/DSS) continue to spread geographically and their incidence continues to increase. They are the most common re-emerging diseases and the most common arbovirus infections. Until safe and effective dengue vaccines are available, the only option to prevent dengue infection is effective control of the mosquito vectors. Most dengue-endemic countries have operational vector control programs. The organization, strategy, methods and effectiveness of these programs will be reviewed briefly in this chapter.

All dengue vectors are members of the *Aedes* subgenus *Stegomyia*. *Aedes aegypti* is the most important vector in almost all dengue-endemic countries. Other species, such as *Ae. albopictus* and *Ae. polynesiensis*, serve as vectors in limited geographical areas. Dengue vector control is thus mainly aimed at *Ae. aegypti*. *Ae. aegypti* is also the vector of yellow fever and chikungunya, and has been described as a vector of West Nile virus in Madagascar.[1]

Vector control operations seek to reduce adult mosquito population densities to a level that does not support transmission, or at least to reduced levels of virus transmission and disease incidence that are acceptable to public health authorities. Reductions in *Ae. aegypti* densities can be brought about by controlling larvae or pupae or adults.

Halstead has asked why modern societies have been unable to control *Ae. aegypti.*[2] He pointed to the increasing magnitude of *Ae. aegypti*–borne diseases, including yellow fever, dengue, dengue hemorrhagic fever and chikungunya, and observed that even though inadequate funds are allocated for the control of this vector, *Ae. aegypti* control programs may well be broken beyond the power of money to fix them. Complicating chronic underfunding were five other factors contributing to the failure of dengue vector control programs:

(1) Desire to find easy solutions.
(2) Degradation of technical and managerial skills.
(3) The increasing scope of the problem.
(4) The shortness of human memory.
(5) Expectation of failure.

Nearly two decades later, the incidence of *Ae. aegypti*–borne diseases continues to increase — further evidence that existing vector control programs are, for the most part, "broken." This was not always the case. In the Americas during the late 1940's, a successful campaign to eradicate *Ae. aegypti* from the entire hemisphere was implemented. This campaign was motivated by the fear of the spread of yellow fever from its sylvatic cycle to the urban cycle resulting in the large epidemics that had once characterized this disease. The eradication strategy was based on the knowledge that in the Americas *Ae. aegypti* virtually always relied on man-made larval habitats such as containers storing water for household use, including water drums and ceramic pots, discarded tires and other cast-off containers. Between 1948 and 1962, the species was eliminated from 21 countries in the region.

This campaign was centered on a closely supervised program in which all actual and potential breeding sites were identified and then treated with insecticides or destroyed. The program was directed by Dr. Fred Soper, the architect of the successful *Anopheles gambiae* eradication program in Brazil. After early success, and once Dr. Soper left the scene, there was a letdown and *Ae. aegypti* gradually and then rapidly reinfested its former range. This was due to several factors: lack of leadership and political will, mounting costs of the eradication programs, changes in socioeconomic and legal conditions, and insecticide resistance.

It has recently been proposed that another attempt be made to eradicate the species in the Americas. A hemispheric plan to achieve the goal was drawn up and published in 1998.[3] As yet no concerted action has been taken to implement this proposal.

Most countries have failed to achieve effective control of *Ae. aegypti*, possibly for many of the reasons listed above. This article will comment on the efficacy of standard methods used in vector control operations and describe new technologies or strategies which may be effective. It will emphasize that the most important factor in achieving successful control of *Ae. aegypti* is a program led by a high caliber administration and staffed by well-trained, -supervised and -motivated personnel.

Existing Control Methods; Control of the Aquatic Stages

Control of the aquatic stages, i.e. the larval and pupal populations, is commonly used. Most of the methods described are applicable to ecological conditions found in Asia and the Americas, where *Ae. aegypti* breeds almost entirely in man-made containers. In Africa *Aedes* larvae may be found in natural habitats such as tree holes, leaf axils and rock pools, as well as in man-made containers.

The Environmental Control of *Ae. aegypti* Larval Habitats

Environmental measures aim to eliminate actual or potential larval habitats. These include clay or ceramic water jars or pots, 55-gallon metal drums and similar containers storing water for household use; breeding is also very common in discarded containers, such as used tires, cans, glass and plastic bottles, and other cast-offs that can fill with rainwater and permit development of larvae. In Latin America, flower-holding containers in cemeteries are a prolific source of *Ae. aegypti*. Preventing the creation of such man-made habitats or disposing of them should not present insurmountable problems. In practice, however, programs attempting to control the vector through environmental measures have frequently failed to reduce vector densities to a point where transmission of the dengue virus is reduced or interrupted.

Environmental control methods have the virtue of not requiring the application of expensive chemical or biological agents to water stored for drinking purposes. At the same time, discarding potential larval habitats improves the environment by disposing of waste containers. It has been postulated that environmental control alone, particularly with the participation of the community, might provide effective control of dengue vector populations, and hence significantly reduce disease transmission.[4–6]

Unfortunately, *Ae. aegypti* control programs relying on environmental measures have so far had limited impact on dengue transmission in urban areas and even less in rural areas. The concept of environmental control appears to be both reasonable and desirable. Why has this approach met with relatively little success in most large scale, operational vector control programs?

In poorer sections of cities in dengue-endemic areas of the developing world, households often have little access to piped water and must depend on obtaining water supplied from neighboring standpipes. Often such a supply does not exist at all. The option of storing of rain or well water in and around houses lessens the dependence on standpipe water and the burden of having to transport water to homes. Even when standpipes exist water is still stored in homes to reduce the effort of carrying water into the house many times a day. Stored rain or well water serves as the main source of breeding of *Ae. aegypti* in most tropical cities.

Even the introduction of adequate water into an area does not necessarily result in cessation of the practice of storing water. In a town in Venezuela, when a piped water system was installed, householders continued to store water in containers because of frequent interruptions in the water supply in the past and a lack of confidence in the reliability of water supply in the future.[7] In many cities in the developing world, water supply is irregular and householders will store water against the possibility of it not being available for periods of time. In Southeast Asian countries, despite modern plumbing, water is traditionally kept in large ceramic water jars in and around the houses, these being periodically filled from taps.

In addition to domestic water storage containers, *Ae. aegypti* breeds in a plethora of waste containers, including discarded tires, empty tin cans, plastic containers, bottles and jars, and old automobile bodies. The number of different containers that serve as *Ae. aeygpti* larval habitats is astonishing. In a 1963 survey of Georgetown, British Guiana, Burton *et al.* found more than 200,000 containers capable of serving as larval habitats for *Ae. aegypti.*[8] In cemeteries, particularly in the Americas, water-filled jars for holding flowers are frequently placed on graves and can produce prodigious numbers of mosquitoes; Barrear-Rodriquez *et al.* found 190,000 flower jars on graves in the main cemetery of Caracas, Venezuela, and estimated that these were capable of supporting the breeding of some 50 million fourth instar mosquito larvae of various mosquito species, including *Ae. aegypti*, on any given day.[9]

A three-year-long survey carried out by Tonn *et al.* in Bangkok, Thailand, showed that there were an average of 6 water storage containers per house in the city, most of them ceramic water jars of 150–180-litre capacity; of these, an average of 2.6 per house were actually positive for *Ae. aegypti* larvae.[10] The city, and its immediate suburbs, with a total population of 2.5 million at the time, was estimated to have some two million containers which could serve as larval habitats for *Ae. aegypti*, of which almost 800,000 were positive for larvae on any given day. The number of adult mosquitoes estimated as emerging from these containers in a 24-hour period was 1,891,670. Since that time the city has grown to almost 8 million. For the cultural reasons described above, the number of containers may have increased even though most homes are now well supplied with piped water. Household water storage in Thailand is not likely to change without intensive educational programs. In Jakarta, Indonesia, a study found an average of 185 water containers per 100 houses, of which 60 were positive for *Ae. aegypti*, giving a container index of 32%. Indoor water jars produced more pupae per house than all other containers combined. The infestation rate of covered containers was significantly higher than that of uncovered containers, perhaps because loose-fitting lids allowed entrance of gravid females to the attractive, darkened interior of the container.[11] In Thailand, outdoor water jars with covers were found to be infested significantly more frequently than those without covers.[12] These findings question the usefulness of recommendations widely made that covers be placed on water jars to prevent mosquito breeding.

Effective control of the aquatic stages of *Ae. aegypti*, where population densities of the vector are as high as those described above, requires either a large, well-staffed, well-trained (and expensive) vector control organization or a degree of participation by the community in the disposal of larval habitats, which is rarely achieved.[13] It is not enough for the community to collect waste containers from the vicinity of their homes. The municipal authorities must also ensure that solid waste which has been collected from homes and deposited at neighborhood collection points is taken to sanitary landfills.

Singapore, by means of an adequately funded vector control organization, has succeeded in reducing *Stegomyia* larval indices from around 25% in the 1960's to < 2% by 1998;[14] despite this, 5183 cases of dengue, including 75 DHF cases and one death, were reported in 1998.[15] Obviously, intensive larval control programs did not interrupt introduction and transmission of the virus Control efforts aimed at reduction of the *Ae. aegypti* larval populations in most other cities of Asia and the Americas have had even less effect on mosquito population densities and virtually none on the incidence of disease transmission. Focks *et al.* concluded that source reduction efforts to control *Ae. aegypti* populations have met with little documented success and not been able to obtain, and sustain, the high level of control necessary to eliminate the threat of dengue transmission.[16]

The failure of most attempts to reduce dengue transmission usually by environmental control methods can be explained by the conclusions of Pontes *et al.* (2000), who estimated that in Brazil, source reduction measures would have to suppress the vector to no more than 1% of a community's houses to avert outbreaks of dengue.[17] This degree of reduction has rarely been achieved or maintained by environmental control programs, in Brazil or elsewhere.

The goal of controlling mosquito populations using environmental measures is a desirable one but is difficult to achieve in most cities or rural areas where sizeable infestations of *Ae. aegypti* are present. In the Americas, ecological conditions have changed since the epoch of the

Ae. aegypti eradication campaigns. The number of waste containers available to support *Ae. aegypti* breeding has greatly increased. Unfortunately, municipal services, including the provision of piped water supply to houses and the disposal of solid wastes, lag far behind human population increases.

Larvicides Used for the Control of *Ae. aegypti*

Many *Ae. aegypti* control programs are based virtually entirely on larvicide applications to containers to control the aquatic stages of the mosquito. An efficient larviciding program should be able to achieve, at least in limited areas, a level of control which would reduce the density of the adult vector population. Ensuring the efficient treatment of the very large numbers of containers found in urban areas, many of which may be small and difficult to find, requires a closely supervised control program, frequent checking of the degree of coverage of the control teams and periodic evaluation of the adult population to determine the impact of the control measures. Both supervisors and control operators must be well trained and motivated.

Inhabitants of dengue-endemic areas now more frequently resist treatments of their water jars than in the past; they feel that the larvicides impart an undesirable taste to the water and are less convinced that the larvicide applications will protect them from dengue. Phantthumachinda evaluated a larviciding field trial in Thailand and observed that the resistance of householders to letting their water jars be treated would prevent adequate coverage of mosquito control.[18] It was noted that periodic, large scale treatments were difficult especially in large cities. Nevertheless, many programs continue to apply larvicides. The following section will briefly review the efficacy of larvicides in common use.

Larvicides added to water that may be consumed by man and animals must have an extremely low degree of mammalian toxicity, both acute and chronic, and be colorless, tasteless and odorless. Only four compounds have met these criteria as set by the International Program for Chemical Safety and the World Health Organization: the organophosphorus compound temephos, the insect growth regulators (IGR's) methoprene and pyriproxyfen, and the biological agent *Bacillus thuringiensis israelensis* (Bti).[19,20] Earlier, larvicides of several chemical groups were available for use in containers. Due to growing vector resistance (Rawlins, 1998; Wirth and Georghiou, 1999; Lee, 1991; de la Cruz, 2001) and newer toxicological standards, the number of available compounds has greatly diminished.[21–24]

The most commonly used larvicides used at present in *Ae. aegypti* control programs are temephos and *Bacillus thuringiensis israelensis*. The IGR methoprene is effective but less widely used. One percent temephos, usually in the form of easily applied sand granules, has been the most widely used larvicide for the control of *Ae. aegypti*. In an early field trial in Thailand, Bang and Tonn (1969) found that 1% temephos sand granules applied at a target dosage of 1 ppm would provide up to two months of effective control of *Ae. aegypti* larvae in ceramic and concrete water jars and somewhat less in clay water jars. Similar persistent control was obtained in many other field trials and in operational use, including in India.[25] In Vietnam, three months of operational control was reported.[26] In Indonesia, an operational field trial of temephos sand granules achieved a 37% reduction in the rates of hospitalization for DHF among a group of 744 children, compared to a 110% increase among 7452 children in a neighboring, untreated area.[27]

Sulaiman *et al.* tested briquettes containing methoprene (Altosid) on *Ae. aegypti* larvae held in plastic containers and controlled adult emergence for 114–122 days posttreatment.[28] This was longer than the control achieved with Bti briquettes, with which they were compared. Methoprene has been successfully used in Australia for the control of *Ae. aegypti* breeding in ornamental bromeliads.[29] While effective and toxicologically acceptable, this compound has not yet come into widespread use against container breeding mosquito species.

Pyriproxyfen has been tested extensively in the field against *Ae. aegypti* and *Ae. albopictus,* and has provided high levels of control for several weeks.[30,31] This toxicity and the potential use of this IGR were reviewed by the International Program for Chemical Safety in 2002, and it was concluded that its use in potable water was safe at concentrations that provide long term control of mosquito larvae but pose no hazards to humans.

The Biological Control of *Ae. aegypti* Larvae

The biological control of mosquito larvae includes use of predators, parasites or pathogens or, in the case of *Bacillus thuringiensis israelensis*, the toxins produced by a bacterial species.

The Use of Fish Predators Against *Ae. aegypti* Larvae

Fish are useful in certain settings for the control of *Ae. aegypti*. Workers in Guangxi on the southern coast of China added *Claris fuscus* to clay water tanks, as required by local law.[32,33] The larval index in tanks without fish was higher than in those with fish. Wang *et al.* (1990) successfully used fish to control *Ae. aegypti* in Taiwan, in a test village where of 6533 containers, 49.83% were used for water storage.[34] In their study, *Poecilia reticulata, Tilapia mossambica* and *Sarotherodon niloticus* were released into household water containers in the Liouchyou areas to control *Ae. aegypti* and *Ae. albopictus,* with a significant but not complete reduction in the number of larvae in jars treated with fish. MacDonald observed that *Poecilia reticulata* placed in water jars in refugee camps on the Thai–Cambodian border did not result in effective control of *Ae. aegypti* larvae.[35] Larvivorous fish were used more successfully in a small field trial in Burma (Myanmar), where *Trichogaster trichopterus* were introduced into containers infested with *Ae. aegypti* larvae. There was complete control for a period of nine months, after which time no more fish remained as they had been killed during the cleaning of the water containers or removed by children.[36] In Nicaragua, 324,394 fish were introduced into barrels, water toughs, wells and smaller water containers, and *Ae. aegypti* larvae were sampled over a five-month period — June to October, 1996. No *Aedes* larvae and pupae were found. The importance of continuing such studies over a longer period was stressed by the organizers of the trial.[37]

Fish may contaminate potable water. Chadee isolated *Citrobacter freudi*, *Escherichia coli* and *Pseudomonas aeruginosa* in guppies and in the water into which guppies had been introduced. Since guppies are frequently put into freshwater wells for *Ae. aegypti* control, the author cautions against the danger of their contaminating the water.[38]

The addition of fish to containers is labor-intensive, especially when much of the breeding is in small containers. Fish achieve better larval control when *Ae. aegypti* breed in larger containers and when the population can be educated not to remove them. Inasmuch as the bacterial larvicide, *Bacillus thuringiensis var. israelensis* (Bti), is not toxic to fish, the integrated control of larvae using both fish and Bti is a promising approach.

Predaceous Mosquito Larvae for the Control of *Ae. aegypti*

Larvae of a number of mosquito species are predators of *Ae. aegypti* larvae. The best-known and most-widely-used are members of the genus *Toxorhynchites*. The larvae of this genus are carnivorous, while adults feed only on plant nectars and therefore are not vectors of disease. Most published studies on *Toxorhynchites* in the control of *Ae. aegypti* describe the number of prey larvae which various instars of the predators are capable of consuming in the laboratory. It is difficult to relate the conclusions of these studies to field conditions, inasmuch as in many of the studies no alternative prey was offered. Larvae of one of several species of *Toxorhynchites* introduced into containers, or laboratory-reared adults released on islands or in small, isolated areas, reduced the density of *Ae. aegypti* populations.[39] Besides the limited efficacy of controlled experimental releases, the difficulty and expense of rearing and introducing large numbers of *Toxorhynchites* adults or larvae into *Ae. aegypti* larval habitats is beyond the capability of most vector control programs and too costly for most programs to sustain.

The Use of Copepod Predators

Cyclopid copepods have been applied as effective predators of mosquito larvae including *Ae. aegypti*. In New Orleans discarded tires were treated effectively with several species of copepods, eliminating both *Ae. aegypti* and *Ae. albopictus*.[40] Brown *et al.* carried out laboratory trials using seven species of *Mesocylops* from northeastern Australia, observing that *Mesocyclops aspericornis* was the most successful predator on *Ae. aegypti*.[41] However, a 1987 field trial in French Polynesia failed to lower the biting rate of adult *Ae. polynesiensis*. It was thought that this was due to the introduced copepods being unable to resist desiccation in land crab burrows in which mosquitoes breed.[42] Field trials in Laos,[43] Australia,[44] Puerto Rico, USA,[45] and Vietnam[46] showed the potential value of these agents in the control of *Ae. aeg*ypti and *albopictus* larvae. In trials carried out with a number of species of copepods in Costa Rica[47] *Mesocyclops thermocyclopoides* survived for 2–5 months in bromeliad leaf axils and for 3–6 months in used tires, and reduced the number of *Ae. aegypti* larvae by 79–99%.

Many factors limit the routine used of copepods as a biological control agent against container breeding *Stegomyia* species. They include the effect of precipitation or lack of it.[47,48] Drinking water storage containers are frequently emptied and washed out and copepods which have been introduced will be destroyed, as was the case in a trial in Mexico where, despite community participation, copepods failed to control *Ae. aegypti*.[48] In a trial in Honduras, householders could not maintain copepods because these were discarded along with the periodic change of water in jars for cleaning. Copepods persisted for long periods in tires or other containers which were not cleaned.[49]

However, Kay *et al.* successfully used copepods to control *Ae. aegypti* in 6 communes consisting of 11,675 households and 49,647 people living in north Vietnam.[50] Because important and sustained vector control achievements have been made in rural Vietnam on a very large scale, they are described in Chap. 12.

The Use of Bti

Bacteria used in the control of mosquitoes include Bti and *Bacillus sphaericus*. Both bacilli produce crystalline proteinaceous toxins during sporolation. These toxic particles when ingested by mosquito larvae are activated by enzymatic action in the alkaline medium of the midgut of the larva. The activated toxins destroy the gut lining. *B. sphaericus* has little action against *Ae. aegypti* and will not be further considered here. Many laboratory and field trials have shown the efficacy of Bti against container-breeding *Stegomyia* (spp.), particularly *Ae. aegypti* and *Ae. albopictus*.[51–54] Bti has been in wide use since 1980 and is very effective against

Ae. aegypti. Activity varies with the dosage rates. A factor limiting the use of Bti in the control of *Ae. aegypti* has been the relatively short time of action of the toxin, often no more than a few days. The short period of effectiveness has prevented wide use in developing countries.[55] However, slow-release granules and briquettes have been developed and these extend efficacy for as long as a month or more.[56] Some slow-release formulations of Bti have provided control for as long as 150 days.[57] Bti can be produced readily in commercial quantities, has a high efficacy, is environmentally safe, cost-effective, and is easily integrated into control programs. It may, in fact, be as effective as temephos sand granule formulations.[58] No resistance to this biological larvicide has appeared in mosquito populations in the field, nor have attempts to select for resistance in the laboratory been successful. It is likely that Bti will provide effective vector control in dengue-endemic countries. If formulations of Bti with sufficient persistence, reasonable cost, and acceptability to householders can be made available, its use should increase. The safety of Bti in potable water has been reviewed by the World Health Organization's International Program for Chemical Safety which found no reports of adverse effects on human health when it was consumed via drinking water or food. Recommendations for safe use will be published as an addendum to the Guidelines for Drinking-Water Quality.[59]

As noted above, Bti can also be incorporated in an integrated program along with fish predators.

Conclusions on the Feasibility of Larval Control

It is apparent that many problems hamper operational control of the aquatic stages of *Stegomyia* species. Among them are the extraordinarily large numbers of actual or potential larval habitats in most urban areas (many of them are small and difficult to locate), the limited number of effective and toxicologically acceptable insecticide compounds available for use in containers that store potable water, insecticide resistance, and the reluctance of communities to accept repeated treatment of their drinking water containers. The most important difficulty is that of obtaining the cooperation of human populations in control programs. If residents of dengue-endemic areas can be persuaded to cooperate in preventing *Aedes* breeding, control of larvae could be the centerpiece in the control of dengue.

Adulticidal Control of *Ae. aegypti*

Ae. aegypti rests and bites both indoors and outdoors in most of the areas in which it is found; because of this, indoor residual sprays applied as for the control of malaria vectors will not necessarily achieve a high level of control of the dengue vector populations. In some field trials in Southeast Asia, indoor residual sprays of pyrethroids have provided a high level of vector control, especially on the thatch and wood surfaces common in homes in this region. Evaluation of the efficacy of residual sprays has been based primarily on bioassays rather than studies of wild adult mosquito populations or dengue viral transmission. Lien *et al.* described the history of the control of *Ae. aegypti* and *Ae. albopictus* in Taiwan during 1987–1992.[60] *Ae. aegypti* was eradicated from Taiwan as a consequence of DDT spraying for the control of malaria vectors.[60] When DDT applications ceased following the eradication of malaria, *Ae. aegypti* returned. The authors endorsed the use of residual applications of the pyrethroid alphacypermethrin for the control of *Ae. aegypti*.

Sulaiman *et al.* applied lambdacyhalothrin and cyfluthrin within houses at doses of 0.01–0.05 mg/a.i.[61] At the higher doses the knockdown rates were high and control of natural populations was sustained for as long as 56 days. In a large study in Taiwan, Lin applied alphacypermethrin as a residual spray during epidemic outbreaks of dengue in 1990–1993 for control of the vector in areas where mosquito population densities were high, but methodological problems make interpretation of results difficult.[62]

Attempts to control *Ae. aegypti* populations by the application of indoor residual spray without also controlling outdoor adult and aquatic populations will not successfully degrade vector populations. Indoor residual adulticides are far too costly to apply routinely in large cities. However, as described below, sequential applications of insecticides by ultralow volume spray equipment have been effective is some areas in achieving considerable reduction in adult mosquito populations.

Space Spray Adulticiding for Control of *Ae. aegypti*; The Use of Thermal Fogs

Space sprays are commonly used in pest mosquito control programs to achieve rapid reduction of pest mosquitoes. They are also commonly applied for the control of mosquito vectors of disease, particularly during epidemics. Globally, the use of space sprays for emergency control of adult *Ae. aegypti* is very prevalent.[63] The two most common methods for this type of applications are by insecticides formulated as thermal fogs or by the application of ultralow volumes of insecticide concentrates. Thermal fogs are produced by equipment which mixes an insecticide at a relatively low concentration with diesel oil or kerosene as a carrier. The mixture is vaporized by being injected into a stream of high-velocity gas, resulting in a thick, highly visible fog. Until the development of ultralow volume cold fogs, thermal fogging was the most commonly used spray method to control *Ae. aegypti*. It remains popular in many countries. The insecticide most commonly used was 5% malathion and, more recently, pyrethroids. Thermal fogging can successfully control biting populations but insecticide persistence is low, generally no more than 3–4 days. Other limitations on thermal fogging are heat decomposition of insecticides and high costs, e.g. purchase and transport of the oil and kerosene diluents. On the other hand, an "advantage" of thermal fogging from the public relations standpoint is the high degree of visibility of the fogging operations, which provides inhabitants with visual evidence (plus smell) that the authorities are acting against the vector populations.[64]

Space Spray Adulticiding for Control of *Ae. aegypti*; The Use of ULV

The ultralow volume (ULV) spray technique applies high concentrations of insecticides in a small droplet liquid formulation, usually less than, 500 ml/he. ULV requires little or no diluent, which reduces the cost of transport and diluent. There are three factors to consider prior to the selection of ULV equipment and insecticides, optimum droplet size, application strategies and environmental impact. It is recommended that ULV equipment should produce a droplet size of 10–15 μ diameter. Droplets less than 5 μ in diameter do not readily settle or impact on mosquitoes. Those larger than 25 μ efficiently kill mosquitoes, but do not drift with wind currents. For aerial applications the droplet size should be less than 50 μ. ULV application equipment may be hand-carried, vehicle-mounted or mounted on aircraft.[65,66]

The WHO Aedes Research Unit (ARU) in Bangkok, Thailand, carried out a series of field trials on aerial and ground applications of ULV. The largest aerial application was conducted a city in northern Thailand with a population of some 50,000 living in an area of 7.75 km^2. Using a C-47 aircraft equipped with a fuselage-mounted spray boom, two ULV applications of 95% malathion were made at an interval of 4 days. The target dosage was 438/ml ha, which was higher than the application rates used in the USA at that time. Pretreatment landing rates of *Ae. aegypti* were 8.6 adults/man-hour and premise indices were 58–95%. The landing rates were reduced by 95 and 99% respectively after each application and reductions remained at 88–89% for the 10-day post-application-period. All ovitraps in the treated area were negative for 4 days after the first application and landing rates of other mosquito species in the area were reduced by 82–97%.[67]

Aerial application of insecticides for mosquito control is costly and requires skilled manpower and equipment often not available in resource-constrained dengue-endemic countries. Accordingly, the ARU thoroughly investigated the ground application of ULV insecticide concentrates in suburban Bangkok villages.

Field trials were carried out in isolated small villages or towns with backpack-portable mist blowers.[68] Six treatments of 856–1363 ml/ha of fenitrothion 8370 ULV concentrate at intervals of 13–69 days provided a high level of control for 7–8 months. Pretreatment landing rates were almost 25/man-hour. Immediately after treatment, landing and oviposition rates were zero. Ten weeks after the first and second treatments, which were spaced at 13 days, the landing rate was only 5.1/man-hour, a 78% reduction. ULV applications were made by spraymen walking along paths among houses and pointing the nozzle toward the houses as they moved, enabling rapid treatment to be made of entire villages. The equipment and insecticide concentrate are readily transportable by vehicle or air to remote areas and can quickly respond to the need for controlling vectors during an epidemic of dengue/DHF. Indoor applications of fenitrothion ULV concentrates by portable mist blowers enabled an immediate kill of adult mosquitoes and provided some residual effect from the droplets of insecticide concentrate on wall surfaces; larvicidal action was also noted. To study the effect of applying ULV fenitrothion concentrates directly into rooms and houses, applications were made in a 20 ha residential area on the outskirts of Bangkok with 1500 homes and a population of about 11,500.[69] The dosage rate was targeted at 0.1 ml/m^3 of room space and each room was treated for a count of about 20 s. Two treatments were made at an interval of 2 weeks. There was an immediate reduction of the population of *aegypti* from a pretreatment landing rate of 25.2 adults/man-hour to below 0.1/man-hour, a reduction of more than 99%. The few females collected on treated premises were all nulliparous. With the exception of one fully gravid female collected indoors 121 days after treatment, no further adults could be captured until 6 months later. Only 8 months after treatment did adult populations show signs of recovery, though landing rates remained at 0.9/man-hour. From 9 to 17 months after treatment, landing rates in the treated area were 69–92% lower than in the untreated area. No oviposition was found for 8 months after treatment and oviposition remained at a much lower level than in the control area for some time thereafter. It would thus appear that the principle of achieving very long term *Ae. aegypti* control using indoor ULV fenitrothion was demonstrated under the ecological conditions of Bangkok.

In the Americas, ULV has been much less successful in controlling *Ae. aegypti*. Aerial and ground applications were carried out in Colombia,[70,71] Puerto Rico, USA,[72] the Dominican Republic,[73] Jamaica,[74,75] Mexico[76] and Panama.[77,78] The poorer results in the Americas as compared to those in Bangkok and elsewhere in Southeast Asia may be attributable to the more closed and tighter house construction in the Americas plus the lower concentrations of insecticides applied in the Americas. Small ULV droplets (2–4 microns) penetrated to the interior in limited amounts, resulting in limited control by ULV applications made from equipment on the street.[78] Insecticide resistance, in some cases, may have affected the efficacy of the applications in American studies. In Mexico, *Ae. aegypti* has become highly endophilic and might be effectively controlled by indoor spraying.[79] In Panama, 75.1% of the adult *Ae. aegypti* rested indoors.[78]

Conclusions: What Are the Options?

Clearly, many problems face the organizers and managers of *Ae. aegypti* vector control programs. The success achieved by *Ae. aegypti* eradication programs in 21 countries of the Americas in the 1950's and 1960's should be an inspiration to modern day vector control teams. Indeed, a plan drawn up in Caracas, Venezuela, in 1997 called for continental eradication of the species.[3] But, in most countries, the methods of the 1950's and 1960's cannot be applied due to changed ecological and social conditions. Industrial products, together with environmental changes in the urban areas of the Americas, have created large numbers of new larval habitats, which did not exist or were uncommon at the time of earlier eradication campaigns in the Americas. Plastic containers, used tires and other waste containers are discarded in enormous numbers into vacant lots and onto streets of the cities and towns in dengue-endemic countries. Environmental and toxicological concerns now limit the types of insecticides that can be used in potable water or as aerial sprays. Insect resistance is developing to a large range of compounds. Changing attitudes now prevent the possibility of exercising the kind of staff discipline commonly applied during the eradication era. If it was technically and politically feasible, eradication of *Ae. aegypti* would indeed resolve the problem of dengue, at least in the Americas. However, for the reasons given here and because of the frequency with which mosquitoes are transported from country to country in aircraft (not to mention the introduction of *Ae. albopictus*), species eradication is no longer a feasible option. Abundant *Ae. aegypti* in suburban and rural areas in close proximity to jungles and to the cities of South America and Africa also poses a risk of urbanization of yellow fever. Urban yellow fever epidemics have recently occurred in Africa.[80] Poor funding and low priority continues to retard the universal supply of piped water to villages and poor urban suburbs in most dengue-endemic countries. Unfortunately, this results in the continued practice of storing water for household use.

It must be emphasized again that until such time as an effective, safe and affordable multivalent vaccine is available, control of dengue transmission must depend on the effective control of its vector(s). This will only be achieved through establishing vector control programs supervised by well-trained professional staff provided with adequate budgetary support for their control operations. Faced with the changing nature of the problems in virtually all areas in which the vectors are found, and their spread to areas in which they have not previously been reported, it is imperative that the nature of vector control activities and the methods and materials also change and adapt. Vector control carried out by poorly trained workmen under supervisors with little knowledge of the bionomics of the vectors, their population dynamics and insecticide susceptibility has yielded poor results and as had little to no impact on dengue viral transmission rates. The recruitment of well-educated, professional entomologists (with an adequate career structure open to them) will enable more judicious use to be made of the techniques and materials which are available.[81] Familiarity with the literature on vector control, and knowledge of earlier field trials and experiences in their operational areas, will add to the improvement of operational control techniques.[13] A higher level of professional training must be given to vector control personnel, enabling them to make more efficient use of available technologies and adapt these to match local ecological conditions. Better mapping of dengue disease and dengue vectors is needed, so that control personnel can identify those areas requiring priority action. Accurate surveillance for the distribution of vector and population densities will identify areas at elevated risk of disease transmission. The status of insecticide susceptibility and resistance in vector populations must be routinely monitored, particularly to those insecticides which are used in the control operations.

There is also an increasing recognition that both informed involvement of the community and education of children must play a greater role in dengue vector control, and this may, at least, reduce the incidence of dengue.[82–84] Valero, reflecting on the inadequacy of the dengue vector control program in Venezuela, concluded that the emergency measures to combat the epidemics had limited effect.[85] Rather, an integrated control program is needed. In practice, environmental improvements must be combined with changes in human behavior affected by sanitary education and the implementation of the necessary laws to support it. This last aspect is fundamental. Much has been written about the role of people in preventing or reducing vector breeding. Often, the authorities provide the community with modest information about the disease, the breeding sites of mosquitoes and control procedures, then abandon them to fend for themselves. As has been noted, there are cases in which the community has collected refuse which may have provided mosquito breeding sites only to find out that the authorities have failed to dispose of the solid waste. People who reside in dengue-endemic epidemic areas have many pressing priorities to survive and may not place a high priority on their role in controlling the vector mosquito that transmits dengue. Often they are confused by admonitions to dispose of larval habitats when some containers are essential for household use.

Greater efforts must be made to coordinate epidemiological risk assessments of dengue epidemics with timely interventions against the vectors, employing whatever methods have been shown to be locally effective, realistically applicable, and affordable and acceptable to the population. At all times, it is essential that continuous efforts are made to reduce the population densities of the vectors, both by encouraging and supporting environmental measures undertaken by the populations of the endemic areas and by improving the vector control (and prevention) services of municipalities and governments.

Finally, two elements critical to success in the war against *Ae. aegypti* are missing — research/education and the use of the private sector. Both of these elements have been ignored virtually everywhere in the world of dengue. Gratz surveyed graduate programs in medical entomology in Asian countries and found a serious deficit in all countries in university-based research on the bionomics and control of *Ae. aegypti.*[81] The result is a severe dearth of research on control of *Ae. aegypti* and governments in dengue-endemic countries who are unable to hire vector control professionals specifically trained in relevant methods. What is more, despite the example and the long term success of mosquito abatement in the Central Valley of California, the model of competitive award of contracts to private vector control companies who are paid with revenues has not been even explored in any country battling dengue.[86,87] The role of the government in this model is to set standards, award and manage contracts, evaluate performance, and possibly set up vector control training courses. The day-to-day work of vector control is performed by employees who can be responsive to the rewards of the private sector.

References

1. Mathiot C, Clerc Y, Rodhain F, Digoutte JP, Coulanges P. [West Nile virus in Madagascar]. *Arch Inst Pasteur Madagascar* 1984;51(1):113–123.
2. Halstead SB. *Aedes aegypti*: Why can't we control it? *Bull Soc Vector Ecol* 1988;13:304–311.
3. PAHO. [Hemispheric plan to expand and intensify efforts to combat *Aedes aegypti*]. *Rev Salud Pan Am* 1998;3:124–130.
4. Bronfman M, Gleizer M. Appearances and reality in community participation: need, excuse or strategy. In: Halstead SB, Gomez-Dantes H (eds.) *Dengue: A Worldwide Problem, A Common Strategy. Proceedings of the International Conference on Dengue and Aedes aegypti Community-Based Control*, Merida, Mexico, July 11–16 1992. Ministry of Health, Mexico, 1992, pp. 63–73.
5. Ault SK. Environmental management: a re-emerging vector control strategy. *Am J Trop Med Hyg* 1994;50S:35–49.
6. Gubler DJ, Clark GG. Community involvement in the control of *Aedes aegypti*. *Acta Trop* 1996;61:169–179.
7. Barrera R, Avila J, Gonzalez-Telez S. Unreliable supply of potable water and elevated *Aedes aegypti* larval indices: a causal relationship? *Am J Mosq Control Assoc* 1993;35:141–148.
8. Burton GJ. Coastal survey of *Aedes aegypti* breeding in British Guiana. *Ann Trop Med Parasitol* 1963;57:446–451.
9. Barrear-Rodriquez R, Machado CE, Bulla L, DRS. Mosquitoes and mourning in the Caracas cemetery. *Antenna* 1982;6:250–252.
10. Tonn RL, Bang YH, Yasuno M. *Water and Mosquito Populations in Bangkok, Thailand*. Report No. WHO/VBC/69.166, World Health Organization, Geneva, Switzerland, 1969.

11. Nelson MJ, Pant CP, Self LS, Usman S. Observations on the breeding habits of *Aedes aegypti* in Jakarta, Indonesia. *Southeast Asian J Trop Med Public Health* 1976;7:424–442.
12. Kittayapong P, Strickman D. Distribution of container-inhabiting *Aedes* larvae (*Diptera*: *Culicidae*) at a dengue focus in Thailand. *J Med Entomol* 1993;30:601–606.
13. Gratz NG. Lessons of *Aedes aegypti* control in Thailand. *Med Vet Entomol* 1993;7:1–10.
14. Chan KL. *Singapore's Dengue Haemorrhagic Fever Control Programme: A Case Study on the Successful Control of Aedes aegypti and Aedes albopictus Using Mainly Environmental Measures as a Part of Integrated Vector Control.* Southeast Asian Medical Information Center, Tokyo, 1985.
15. Ooi EE, Goh KT, Gubler DJ. Dengue prevention and 35 years of vector control in Singapore. *Emerg Infect Dis* 2006;12:887–893.
16. Focks DA, Brenner RJ, Hayes J, Daniels E. Transmission thresholds for dengue in terms of *Aedes aegypti* pupae per person with discussion of their utility in source reduction efforts. *Am J Trop Med Hyg* 2000;62(1):11–18.
17. Pontes RJ, Freeman J, Oliveira-Lima JW, Hodgson JC, Spielman A. Vector densities that potentiate dengue outbreaks in a Brazilian city. *Am J Trop Med Hyg* 2000;62:378–383.
18. Phanthumachinda B. Pilot studies on community participation: *Aedes aegypti* control in Phanus Nikyom district, Chonburi province, Thailand. *Dengue Bull* 1984;10:35–41.
19. WHO. *14th Report of the Expert Committee on the Safe Use of Pesticides.* Tech Report Series, No. 813, World Health Organization, Geneva, Switzerland, 1991.
20. WHO. *Report of the Informal Consultation on the Safety of Microbial Pest Control Agents.* Report No. WHO/PCS 93.47, World Health Organization, Geneva, Switzerland, 1993.
21. Rawlins SC. Spatial distribution of insecticide resistance in Caribbean populations of *Aedes aegypti* and its significance. *Rev Panam Salud Publica* 1998;4(4):243–251.

22. Wirth MC, Georghiou GP. Selection and characterization of temephos resistance in a population of *Aedes aegypti* from Tortola, British Virgin Islands. *J Am Mosq Control Assoc* 1999;15:315–320.
23. Lee HL. Esterases activity and temephos susceptibility in *Aedes aegypti* larvae. *Mosq Borne Dis Bull* 1991;2:61–66.
24. de la Cruz AM, Mesa A, San Martin JL. The community and the control of *Aedes aegypti*: perception and behavior regarding temephos larvicide. *Rev Cubana Med Trop* 2001;53:44–47.
25. Geevarghese G, Dhanda V, Rao PNR, Deobhankar RB. Field trials for the control of *Aedes aegypti* with Abate in Poona city and suburbs. *Indian J Med Res* 1977;65:466–473.
26. Que ND, Rung DV, Chow CY. *Aedes* mosquito surveillance in the Republic of Vietnam. *Southeast Asian J Trop Med Public Health* 1974;5:569–573.
27. Lubis I, Suharyono W. Reduction in dengue transmission after mass treatment of temephos SG 1 ppm in Indonesia. *Dengue Bull* 1985;11:15.
28. Sulaiman S, Jeffery J, Sohadi AR. Residual efficacy of Altosid and bactimos briquets for control of dengue/dengue haemorrhagic fever vector *Aedes aegypti* (L). *Mosq Borne Dis Bull* 1991;8(1):123–126.
29. Ritchie SA, Broadsmith G. Efficacy of ALTOSID pellets and granules against *Aedes aegypti* in ornamental bromeliads. *J Am Mosq Control Assoc* 1997;13(2):201–202.
30. Nyar JK, Ali A, Zaim M. Effectiveness and residual activity comparison of granular formulations of insect growth regulators pyriproxyfen and s-methoprene against Florida mosquitoes in laboratory and outdoor conditions. *J Am Mosq Control Assoc* 2002;18:196–201.
31. Yapabandara AM, Curtis CF. Laboratory and field comparisons of pyriproxyfen, polystyrene beads and other larvicidal methods against malaria vectors in Sri Lanka. *Acta Trop* 2002;81:211–223.
32. Wu N, Wang S. Vector control for the prevention of dengue fever in Guangxi, People's Republic of China. *Dengue Bull* 1985;11:41–45.
33. Wu N, Wang SS. Fish as front-line health workers. *World Health Forum* 1988;9:220.
34. Wang CH, Hwang JS, Lay JR. [Preliminary study on the biological control of dengue vectors by fish in Liouchyou Prefecture, Pingtung County, Taiwan]. *Kao-Hsiung I Hsueh Ko Hsueh Tsa Chih (Kaohsiung J Med Sci)* 1990;6: 382–388.

35. MacDonald MB. Fighting a new disease: issues in community mobilization for controlling DHF in dispaced persons camps along the Thai–Cambodian border. *Southeast Asian J Trop Med Public Health* 1990;21:681–682.
36. Aung H, Paing M, Oo LL. The efficacy of mosquito fish *Trichogaster trichopterus* for control of *Aedes aegypti* in a community, Yangon. *Myanmar Health Sci Res J* 1991;3:36–40.
37. Miranda MM, Mendoza C, Guadamuz J, Perez R. [Evaluation of the predatory capacity of various genera of larvivorous fish and their use in the control of the aquatic phases of dengue and malaria vectors]. *Rev Nicar Entomol* 1990;50:47–54.
38. Chadee DD. Bacterial pathogens isolated from guppies (*Poecilia retulata*) used to control *Aedes aegypti* in Trinidad. *Trans R Soc Trop Med Hyg* 1992; 86:693.
39. Gerberg EJ. Sequential biocontrol application in the use of *Toxorhynchites* spp. In: Laird M, Miles JA (eds.) *Integrated Mosquito Control Methodologies II*. Academic Press, London, 1985.
40. Marten GG. Evaluation of cyclopoid copepods for *Aedes albopictus* control in tires. *J Am Mosq Control Assoc* 1990;6:681–688.
41. Brown MD, Kay BH. Evaluation of Australian *Mesocyclops* (*Cyclopoida*: *Cyclopidae*) for mosquito control. *J Med Entomol* 1991;28:618–623.
42. Lardieux F, Riviere F, Sechan Y, Kay BH. Release of *Mesocyclops aspercornis* for control of larval *Aedes polynesiensis* in land crab burrows on an atoll of French Polynesia. *J Med Entomol* 1992;29:571–576.
43. Jennings CD, Bounlay P, Bounsoum S, Kay BH. *Aedes aegypti* control in the Lao People's Democratic Republic with reference to copepods. *Am J Trop Med Hyg* 1995;53:324–330.
44. Russell BM, Muir LE, Weinstein P, Kay BH. Surveillance of the mosquito *Aedes aegypti* and its biocontrol with the copepod *Mesocyclops aspercornis* in Australian wells and gold mines. *Med Vet Entomol* 1996;10:155–160.
45. Rawlins SC, Martinez R, Wiltshire S, Clarke D, Prabhakar P, Spinks M. Evaluation of Caribbean strains of *Macrocyclops* and *Mesocyclops* (*Cyclopoida*: *Cyclopidae*) as biological control tools for the dengue vector *Aedes aegypti* *J Am Mosq Control Assoc* 1997;13(1):18–23.
46. Vu SN, Nguyen TY, Kay BH, Marten GG, Reid JW. Eradication of *Aedes aegypti* from a village in Vietnam, using copepods and community participation. *Am J Trop Med Hyg* 1998;59(4):657–660.

47. Schaper S. Evaluation of Costa Rican copepods (*Crustacea*: *Eudecapoda*) for larval A*edes aegypti* control with special reference to *Mesocyclops thermocylopoides. J Am Mosq Control Assoc* 1999;15:510–519.
48. Gorrochotegui-Escalante N, Fernanddez-Salas I, Gomez-Dantes H. Field evaluation of *Mesocyclops longisetus* (*Copepoda*: *Cyclopoidea*) for the control of larval *Aedes aegypti* (*Diptera*: *Culicidae*) in northeastern Mexico. *J Med Entomol* 1998;35:699-703.
49. Marten GG. *Honduras: Use of Cyclopoids for Aedes aegypti Control in El Progresso, October 1990–December 1992.* VBCk Report No. 81329, USAID Washington DC, 1993.
50. Kay BH, Nam VS, Tien TV, Yen NT, Phong TV, Diep VT *et al*. Control of *Aedes* vectors of dengue in three provinces of Vietnam by use of *Mesocyclops* (*Copepoda*) and community-based methods validated by entomologic, clinical and serological surveillance. *Am J Trop Med Hyg* 2002;66:40–48.
51. de Barjac H. [Toxicity of *Bacillus thuringiensis* var *israelensis* for larvae of *Aedes aegypti* and *Anopheles stephensi*]. *C R Acad Sci Hebd Seances Acad Sci D* 1978;286:1175–1178.
52. Hougard JM, Duval J, Escaffre H. Evaluation en milieu naturel de l'activite larvicide d'une formulation de *Bacillus thuringiensis* H 14 sur *Aedes aegypti* dans un foyer epidemique de fievre jaune en Cote d'Ivoire. *Cah ORSTOM Ser Entomol Med Parasitol* 1985;23:235–240.
53. Lee HL, Pe TH, Cheong WH. Laboratory evaluation of the persistence of *Bacillus thuringiensis* var *israelensis* against *Aedes aegypti* larvae. *Mosq Borne Dis Bull* 1986;2:61–66.
54. Lee HL, Cheong WH. Field evaluation of the efficacy of *Bacillus thuringiensis* H-14 for the control of *Aedes* (*Stegomyia*) *albopictus*. *Mosq Borne Dis Bull* 1987;3:57i–i63.
55. Mulla MS. Biological control of mosquitoes with entomopathogenic bacteria. Proceedings of the IV National Vector Symposium, Taichung, Taiwan. *Chin J Entomol* 1991;6:93–103.
56. Novak RJ, Gubler DJ, Underwood D. Evaluation of slow-release formulations of temephos and *Bacillus thuringiensis* var. *israeliensis* for the control of *Aedes aegypti* in Puerto Rico. *J Am Mosq Control Assoc* 1985;18:196–201.
57. Ali M. Comparative efficacy of temephos (Abate 1% SG) and *Bacillus thuringiensis israelensis* tablets (ABG-6499) against *Aedes aegypti* in potable water. In: *13th European SOVE Meeting*, Antalya, Turkey, 2000.

58. Rivera P, Lopez M, Valle S, Lopez O, Espinoza P. Evaluation of the larvicidal impact of culinex (Bti H-14, tablet) and temephos 1% against *Aedes aegypti* under simulated natural condiitions. *J Am Mosq Control Assoc* 1997;13:125.
59. WHO. *Guidelines for Drinking-Water Quality*, 3rd ed. First addendum to Vol. 1. World Health Organization, Geneva, Switzerland, 2006.
60. Lien JC, Wu YC, Huang HM, Chung CL, Yueh IY, Lu LC. Survey and control of dengue fever vectors *Aedes aegypti* and *Aedes albopictus* in Taiwan during 1987–1992. In: Halstead SB, Gomez-Dantes H (eds.) *Dengue: A Worldwide Problem, A Common Strategy. Proceedings of the International Conference on Dengue and Aedes aegypti Community-Based Control*, Merida, Mexico, July 11–16, 1992. Ministry of Health, Mexico, 1992.
61. Sulaiman S, Karim MK, Omar B, Omar S. Field evaluation of lambbda-cyhalothin and cyfluthrin against dengue vectors in an endemic area in Malaysia. *J Florida Mosq Control Assoc* 1993;64:26–29.
62. Lin TH. Surveillance and control of *Aedes aegypti* in epidemic areas of Taiwan. *Kaohsiung J Med Sci* 1994;10:S88–S93.
63. Gratz NG. Emergency control of *Aedes aegypti* as a disease vector in urban areas. *J Am Mosq Control Assoc* 1991;7:353–365.
64. Pant CP. Space sprays used in mosquito vector control. In: Laird M, Miles JW (eds.) *Integrated Mosquito Control Methodologies*. Academic, London, 1983, pp. 37–48.
65. Mount GA. Ultra-Low Volume Application of Insecticides for Vector Control. Report No. VBC 85.919, World Health Organization, Geneva, 1985.
66. Mount GA. A critical review of ultralow-volume aerosols of insecticide applied with vehicle-mounted generators for adult mosquito control. *J Am Mosq Control Assoc* 1998;14:305–334.
67. Lofgren CS, Ford HR, Tonn RJ, Jatanasen S. The effectiveness of ultra-low volumn applications of malathion at a rate of 6 US fluid ounces per acre in controlling *Aedes aegypti* in a large-scale field test at Nakhon Sawan. Thailand. *Bull World Health Organ* 1970;42:15–25.
68. Wirat S, Pant CP. *Sequential Application of ULV Sumithion for Sustained Control of Aedes aegypti: Use of Fontan and a Backpack Portable Mist Blower.* Report No. WHO/VBC/73.432, World Health Organization, Geneva, 1973.

69. Pant CP, Mathis HL, MJN, Phanthumachinda B. A large-scale field trial of ultra-low volume fenitrothion applied by a portable mist blower for the control of *Aedes aegypti. Bull World Health Organ* 1974;51:409–415.
70. Motta Sanchez A, Tonn RJ, Uribe LJ, Calheiros LB. Comparacion de la eficacia de varios metodos de aplicacion de insecticidas para el control o la erradicacion del *Aedes aegypti* en Colombia. *Bol OSP* 1978; 84:24–37.
71. Uribe LJ, Garrido GC, Nelson M, TInker ME, Moquillaza J. Aerial ULV application trial of malathion against *Aedes aegypti* in a city of Colombia. *PAHO Bull* 1984;18:43–57.
72. Fox I, Specht P. Evaluating ultra-low volume ground applications of malathion against *Aedes aegypti* using landing counts in Puerto Rico, 1980–84. *J Am Mosq Control Assoc* 1988;4:163–167.
73. Perich MJ, Tidwell MA, Williams DC, Sardelis MR, Pena CJ, Mandeville D *et al.* Comparison of ground and aerial ultra-low volume applications of malathion against *Aedes aegypti* in Santo Domingo, Dominican Republic. *J Am Mosq Control Assoc* 1990;6(1):1–6.
74. Moody C, Bowen-Wright C, Murray J, Castle T, Barrett MW, Dunkley G *et al.* Emergency control of vectors during a dengue outbreak in Jamaica, 1977. In: *Dengue in the Caribbean*, 1977; 1979. PAHO, Jamaica, 1979, pp. 87–92.
75. Castle T, Amador M, Rawlins S, Figueroa JP, Reiter P. Absence of impact of aerial malathion treatment on *Aedes aegypti* during a dengue outbreak in Kingston, Jamaica. *Rev Panam Salud Publica* 1999;5(2):100–105.
76. Perez Gonzalez JL, Fernandez Salas I, Flores Leal A, Suarez A. Comparison of cold (ULV) versus thermal (fog) indoor spraying to control resting female *Aedes aegypti* in Monterrey, northeastern Mexico. *J Am Mosq Control Assoc* 1997;13:123.
77. Echevers G, Lima MJ, Franco RM, Calheiros LB. Nebulizaciones terrestres con malation a volumen ultrareducido (ULV) en Panama. *Bol Of Sanit Panam* 1975;75:405-12.
78. Perich MJ, Davila G, Turner A, Garcia A, Nelson M. Behavior of resting *Aedes aegypti* (*Culicidae*: *Diptera*) and its relation to ultra-low volume adulticide efficacy in Panama City. *Panama J Med Entomol* 2000;37(4):541–546.
79. Galvan E, Fernandez Salas I, Flores Leal A. House exit and entrance behavior and physiology of female *Aedes aegypti* in Monterrrey City, northeastern Mexico. *J Am Mosq Control Assoc* 1997;13:121–122.

80. Miller BR, Monath TP, Tabachnik WJ, Ezike VI. Epidemic yellow fever caused by an incompetent mosquito vector. *Trop Med Parasitol* 1989;40:396–399.
81. Gratz NG. Education and employment of medical entomologists in *Aedes aegypti* control programmes. *Kaohsiung J Med Sci* 1994;10:S19–S27.
82. Chiaravalloti VB, Morais MS, Ciaravalloti Neto F, Conversani DT, Riorin AM, Barbosa AA *et al*. [Evaluation of compliance with dengue fever prevention: the case of Catanduva, Sao Paulo, Brazil]. *Cad Saude Publica* 2002;18:1321–1329.
83. Madeira NG, Macharelli CA, Pedras JF, Delfino MS. Education in primary school as a strategy to control dengue. *Rev Soc Bras Med Trop* 2002;35:221–226.
84. Tauil PL. [Critical aspects of dengue control in Brazil]. *Cad Saude Publica* 2002;18:867–871.
85. Valero N. [Toward integrated dengue control]. *J Invest Clin* 2002;43:141–144.
86. Challet GL. Mosquito abatement district programs in the United States. *Kaohsiung J Med Sci* 1994;10:S67–S73.
87. Clarke JL. Privatization of mosquito control services in urban areas. *Kaohsiung J Med Sci* 1994;10:S74–S77.

12

Biological Control of Dengue Vectors: Promises from the Past

Brian H. Kay

Natural control of mosquito vectors of dengue may be introduced by adding mass-produced agents either into existing natural or constructed environments or into man-made containers breeding *Aedes* vectors of the dengue viruses. In contrast to agriculture, which considers economic thresholds and crop damage, goals for mosquito control should be to lower populations to levels necessary to break the transmission cycle. Unfortunately, in the case of dengue, we have little idea of what these levels should be. In Trinidad, Focks and Chadee[1] estimated that in most of their 16 study sites, control efficacy was likely to be > 99% reduction in pupal populations. More recent computer modelling simulations of Focks and colleagues[2] suggest a more variable control efficacy of > 90% population reduction for some Caribbean, Central American and South-east Asian localities. In Fortaleza, Brazil, the critical vector prevalence permitting dengue virus transmission was equated with a household index of 1%,[3] but since this index indicates prevalence, the actual adult productivity of this 1% may vary widely according to the container type and its situation.

As the level of suppression required depends on demographic factors such as human population density and dengue immunity plus climate, it seems prudent at this stage to suggest that any contemplated biological control agent should be able to reduce *Aedes* vector populations by at least 95%, the minimum level expected of insecticidal methods. It should be noted that incomplete control may reduce density-dependent mortality factors and actually increase the numbers of emerging adults,[4] as noted in trials with tree-hole breeding *Ae. sierrensis* and the ciliate parasite *Lambornella clarki*.[5]

The published literature is full of reports on potential biological control agents for mosquitoes, mainly involving aquatic stages. Given that in 1964 Jenkins annotated over 400 pathogens, parasites and predators, it is surprising that most still remain in the "promising" category and relatively few field trials have been done.[6] Why is this?

In 1974, Chapman[7] reviewed invertebrates including external and internal ciliates (*Tetrahymena* being an example of the latter), microsporidia (*Edhazardia aedis*), bacteria (*Bacillus sphaericus*), Proteobacteria (rickettsiae), *Wolbachia pipientis*, various virus groups, fungi (*Lagenidium giganteum*), nematode worms (*Romanomermis* spp.), as well as plants which produce toxins or act as predators. Despite some mass production problems for field release, shelf life problems, limited temperature ranges and equivocal field results, Chapman remained generally optimistic about their potential. However, that was over 30 years ago.

Of the numerous agents reviewed, only some of those listed above have had sufficient longevity to enter the modern literature, but sometimes for different reasons. Maternally transmitted *Wolbachia* symbionts cause cytoplasmic incompatibility in 15–20% of insect species, but for medical entomology it is best known in *Culex pipiens* complex mosquitoes.[8,9] Rickettsial-like micro-organisms have been known for over 70 years, as *Wolbachia pipientis* was described by Hertig in 1936.[10] After mass releases of incompatible *Wolbachia*-infected males obtained from Fresno, California, monsoonal rain disrupted an attempted eradication of *Culex pipiens fatigans* (= *Cx quinquefasciatus*) at Okpo, Myanmar, between February and May 1967.[11] There is no doubt that that the withdrawal of WHO field units from Asia in the mid-1970s, stimulated by emotive accusations of development of biological warfare agents in New Delhi,[12] had a prolonged dampening effect on some of this promising technology, but interestingly enough there has been a recent rejuvenation of interest in strategies involving the sterile male approach. In California, a virulent *Wolbachia* strain was discovered from *Drosophila melanogaster*, which had the ability to shorten adult lifespan.[13] This phenotype, which exhibits an inability to control replication in nervous and muscle tissue, is known as "popcorn" (*w*MelPop) and offers another potential approach. The least complicated strategy is to introduce *w*MelPop to vector species via mosquito cell lines and by microinjection into embryos to create stabilized infected lines, and suitable for backcrossing into local target vector populations. Such collaborative work has commenced, led by Scott O'Neill at the University of Queensland, with funding under the banner of the Bill and Melinda Gates Grand Challenges in Global Health, and *w*MelPop lines have been successfully established in *Ae. aegypti*.[14] The objective is to curtail survival to the point where most adult *Ae. aegypti* females will perish prior to the period required for extrinsic incubation of dengue. *Wolbachia* are also being investigated as possible tools for driving genes into vector populations[15,16] for disease prevention, but this approach would fall under the highly charged area of genetically modified organisms. Since insect survival is a geometrically related parameter in the mathematics of pathogen transmission, the interruption of vector survival may have a far greater utility than direct approaches to altering genes affecting susceptibility. Since *Wolbachia*-infected mosquitoes are common and ubiquitous,[14,17] it is hoped that governments will view *w*MelPop-infected vectors as having just another strain of *Wolbachia.* Thus it may qualify as a biological control agent, and a local one at that. One of the reasons why the cyclopoid predator *Mesocyclops* (see later) was accepted so readily in Vietnam was that pre-existing local *Mesocyclops* were found in every village survey[18] and the process was promoted as one of social expansion of something beneficial into non-infected jars and tanks, rather than introduction of something new, foreign and strange.

Baculoviruses (formerly nuclear polyhedrosis viruses) have gained some prominence as vectors in genetic manipulation, as have densoviruses (*Parvoviridae*) such as the *Aedes aegypti* densovirus, which has been developed into an expression and transducing vector in mosquitoes.[19] There is interest in densoviruses in their own right, as potential biological control agents as mortality of *Aedes albopictus* larvae exceeded 97% after 21 days after infection with C6/36 DNV.[20] Furthermore, data from this study suggested an inverse relationship between the densovirus (C6/36DNV) and the titre of the coincidental dengue 2 infection in the mosquitoes.

Tetrahymena pyriformis has been used as a synergistic bioencapsulation tool for *Bacillus thuringiensis israelensis*,[21] as this aquatic protozoan can accumulate up to 240 parasporal crystalline inclusions (the delta endotoxins of *Bt*) per cell in its food vacuoles. This originally was done to enhance a short residual persistence and a narrow spectrum of activity in conventional *Bacillus* while avoiding the regulatory restrictions associated with recombinant organisms. However, significant progress with DNA recombinant technology led to the cloning of the *Bti* genes into other organisms,[22] such as cyanobacteria, gram-negative bacteria and Bacillus *sphaericus*, to produce potential products which theoretically should be effective against a broader range of mosquito genera and blackflies (*Simuliidae*). However, transgenic products such as these still cause anxiety, are time-consuming to construct, and are costly. There have been questions raised about the maintenance of appropriate dosage in such organisms and whether attenuation might occur in the natural environment.

Some of those biological control candidates are now more correctly referred to as biorational insecticides or biopesticides. The juvenile mosquito hormone mimic s-methoprene inhibits emergence from pupae, may cause some physiological malformation of adult genitalia, and larval death at higher dosages, and is widely used at low rates of 8 ppb in saltmarsh and freshwater habitats in Australia with little apparent acute toxicity to non-target organisms.[23,24] For dengue control which is associated mainly with artificial rather than natural habitats, potential toxicity issues relate mainly to drinking water quality. However, *Bacillus* and s-methoprene are well accepted in this regard, and have minimum effect on non-target organisms. Recently another juvenile hormone mimic, pyriproxyfen as chips, has been evaluated for usage in water jars in Vera Cruz, Mexico, with unconvincing results.[25]

Bacillus thuringiensis israelensis is used extensively in broad-scale control in Australia[26] and elsewhere and, as with *B. sphaericus*, is widely accepted as being affordable, environmentally sound products. The former is generally regarded as being useful against *Ae.*

aegypti,[27,28] but *B. sphaericus* was 220–3500 times less efficacious.[29] In the environment, however, *B. thuringiensis* is short-lived, particularly in polluted habitats, although there have been some claims of recycling. In contrast, *B. sphaericus* recycles in larval cadavers, which seem to contain nutrients for vegetative multiplication and toxin synthesis, associated with sporulation,[30] and it is more closely associated with use in polluted water and *Culex quinquefasciatus* control[28] than in predominantly clean water, generally associated with species such as *Ae. aegypti*.

Fungal diseases of insects have been known since 1834, and the concept of using them as biocontrol agents was suggested independently by Pasteur and by Le Conte in 1874.[31] As early as 1879, the fungus *Metarhizium anisopliae* could be mass-produced, and in their extensive review of fungi of potential importance for biological control of agricultural and medically important arthropods, Hall and Papierok[31] noted a resurgence of interest through the 1970s and earlier, because of the growing problem of insecticide resistance. They noted over 400 species of entomogenous fungi with those from the classes deuteromycetes (e.g. *Metarhizium*, *Beauvaria*, *Culicinomyces*, *Tolypocladium*), oomycetes (*Lagenidium*, *Leptolegnia*) and chytridiomycetes (*Coelomomyces*) perhaps showing the greatest applicability for mosquito control. But what happened after that? The authors suggested that shortcomings in bioassay systems, formulation and storage contributed to their failure to develop to their full potential, coupled with shortfalls in the ecological, fundamental and developmental understanding.

Lagenidium giganteum (Lagenex®)[28] is now commercially available for floodwater *Aedes* control, but it seems that alternative entomopathogenic fungi such as *Beauvaria bassiana* and *Metarhizium anisopliae* are only now receiving serious contemporary consideration for vector control of the medically important pests, perhaps stimulated by the presence of commercial products in agriculture for locust and grasshopper control.[32] The search for biopesticides has also intensified due to raised environmental consciousness with respect to not only human health but also inactivity against non-target organisms. There also have been substantial advances in formulation, storage and operational production of entomopathogens.

In the laboratory, Blanford and collaborators[33] demonstrated that spore suspensions of *Beauvaria bassiana* used as surface sprays could reduce the numbers of *Anopheles stephensi* able to transmit malaria (a *Plasmodium chaubaudi* model) by approximately 80-fold, and in Tanzania, applications of *Metarhizium anisopliae* to suspended cotton sheets hung inside huts resulted in 23% infection of resting *Anopheles gambiae* and a potential 75% reduction in transmission of malaria.[34] Similar to the theory behind the release of life shortening strains of *Wolbachia*, this approach to reducing the daily probability of daily survival is now being investigated as a method of disrupting the transmission dynamics of dengue, as *Ae. aegypti* primarily rests indoors.

Up until 1983, some other entomopathogens were trialled in separate and integrated format in Pacific countries. Such studies are documented by Marshall Laird (a champion of biocontrol) and James Miles in volume 2 of their informative books on integrated mosquito control methodologies.[35] In Tuvalu, the mermithid worm *Romanomermis culicivorax*[28] produced uneven infection rates but some control; however, integrated use with Tecknar (*B. thuringiensis israelensis*), Altosid (s-methoprene) and later the adulticide bendiocarb against *Ae. aegypti* proved efficacious when coupled with community source reduction campaigns. In Western Samoa, Pillai attempted to control five mosquito species in ground pools and artificial containers with *Romanomermis* but infection rates varied, and recycling was negligible. When inoculated into *Pandanus* leaf axils at Apia, recycling did not occur, in contrast to Laird's earlier experiments in Tokelau, where the nematode persisted at low levels for up to 35 months. In New Zealand, *Coelomomyces opifexi* failed to control *Aedes australis*. The champions who carried out these experiments have long retired, and results were patchy to say the least, so possibly one could conclude that these potential agents were unsuitable for long term consideration.

Fish continue to be used as biological control agents in man-made containers, and their effectiveness has been reviewed elsewhere.[36] In northern Vietnam, fish were usually found in up to 20% of concrete tanks and other containers,[18] whereas on Small Liu-Chiu Isle, Taiwan, only 32% of residents were willing to try larvivorous fish in their water containers.[37] Towards the end of the 1980–1981 outbreak of dengue fever in southern China, Chinese catfish, *Claris fuscus*, were introduced into household water containers, mainly large jars.[38] The program was enforced by weekly inspection and those without fish were required to replace them or be fined. *Ae. aegypti* reductions over the ensuing four years were good, despite the fact that some residents objected to the oil exuded and fish faeces, particularly when the fish grew larger.

Several unsuccessful field trials with the mosquito predator *Toxorhynchites* have been documented, as have its characteristics which make it unsuitable for serious consideration in dengue vector control programs.[36,39] The studies, including mass release of *Toxorhynchites amboinensis* in French Polynesia,[40] showed 24–75% control efficacy of *Ae. aegypti* and *Ae. polynesiensis* in surveys of various container types, including tyres, drums, ovitraps and tree holes, and *Tx. amboiensis* became well established on wet forested islands, especially in tree holes. However, when *Toxorhynchites* were released to protect the international airport at Faaa, only 1–2.6% became established in the drier urban environment and they disappeared after a year, probably dispersing into the lusher surrounding hillside habitat. Rainfall was also found to be a major limiting factor: dryness limited the supply of *Aedes* hatchlings for this obligate predator, whereas the 50–60 rain days per year (over 10 mm) caused overflowing of containers and spillage of the highly hydrofugal eggs. Apart from the unsuitability of these k-strategists with long life cycles, the cost of producing each *Toxorhynchites* was estimated at US$0.12 per adult in 1985 prices.

Various aquatic beetles (particularly *Dytiscidae* and *Hydrophylidae*) and bugs (especially *Notonectidae*) proved to be effective predators of mosquito immatures, but only in small pools, stock troughs or experimental tubs.[39,41] In French Polynesia,[40] a local notonectid *Anisops tahitiensis*, odonate nymphs of the genera *Tramea*, *Pantala* and *Anax* species, a planarian *Dugesia tahitiensis*, the copepod, *Mesocyclops aspericornis* and the introduced fish, *Poicilia reticulata*, were evaluated by laboratory study. Those carried out on *Mesocyclops* were the forerunner to its successful use in Vietnam and elsewhere, but various biological, ecological and behavioural attributes limited the utility of some of the other candidates. In Vietnam, the naturally occurring corixid bug *Micronecta quadristrigata* was common in a variety of containers and was shown to be exerting a significant mosquito control effect when present.[18] The problem with these known predators lies with mass production and in controlling their dispersal, as they are free to fly away to preferred permanent and semi-permanent natural habitats. Limitations on food resources in artificial containers may also limit numbers of obligate and larger predators, although the corixid *M. quadristrigata* is an omnivore. Dragonfly naiads are considered to have limited potential as broad-scale control agents, because rearing of the young naiads is labour intensive. In Yangon, Myanmar, Sebastian and others reared *Crocothemis servilia* naiads for 21 days from the eggs of wild-caught females but had to provide large numbers of mosquito larvae as food. Monthly inundative releases were done at 800 *Ae. aegypti* breeding sites (mainly metal drums and jars) from June to September 1979 and suppression was effective.[42] Given this apparent success with a local resource, one could ask why this methodology was not exploited further. Did the champions give up or retire, were funds for extension unavailable, or was there an underestimation of what was needed to achieve translation into community practice?

Although there may seem to be a variety of possible explanations for the demise of some potential biological control candidates, it also requires ecological understanding of the agent itself and delivery into the correct habitat niche. However, one can sympathize with Laird and Miles' 1984 sentiments[35] in their preface on integrated methodologies and biological control: "For after the best part of a century of unremitting entomological study, *Ae. aegypti* is the world's most studied and best known mosquito. Short of actually investigating the species to death, its suppression via the methodology of established safety now proposed awaits prompt and positive action by the governments and international Specialized Agencies concerned." In 2008, we still wait, despite several World Health Assembly resolutions on dengue control from 1992 onwards.

One could have asked the same question about *Mesocyclops* and other cyclopoid copepods in 1980. Although some copepod species had been recorded as accidental predators of mosquito larvae (colonized *Ae. albopictus*) since 1957,[43] the story first unfolded in 1981, when Rivière and Thirel published data resulting from inadvertent inoculation of ovitraps with creek water containing *M. aspericornis*.[44] One wonders what would have happened if the water source for the ovitraps in the front and back yards at the Paea laboratory of the Institut Malarde had been the same, and both derived from tap water. In 1984, Suarez and colleagues published data from Colombia which indicated that *M. aspericornis* could possibly be used for control of *Ae. aegypti* larvae,[45] and later, in collaboration with Gerald Marten, noted that *Anopheles albimanus* immature populations were reduced in natural habitats where *Mesocyclops* also occurred.[46] In 1987, results of the first broad-scale field trials in French Polynesia were published indicating that *M. aspericornis* could be inoculated successfully not only into artificial habitats (tyres, drums and ovitraps) but also into natural ones such as tree holes and land crab burrows to reduce *Ae. aegypti* and *Ae. polynesiensis*.[47] By 2000, there had been successful field trials in the USA, Mexico, Honduras, Costa Rica, Colombia, Brazil, Puerto Rico, Laos, Thailand and Vietnam. The progress in utilizing *Mesocyclops* in community-based programs has been encouraging, with the first recorded eradication of *Ae. aegypti* without the use of an insecticide recorded at Phan-boi village, 30 km outside of Hanoi, in 1998.[48] *Mesocyclops* and the preventive strategy was incorporated into the National Dengue Plan by 1998 and basic training on its usage expanded to 26 provinces by 2000,[18] and to 41 thereafter. By 2000 in northern Vietnam, total control of *Ae. aegypti* had been achieved in 4 of 6 communes in Hung Yen, Haiphong and Nam Dinh provinces (phase 1) with 11,675 households and 49,647 people.[49] By 2003, similar results had been achieved in community programs in 3 communes in Quang Nam, Quang Ngai and Khanh Hoa in central Vietnam, involving 5913 households and 27,167 people.[50] The development of the project design in the context of local ownership and participation paid huge dividends as after project funding ceased after 2000 (except for a small allocation for the untreated control commune at Xuan Tien), as locally directed and funded programs expanded to 37 additional communes with 32 eradications by 2005.[51] Although the overall result of the elimination of dengue transmission is probably self-evident, supporting clinical and serological surveillance demonstrated that this had occurred from 2002. This strategy has also been applied successfully in southern Vietnam.

Why did this become such a spectacular success? Apart from the presence of a suitable candidate biological control agent that proved to be present in every village that we have surveyed to date, and the development of quantitative sampling methods to describe risk, the key factor probably lies in committed leadership and the development of genuine partnerships, from commune to national level. Although elements of our strategy can be transferred broadly into other localities with an *Ae. aegypti* problem, *Mesocyclops* is a tool suited to large water storage containers and therefore is broadly applicable to countries which do not have a reliable and safe piped water supply. Cambodia is such a country but our attempts to establish a program there were thwarted because of conflicting and competing interests, and our inability to find a national leader who was willing to take responsibility. As scientists, we have by necessity to be involved in local capacity building, and once the technology has been developed, be willing to effect a gradual transference (along with its power) to local leaders (or budding ones), collaborators and communities. Our novel and cost-effective strategy is so entrenched in some regions of Vietnam that "*Meso*" is a subject of plays, poems and songs, and a football cup in two districts.

In the Laird and Miles book,[35] Bohmfalk suggested that microbial pesticides presented a catch-22 in that in the USA at least, the potential marketers are those which already are responsible for conventional insecticides. Nevertheless, *Bacillus* and s-methoprene products are now well established, accepted and broadly marketed. The mosquito and vector control industry has some resistance with respect to change. With biological control candidates that promise a long residual life and self-driving mechanisms, this also could be perceived to represent an economic threat to present marketers. It was suggested at one stage that *Mesocyclops* could be marketed to make it worth more, but we chose the route of empowering communities.

Over 60 species of *Mesocyclops* occur throughout the world but particularly in the tropics and subtropics, where dengue viruses are a problem. *Mesocyclops* range in length from 0.7 to 1.5 mm, roughly similar to that of newly emerged *Aedes*, *Anopheles* and *Culex* larvae (0.7–1.2 mm). Most *Mesocyclops* are effective predators of *Aedes* and *Anopheles* larvae but, for reasons not fully understood, seldom of *Culex* species.

Some reasons for their success as biological control agents are: (i) they are classic r-strategists which feed on algae and a range of smaller micro-organisms such as protozoans and rotifers; (ii) they readily feed or engage in killing activity of 20–40 first or second instars per day, but in the absence of mosquito larvae resort to more generalized feeding; (iii) their reproductive rate is high, with 50–70 eggs produced every 3 days, depending on species, nutrition and temperature; and (iv) one can presume that most, if not all, species have some capacity to survive desiccation by entering dormancy, a desirable feature for ephemeral sites. They are commonly found in various lentic bodies, but particularly eutrophic ones, which provide an abundance of nutrients to support large populations. Thus the multitude of green-tinged lakes and ponds throughout Indo China in particular, provide a rich local natural resource, and at minimal cost. The fact that suitable *Mesocyclops* stocks are available locally obviates potential problems associated with translocation of exotic species, and also the costs of mass production.

Mesocyclops are easily mass-cultured by starting colonies using single females, preferably gravid, and placing them in a culture containing boiled wheat seeds, and algae (e.g. *Euglena*, *Chilomonas*) and protozoans (*Paramecium*). The New Orleans Mosquito Control Board have established an elegant system with racks of 2.6 × 1.3 × 0.1 m production trays for both *Mesocyclops* and *Macrocyclops*, producing 20,000–30,000 per tray every 2 months. Newly hatched brine shrimp (*Artemia*) can be used as supplementary food for holding stocks prior to release. At the National Institute of Hygiene and Epidemiology, Hanoi, plastic 150-litre garbage bins are filled with water (chlorine-free), and approximately 50 g of boiled wheat seed and 140,000 *Paramecium* are added to produce 4500 *Mesocyclops* 21 days later.[18] In practice, however, mass culture can be circumvented at Vietnamese sites by inoculating 100–200 *Mesocyclops* into approximately 20 selected concrete tanks and wells, the latter preferably public ones. Stocks build up, despite some being redistributed prematurely by villagers in daily water collection. These latter methodologies are particularly suitable for developing countries, as is mass field collection and redistribution into container breeding sites, after a washing process.

Mesocyclops have been trans-shipped in New Orleans on moist polyester sponges, whereas in Vietnam this technique has been refined by packing 20 × 15 × 10 cm foam pieces hollowed in the centre into 24 × 14 × 10 cm lunch boxes, and sending them via the mail system. Each box contains 20,000 copepods which can survive for as long as 30 days. On receipt, health authorities or schoolchildren distribute the foam pieces into target habitats.[18] Rivière and others developed a system where *Mesocyclops* could be sprayed using a backpack sprayer.[47]

Although predacious copepods offer a simple, low-cost technology for *Aedes* control, they generally are applicable for control in large water storage containers, wells, some drums, and sometimes vases and ornamental ponds. Where large water storage containers are the source of the greatest *Aedes* productivity, they provide a long-term and important control option. There have been problems in some situations where residents frequently wash their water containers and inadvertently throw their stocks out. Since *Mesocyclops* take 3–4 weeks to populate suitable sites, control effects are not immediate as larger instars will not be killed. This problem, and for those sites also containing *Culex*, can be solved with concomitant use of *Bacillus thuringiensis*. *Mesocyclops* should not be used in the West African countries which still have a *Dracunculus* problem.

So what of the future? The containers to which we are applying or seeking to apply biological control agents are often nutrient-deprived and so natural density-dependent mortality already operates on species such as *Ae. aegypti*. It is noted,however, that the smaller size of copepods predisposes survival with fewer nutrients. Failed control with copepods is more likely to occur through inappropriate habitat selection, overwashing of containers or incorrect selection of, for example, the nonpredatory *Thermocyclops* as the local tool. However, in Vietnam, I never fail to be impressed by the level of knowledge of the thousands of schoolchildren and health volunteers assisting with local programmes. On the international scale, it comes down to the level of promotion that bodies such as WHO are willing to do.

It has long been known that waters from crowded sites are less attractive to gravid female mosquitoes, as are waters containing entomopathogenic digenean flatworms,[52] notonectids and possibly fish. In contrast, in Mexico, Torres-Estrada and colleagues have demonstrated that water with *Mesocyclops*, or previously containing them, is more attractive to ovipositing *Ae. aegypti*.[53] This increases the utility of copepods as biological control tools and this (or the inverse) may apply to others as a byproduct of such organisms. Fuller elucidation of the attraction and repulsion of gravid female mosquitoes could provide insight into appropriate choice and more informed use of biological control agents.

There are undoubtedly other biological control agents which await a champion. This chapter has demonstrated that such development has been a stop–start affair, if we consider that some evaluations have spanned the millennium. Not forgetting the understated usefulness of native fish for dengue vector control, the strategy behind the broad-scale success of *Mesocyclops* should be more widely employed, and entomopathogenic fungi and popcorn *Wolbachia* deserve to be fully trialled, but this should be done with consideration of future operational partnerships.

References

1. Focks DA, Chadee DD. Pupal survey: an epidemiologically significant surveillance method for *Aedes aegypti* — an example using data from Trinidad. *Am J Trop Med Hyg* 1997;56:159–167.
2. Focks DA, Brenner RJ, Hayes J, Daniels E. Transmission thresholds for dengue in terms of *Aedes aegypti* pupae per person with discussion of their utility in source reduction efforts. *Am J Trop Med Hyg* 2000;62:11–18.
3. Pontes RJ, Freeman J, Oliveira-Lima JW, Hodgson JC, Spielman A. Vector densities that potentiate dengue outbreaks in a Brazilian city. *Am J Trop Med Hyg* 2000;62:378–383.
4. Service MW. Population dynamics and mortalities of mosquito preadults. In: Lounibos LP, Rey JR, Frank JH (eds.) *Ecology of Mosquitoes: Proceedings of a Workshop*. Florida Medical Entomology Laboratory, University of Florida, Tallahassee, 1985, pp. 185–201.
5. Washburn JO, Mercer DR, Anderson JR. Regulatory role of parasites: impact on host population shifts with resource availability. *Science* 1991;253: 185–188.
6. Jenkins DW. Pathogens, parasites and predators of medically important arthropods. *Bull World Health Organ* 1964;30:1–150.
7. Chapman CC. Biological control of mosquito larvae. *Ann Rev Entomol* 1974;19:33–59.
8. O'Neill Sl, Hoffmann AA, Werren JH (eds.). *Influential Passengers: Inherited Microorganisms and Arthropod Reproduction*. Oxford University Press, 1997.

9. O'Neill SL. *Wolbachia pipientis*: symbiont or parasite? *Parasitol Today* 1995;11:33–59.
10. Hertig M. The rickettsia, *Wolbachia pipientis* (gen. et sp. nov.) and associated inclusions in the mosquito, *Culex pipiens*. *Parasitology* 1936;28: 453–486.
11. Laven H. Eradication of *Culex pipiens fatigans* through cytoplasmic incompatibility. *Nature* 1967;216:383–384.
12. Powell K, Jayaraman KS. Mosquito researchers deny plotting secret biowarfare test. *Nature* 2002;419:867.
13. Min KT, Benzer S. *Wolbachia*, normally a symbiont of *Drosophila*, can be virulent, causing degeneration and early death. *Proc Natl Acad Sci USA* 1997; 94:10792–10796.
14. Cook PE, McMeniman CJ, O'Neill SL. Modifying insect population age structure to control vector-borne disease. In: Aksoy S (ed.) *Transgenesis and the Management of Vector-Borne Disease*. Landes Bioscience, Austin, Texas, USA, 2007.
15. Brownstein JS, Hett E, O'Neill SL. The potential of virulent *Wolbachia* to modulate disease transmission by insects. *J Invertebr Pathol* 2003;84:24–29.
16. Rasgon JL, Scott TW. Impact of population age structure on *Wolbachia* transgene driver efficacy: ecologically complex factors and release of genetically modified mosquitoes. *Insect Biochem Mol Biol* 2004; 34:707–713.
17. Kittayapong P, Baisley KJ, Baimai V, O'Neill SL. Distribution and diversity of *Wolbachia* infections in Southeast Asian mosquitoes (*Diptera*: *Culicidae*) *J Med Entomol* 2000;37:340–345.
18. Nam VS, Yen NT, Holynska M, Reid JW, Kay BH. National progress in dengue vector control in Vietnam: survey for *Mesocyclops* (*Copepoda*), *Micronecta* (*Corixidae*), and fish as biological control agents. *Am J Trop Med Hyg* 2000;62:5–10.
19. Ward TW, Jenkins MS, Afanasiev BN, Edwards M, Duda BA, Suchman E, Jacobs-Lorena M, Beaty BJ, Carlson JO. *Aedes aegypti* transducing densovirus pathogenesis and expression in *Aedes aegypti* and *Anopheles gambiae* larvae. *Insect Mol Biol* 2001;10:397–405.
20. Wei W, Shao D, Huang X, Li J, Chen H, Zhang Q, Zhang J. The pathogenicity of mosquito densovirus (C6/36DNV) and its interaction with dengue type 2 in *Aedes albopictus*. *Am J Trop Med Hyg* 2006;75: 1118–1126.

21. Manasherob R, Ben-Dov E, Margalit J, Zaritsky A, Barak Z. Raising activity of *Bacillus thuringiensis* var *israelensis* against *Anopheles stephensi* larvae by encapsulation in *Tetrahymena pyriformis* (*Hymenostomatida*: *Tetrahymenidae*). *J Am Mosq Control Assoc* 1996;12:627–631.
22. Kucinska J, Lonc E, Rydzanicz K. Transgenic bioinsecticides inimical to parasites but imical to environment. *Wiad Parazytol* 2003;49:11–20.
23. Brown MD, Watson TM, Green S, Greenwood JG, Purdie D, Kay BH. Toxicity of insecticides for control of freshwater *Culex annulirostris* (*Diptera*: *Culicidae*) to the non-target shrimp, *Caradina indistincta* (*Decapoda*: *Atyidae*). *J Econ Entomol* 2000;93:667–672.
24. Brown MD, Carter J, Thomas D, Purdie DM, Kay BH. Pulse-exposure effects of selected insecticides to juvenile Australian crimson-spotted rainbowfish (*Melanotaenia duboulayi*). *J Econ Entomol* 2002;95: 294–298.
25. Kroeger A, Lenhart A, Ochoa M, Villegas E, Levy M, Alexander N, McCall PJ. Effective control of dengue vectors with curtains and water container covers treated with insecticide in Mexico and Venezuela: cluster randomized trials. *Br Med J* 2006;332:1247–1252.
26. Russell TL, Kay BH, Ryan PA. Mosquito control and *Bacillus thuringiensis* var. *israelensis* use: an Australian perspective. In: *Biotechnology of Bacillus thuringiensis*, Vol. 5. Science and Technics, Hanoi, 2005, pp. 57–72.
27. Russell TL, Brown MD, Purdie DM, Ryan PA, Kay BH. Efficacy of Vectobac (*Bacillus thuringiensis* var. *israelensis*) formulations for mosquito control in Australia. *J Econ Entomol* 2003;96:1786–1791.
28. Lacey LA, Undeen AH. Microbial control of black flies and mosquitoes. *Ann Rev Entomol* 1986;31:265–269.
29. Brown ID, Watson TM, Carter J, Purdie DM, Kay BH. Toxicity of VectoLex CG (*Bacillus sphaericus*) to Australian mosquitoes and selected non-target species. *J Econ Entomol* 2004;97:51–58.
30. Becker N, Zgomba M, Petric D, Beck M, Ludwig M. Role of larval cadavers in recycling processes of *Bacillus sphaericus*. *J Am Mosq Control Assoc* 1995;11:329–334.
31. Hall RA, Papierok B. Fungi as biological control agents of arthropods of agricultural and medical importance. *Parasitology* 1982;84:205–240.
32. Thomas MB, Kooyman C. Locust biopesticides: a tale of two continents. *Biocontrol News Inf* 2004;25:47N–51N.

33. Blanford S, Chan BHK, Jenkins NE, Sim D, Turner RJ, Read AF, Thomas MB. Fungal pathogen reduces potential for malaria transmission. *Science* 2005;308:1638–1641.
34. Scholte E-J, Ng'habi K, Kihonda J, Takken W, Paaijmans K, Abdulla S, Kileen GF, Knols BGJ. An entomopathogenic fungus for control of adult African malaria mosquitoes. *Science* 2005;308:1641–1642.
35. Laird M, Miles JW (eds.). *Integrated Mosquito Control Methodologies*, Vol. 2. Academic, London, 1985.
36. Lacey LA, Orr BK. The role of biological control of mosquitoes in integrated vector control. *Am J Trop Med Hyg* 1994;50:97–115.
37. Wang CH, Chang NT, Wu HH, Ho CM. Integrated control of the dengue vector *Aedes aegypti* in Liu-Chiu village, Ping-Tung county, Taiwan. *J Am Mosq Control Assoc* 2000;16:93–99.
38. Wu N, Wang SS, Han GX, Xu RM, Tang GK, Qian C. Control of *Aedes aegypti* larvae in household water containers by Chinese cat fish. *Bull World Health Organ* 1987;65:503–536.
39. Bay EC. Predator-prey relationships among aquatic insects. *Ann Rev Entomol* 1974;19:441–453.
40. Riviere F, Sechan Y, Kay BH. The evaluation of predators for mosquito control in French Polynesia. *Arbovirus Res Australia* 1986;4:150–155.
41. Chesson J. Effect of notonectids (*Hemiptera*: *Notonectidae*) on mosquitoes (*Diptera*: *Culicidae*): predation or selective oviposition? *J Econ Entomol* 1984;13:531–538.
42. Sebastian AA, Sein MM, Thu MM, Corbet PS. Suppression of *Aedes aegypti* (*Diptera*: *Culicidae*) using augmentative release of dragonfly larvae (*Odonata*: *Libellulidae*) with community participation in Yangon, Myanmar. *Bull Entomol Res* 1990;80:223–232.
43. Bonnet D, Mukaida T. A copepod predations on mosquito larvae. *Mosq News* 1957;17:99–100.
44. Riviere F, Thirel R. La predation du copepode *Mesocyclops leuckarti pilosa* (Crustacea) sur les larves de *Aedes aegypti* et *Ae. (St.) polynesiensis*: essais preliminaires d'utilization comme agent lutte biologique. *Entomophaga* 1981;26:427–439.
45. Suarez MF, Ayala D,Nelson MJ, Reid JW. Hallazgo de *Mesocyclops aspericornis* (Daday) (*Copepoda*: *Cyclopidae*) depredator de larvas de *Aedes aegypti* en Anapoima-Colombia. *Biomedica* 1984;4:74–76.

46. Marten GG, Astaiza R, Suarez MF, Monje C, Reid JW. Natural control of larval *Anopheles albimanus* (*Diptera*: *Culicidae*) by the predator *Mesocyclops* (*Copepoda*: *Cyclopoida*). *J Med Entomol* 1989;26:624–627.
47. Riviere T, Kay BH, Klein J-M, Sechan Y. *Mesocyclops aspericornis* (*Copepoda*) and *Bacillus thuringiensis* var *israelensis* for the biological control of *Aedes* and *Culex* vectors (*Diptera*: *Culicidae*) breeding in crab holes, tree holes and artificial containers. *J Med Entomol* 1987;24:425–430.
48. Nam VS, Yen NT, Kay BH, Marten GG, Reid JW. Eradication of *Aedes aegypti* from a village in Vietnam, using copepods and community participation. *Am J Trop Med Hyg* 1998;59:657–660.
49. Kay BH, Nam VS, Tien TV, Yen NT, Phong TV, Diep VT *et al.* Control of *Aedes* vectors of dengue in three provinces of Vietnam by use of *Mesocyclops* (*Copepoda*) and community-based methods validated by entomologic, clinical and serological surveillance. *Am J Trop Med Hyg* 2002;66:40–48.
50. Nam VS, Yen NT, Phong TV, Ninh TU, Mai LQ, Lo LV, Nghia LT, Bektas A, Briscombe A, Aaskov JG, Ryan PA, Kay BH. Elimination of dengue by community programs using *Mesocyclops* (*Copepoda*) against *Aedes aegypti* in Central Vietnam. *Am J Trop Med Hyg* 2005;72:67–73.
51. Kay BH, Nam VS. New strategy against *Aedes aegypti* in Vietnam. *Lancet* 2005;365:613–617 (with editorial comment).
52. Zahiri N, Rau ME. Oviposition attraction and repellency of *Aedes aegypti* (*Diptera*: *Culicidae*) to waters from conspecific larvae subjected to crowding, confinement, starvation, or infection. *J Med Entomol* 1998;35:782–787.
53. Torres-Estrada JL, Rodriguez MH, Cruz-Lopez L, Arredondo-Jiminez JI. Selective oviposition by *Aedes aegypti* (*Diptera*: *Culicidae*) in response to *Mesocyclops longisetus* (*Copepoda*: *Cyclopoida*) under laboratory and field conditions. *J Med Entomol* 2001;38:188–192.

13

Mosquito Control: Behavioral and Community Interventions

Peter Winch, Elli Leontsini and Linda S. Lloyd

> *"We cannot overstate the role of behavioral science in our effort to get ahead of the curve with emerging infections. Having the science of laboratory technology to control infectious diseases is not enough, unless we can influence people to behave in ways that minimize the transmission of infections and maximize the efforts of medical interventions."*[1]

Dengue is in many ways the prototype of a "man-made disease."[2] The larval habitats of its principal vector, *Aedes aegypti*, are artificial containers found in and around the home in most settings where transmission is occurring. Promotion of elimination or control of larval habitats at the household level through partnerships between community groups, municipalities and the private sector is one of the cornerstones of a successful integrated strategy for preventing dengue and dengue hemorrhagic fever.[3–7] Control of larval container habitats is only one of several sets of behaviors that need to be promoted, but it has received the most attention and effort. Other important sets of behaviors include self-protection from mosquito bites, and early and appropriate treatment-seeking and home care for suspected cases of dengue. Appropriate treatment-seeking and care is an important element of current efforts to reduce the impact of malaria, but few studies have been conducted in relation to dengue. Notable exceptions include a study in Thailand[8] and a study in Cambodia.[9] The current discussion, however, will focus on control of larval container habitats. In practice, behavioral and community interventions are inseparable. In the following pages we will present and discuss the behavioral and community issues separately, before concluding by presenting a vision for integrated programs.

Conventional Approaches to Behavior Change Interventions

A starting point for the design of behavior change interventions is often a model that summarizes the factors that affect whether people do or do not perform a specific behavior. A communication strategy is then developed that is directed at modifying these factors as a way of promoting the behavior in question. The Health Belief Model, developed by a group of social psychologists in the U.S. Public Health Service in the 1950's,[10] is the model that has been most commonly applied to dengue control.[11] Vector control personnel, nevertheless, seldom are aware that they are using a model at all, much less that the model has this specific name. According to the original version of the Health Belief Model, people are likely to carry out a treatment or preventive behavior if they believe they are susceptible to the disease, believe the disease is serious, believe the treatment or preventive method is effective, consider the treatment or preventive method to be affordable, and receive a prompt or cue telling them when to carry out the behavior. Later, other constructs such as self-efficacy were added to the model.[12] If we apply the original model to elimination of used tires stored in the yard or patio around the house, a typical behavior change strategy consists in dissemination of four key messages: (1) you are susceptible to dengue and dengue hemorrhagic fever, (2) dengue and DHF are serious diseases, (3) elimination of used tires is an effective way of preventing dengue and DHF, and (4) elimination of tires costs little in terms of time, effort or money. We might also give people a reminder or cue to eliminate tires at the beginning of the transmission season, or when a dengue epidemic is anticipated.

While proven effective as a framework for developing interventions for other health problems, the Health Belief Model, and other behavior change models that have been proposed, have been less successful as frameworks for developing interventions to promote dengue prevention. Numerous studies have demonstrated that increasing awareness of the seriousness of dengue and DHF and prevention methods does not translate into behavior change as indicated by decreases in larval indices.[13–18] A number of reasons have been identified for the limitations of these models as frameworks for the design of interventions. Among the principal reasons are the following:

Dengue and DHF not viewed as serious. While educational messages stress the seriousness of dengue and DHF, they frequently are not believed. This is particularly the case in the Americas, where large epidemics of DHF have yet to occur outside of Cuba. First, personal experience with "classical" dengue fever leads many to conclude that dengue is a minor problem, at least for them, similar in both symptoms and severity to influenza.[19,20] Second, dengue is frequently overshadowed by HIV/AIDS, and seen as trivial in comparison. Both dengue and HIV/AIDS are common in urban areas undergoing rapid social change.[7] A number of countries now have significant levels of both dengue and HIV transmission, such as Thailand and Vietnam in Asia, or Brazil and the countries of the Caribbean Basin in the Americas.

Complex nature of the behaviors being promoted. Public health officials have frequently underestimated the complexity of the behaviors being promoted. General messages applicable to all containers, such as "eliminate or cover all containers on your property," sound reasonable in the abstract, but it often is not clear how to apply such recommendations to specific container types. Taking tires as an example, householders may find both elimination and covering to be difficult to perform. People frequently do not want to eliminate tires because they potentially can be sold or used at some future date, and indeed there is a worldwide market in sale of used tires, which has been implicated as one of the causes of the global spread of *Aedes albopictus* in recent years.[21,22] Exhortations to throw out tires with other refuse may not resonate with householders if tires are not defined as a type of "refuse."[23] When householders do try to eliminate tires from their property, they may find that they are not picked up in the regular refuse collection, or that they are picked up by a neighbor and then stored in his/her yard.[23] If tires are cut up into pieces so that they may no longer serve as a container habitat, the scraps of tires may still contain viable eggs that can serve to disperse mosquitoes.[24] Emptying tires that contain water is a difficult undertaking, especially for truck tires. Stacking tires and covering them with a plastic sheet or tarpaulin may create new larval habitats in the depression where the sheet or tarpaulin rests on the top tire, and storage of tires inside the house is seldom an option.

Large number of behaviors to be promoted. There is tremendous variety in the types of containers where *Aedes* larvae can be found, particularly in complex urban environments where much dengue transmission is occurring.[7,25] For many of these container types, specific behavioral recommendations will need to be formulated that are both feasible and acceptable to the householder. Even for one container type, such as potted plants, specific recommendations may be needed for different types of plants.[26] Furthermore, behavior change specialists have found that effective promotion of one "behavior" such as administering oral rehydration therapy to a child with diarrhea requires breaking down this "behavior" into observable and measurable component steps or behaviors and determining how to explain and promote each one.[27] Thus the recommended "behavior" for each type or subtype of container may need to be broken down into a series of steps or subbehaviors in order to develop an effective behavior change strategy. The implications of this are that a dengue control program may need to promote dozens of specific behaviors in order to be effective. At the same time, units of ministries of health responsible for vector control rarely have the specialists in health education and behavior change necessary to carry out such a complex undertaking.

Concern about the effectiveness of larval source reduction. There is no doubt that larval source reduction is efficacious: if all of the container habitats of a container-breeding mosquito are eliminated or controlled, the production of adult mosquitoes should cease. This has been demonstrated in the past by the vertically organized eradication programs. There is concern, however, that even under ideal circumstances voluntary control of larval container habitats may be an ineffective control measure because of the high level of coverage required. In other words, even if most people control most of the larval habitats on their property most of the time, which is more than most community-based programs have been able to achieve, there may be no demonstrable impact on transmission. This issue was quantified in a study on transmission thresholds for dengue in terms of *Aedes aegypti* pupae per person.[28] Threshold levels for transmission were found to range from 0.5 to 1.5 pupae per person at 28°C, while observed levels of pupae per person ranged from 0.3 to greater than 60. For a site in Reynosa, Mexico, the authors calculated that 9 out of every 10 containers would need to be eliminated, while in Trinidad the figure was 24 out of every 25 containers.[28]

The Emerging Paradigm for Behavioral Interventions

The previous section postulated that dengue control programs have been trying to do the impossible: effectively promote a large number of complex and time-consuming behaviors to be carried out in and around the home in order to eliminate 90% of the container habitats, to prevent a disease that most people do not view as an important public health problem, at least in interepidemic periods. In response to this situation, a new paradigm for behavioral interventions to promote control of larval container habitats has emerged in recent years. The components of the paradigm are: (1) prioritization of container types based on their productivity; (2) in-depth formative research, in collaboration with entomologists to develop feasible, acceptable and effective behavioral recommendations; (3) integration of physical, biological and chemical control methods to increase the effectiveness of behavioral interventions; (4) evaluation of both entomological and behavioral outcomes.

Prioritization of container types

Given the impossibility of effectively addressing all container types, control programs are increasingly examining the productivity of different container types, and giving priority to the most productive containers in behavior change interventions.[29] The simplest measure of productivity is the proportion of any type of container that contain larvae. Pupal measures, such as average number of pupae per container or per hectare, have been proposed as superior and practical ways to prioritize containers in control programs.[30,31] Pupal measures have been used in several studies and programs to prioritize containers.

A study in Mérida, Yucatán, Mexico, illustrates the application of this approach. Background research between 1989 and 1994 revealed high levels of knowledge about dengue and dengue hemorrhagic fever, transmission of the virus, and breeding sites of the mosquito vector.[32,33] However, entomological and observational studies revealed little or no adoption of control strategies recommended by the Ministry of Public Health to prevent the breeding of *Aedes aegypti* in water-holding containers.[33] Formative research using qualitative and observational methods examined the role various containers play within the domestic setting,[20,32] and identified the function or use of the container as a key construct which determined not only the type of control strategies which might be appropriate for the container but also the level of effort household members would be willing to commit to specific actions.[32,34] The functional categories identified were: water storage, animal water dishes, vases and aquatic plants, other uses, no specific use, trash and tires. A pupal productivity study (container productivity × container abundance) was then carried out to determine the relative contribution of each functional category to adult *Aedes* production. The five most productive functional categories (water storage, animal water dishes, vases and aquatic plants, tires, and containers without a specific use) were selected for a subsequent behavior change intervention.

In-depth formative research

Once the container types that will be the focus of the intervention have been identified, there is a phase of formative research in which the content or "form" of the behavior change intervention is defined in more detail.[35] A mix of qualitative and quantitative methods is used. Questions that formative research addresses include: Who is responsible for looking after or making decisions about the elimination of the container type in question, how is the container used, which control measures might be acceptable, and comparison of families who do and do not contract dengue.[36] Close collaboration between social scientists and entomologists is needed to identify behaviors which are both acceptable to the householder and effective in eliminating mosquito eggs or larvae. Mark Nichter of the University of Arizona coined the term "ethnoentomology" to describe this collaboration.

There is growing recognition that it is not sufficient to simply define the behavior to be promoted, even if this is done in collaboration with the community. The researchers work to identify feasible methods of controlling larval container habitats and implement them through behavioral trials with a small number of households. After identifying the successful practices, researchers work with the community to develop methods to promote these practices throughout the population. This approach has been described in most detail for trials of new weaning foods for children.[37]

Behavioral trials have been used in studies in Honduras and the Dominican Republic. When manual cleaning was found to be ineffective in eliminating mosquito larvae from water storage containers in a community-based control program in El Progreso, Honduras, it was decided to develop and evaluate an improved method of removing mosquito eggs based on commonly available materials. Chlorine bleach and detergent are routinely used by householders in the city to clean concrete washbasins and metal drums, the two most important larval habitats in terms of pupal productivity. In collaboration with entomologists, the efficacy of these materials in eliminating eggs, larvae and pupae of *Ae. aegypti* was assessed first under controlled conditions in an insectary, and then in community trials when applied by householders, and the bleach was found to be highly ovicidal.[38] It was decided to promote the application of a combination of chlorine bleach and detergent to the walls of washbasins and drums as a method for eliminating eggs. The bleach maintained its ovicidal properties when mixed with detergent, and the detergent gave the mixture consistency so that it could be applied as a thin film to the walls. A field trial of the method was conducted in 13 periurban neighborhoods.[39] In the first postintervention survey, little or no impact was discernable on mosquito larvae and pupae in water storage containers. The method was then modified by increasing the recommended quantities of bleach and detergent and simplifying the instructions. In the second postintervention survey, knowledge of the steps and their order increased further; the intervention neighborhoods had significantly fewer algae on washbasin walls, an indicator of more effective cleaning; and numbers of pupae and third and fourth instar larvae were significantly lower than in untreated neighborhoods.

In the Dominican Republic, the methodology used was coined as Negotiation of Improved Practices (Negociación de Prácticas Mejoradas or NEPRAM).[40] The reliability of the piped water supply is very poor, and water pressure is low — factors which have been associated with high levels of larval infestation in other countries.[41] The main containers for storing water are large metal drums, frequently lined with a layer of concrete and being in constant use. Initially it was thought that one of two designs of improved lids made from locally available materials would be accepted by householders and would hermetically seal the drums from mosquitoes. Trials in households demonstrated, however, that people found both designs of lids inconvenient and difficult to use correctly.[40] An additional control method, dabbing small amounts of undiluted chlorine bleach directly on the sides of even half-empty drums, was much better accepted. People appreciated the cleaner appearance of the sides of the drum and of the water.

Integration of physical, biological and chemical control methods

In the past, physical, biological and chemical control methods for container-breeding mosquitoes such as larvicides have been developed for and marketed to ministries of health and local vector control services, but not to householders. Vector control specialists are increasingly of the opinion that a major target for technologies to control mosquito larvae should be the household. One reason for this is the limited effectiveness of many of the physical methods for maintaining container habitats free of larvae, such as cleaning the container or changing the water. A second reason is that household surveys in every country where they are conducted demonstrate that households dedicate significant proportions of their income to vector abatement, most commonly aerosol insecticides and mosquito coils.

Technologies that have been proposed for use at the household level include improved lids or insecticide-treated covers for water storage containers,[42,43] nets to sweep water storage containers for larvae,[44] application of biological larvicides such at BTI to water storage containers,[45] and community application of copepods to containers.[46,47]

Evaluation of both entomological and behavioral outcomes

Surveillance of behavioral outcomes is necessary for monitoring progress toward program objectives. Behavioral surveillance is often not carried out by programs. Traditional entomological indicators such as the House (Premise) or Breteau Indices are monitored, and inferences about human behavior are made based on them. The concept of behavioral surveillance usually refers to "behavioral risk factor surveillance," where "the objective is simply to collect data on those risk factors that have been causally linked to or associated with" the diseases of interest.[48] Applying this definition to dengue fever/DHF prevention and control, behavioral surveillance might be defined as the monitoring of behaviors in individuals that (a) increase or decrease the risk of adult *Aedes* mosquitoes being produced on their property, (b) increase or decrease the risk that they or other members of their family will be bitten by *Aedes* mosquitoes, and (c) increase or decrease the risk of adverse outcomes when they or other members of their household contract a dengue virus infection. Specific behaviors that might be monitored include water storage practices, acquisition, storage and disposal of containers such as tires and metal cans, practices aimed at preventing mosquito bites such as insecticide and repellent use, and case management of febrile illnesses.

Evaluation of individual behavior toward larval container habitats is becoming more sophisticated. In a study in El Progreso, Honduras, researchers found that traditional indices were not sufficiently sensitive to changes in human behavior, as they did not differentiate between containers in which all the immature stages are present and those which hold only first- and second-instar larvae.[49] It was not considered essential to prevent all larval development to limit transmission of pathogens by the adults;

if the *Ae. aegypti* in the containers only manage to develop to young larvae before the containers are cleaned, then control of the vector in these containers will be effective. An index was developed as a summary measure of the degree of infestation of a washbasin by *Ae. aegypti*. It was the sum of four variables assessed in the survey: presence of any immature stages (larvae and/or pupae); presence of pupae; detection of third–fourth-instar larvae in a five-dip sample; and a log transformation of the number of larvae recovered.[49] Such indices can make control programs more efficient by targeting the greatest efforts at households with higher levels of infestation.

This approach is a reasonable starting point, but it is limited to individual and household behaviors. While behavior change at this level is certainly necessary, it is rarely sufficient and may not even be the most important set of behaviors to examine. A number of examples now exist for models of behavior change that include not only the individual level, but also laws, regulations and policies and other factors operating at the community, state and national levels that facilitate individual behavior change.[50–52] Taking tires as an example, a comprehensive approach to behavioral surveillance might look not only at whether householders have unprotected tires on their property, but also at the number of tires collected during community cleanup campaigns, the proportion of labeled "sentinel tires" that are taken away during routine refuse collection, the number of tires on public lands such as roadsides, the degree to which existing laws about tire disposal/recycling are implemented, and tire storage in private businesses where tires are bought and sold.

Role of the Community

The preceding section has examined the question of behavior: What actions do we want people to take, and how do we monitor whether they are actually performing them? Beyond this there is the question of assignment of responsibility for activities to promote and facilitate these behaviors at the community level. Which organizations will explain to people why and how to control larval container habitats in and around their homes, and mobilize people to control other container habitats found outside the home?

The traditional response to this question has been the vertically organized vector control programs housed within ministries of health. William Crawford Gorgas pioneered the vertical approach to the control of mosquito larval habitats in Havana, Cuba, in 1899. This approach to vector control subsequently became the basis of successful programs for many vector-borne diseases, including dengue. Among people living in communities that benefited from these vertical programs, many came to view vector control as the exclusive responsibility of the government.[53] The use of technologies such as spraying equipment and the limited opportunities for local input into program design and evaluation contributed to the impression that only governments have the capability to control vector control activities.[53]

As *Ae. aegypti* densities have continued to increase over the past 30 years, vector control professionals have increasingly recognized that assigning responsibility for vector control exclusively to the government was a recipe for failure due to shrinking government budgets for vector control, expanding urban populations,[7] and the passive role communities frequently play in the presence of vertical programs.[18] A number of authors have advocated an expanded role for communities in control of *Aedes* mosquitoes and other disease vectors.[3–5,54,55] This call for an expanded role of communities in vector control has coincided with the growth of the private sector and civil society, including nongovernmental organizations, and debates about the role of civil society in advocating improvements to health, and in implementing health interventions.[56]

Efforts to promote community participation in dengue prevention and control face a number of practical difficulties. The first and most immediate is defining the scope of the participation. In programs where the scope of participation is narrowly defined and limited primarily to *Aedes* mosquitoes, it usually proves impossible to sustain community interest, as dengue is seldom a community priority during interepidemic periods. At the opposite end of the spectrum are programs where the scope of participation is broad. The ultimate objective may be not only the control of dengue but also the development of the community as a whole with an emphasis on self-reliance and planning in response to needs and priorities defined by the community — an approach sometimes referred to as the community development approach.[53,57] While community development approaches are advocated by many authors, they often prove incompatible with centralized government policies and programs.[58] Furthermore, they may have little impact on mosquito densities if communities focus on their defined priorities, such as employment, housing and education, rather than specific actions related to container habitats, such as elimination of tires.

The emerging paradigm for the role of the community in dengue prevention and control does not ascribe a dominant role to either the government or the community. Instead, a broad spectrum of groups and organizations are involved, each with specific roles to play.[5,55] These include the government vector control services, municipal governments, nongovernmental organizations, faith-based organizations, schools, grassroots community groups and private companies.[3,4] In such a program, the role of government vector control services becomes one of coordination, oversight, monitoring and regulation rather than implementation of vector control measures by government employees. Many governments are unprepared for such a role, and the transition to this form of collaboration with civil society will be an extended one.[56]

Conclusions

Conventional approaches to involving individuals in control of the larval container habitats of dengue vectors have proven ineffective, leading to a search for new approaches. This chapter has described an emerging paradigm for behavioral change interventions that consists of (1) prioritization of container types based on their productivity; (2) in-depth formative research in collaboration with entomologists to develop feasible, acceptable and effective behavioral recommendations; (3) integration of physical, biological and chemical control methods to increase the effectiveness of behavioral interventions; (4) evaluation of both entomological and behavioral outcomes, as well an emerging paradigm for community involvement in vector control that calls on the government to assume the role of coordination and oversight in a coalition of groups and organizations, each with specific functions related to vector control.

References

1. Satcher D. Emerging infections: getting ahead of the curve. *Emerg Infect Dis* 1995;1:1–6.
2. Gillett JD. The behaviour of *Homo sapiens*, the forgotten factor in the transmission of tropical disease. *Trans R Soc Trop Med Hyg* 1985;79:12–20.
3. Gubler DJ. *Aedes aegypti* and *Aedes aegypti*-borne disease control in the 1990s: top down or bottom up? Charles Franklin Craig Lecture. *Am J Trop Med Hyg* 1989;40:571–578.
4. Gubler DJ, Clark GG. Community-based integrated control of *Aedes aegypti*: a brief overview of current programs. *Am J Trop Med Hyg* 1994;50:50–60.
5. Gubler DJ, Clark GG. Community involvement in the control of *Aedes aegypti*. *Acta Trop* 1996;61:169–179.
6. Halstead SB. Dengue in the health transition. *Gaoxiong Yi Xue Ke Xue Za Zhi* 1994;10(Suppl):S2–14.
7. Knudsen AB, Slooff R. Vector-borne disease problems in rapid urbanization: new approaches to vector control. *Bull World Health Organ* 1992;70:1–6.
8. Okanurak K, Sornmani S, Mas-ngammueng R, Sitaputra P, Krachangsang S, Limsomboon J. Treatment seeking behavior of DHF patients in Thailand. *Southeast Asian J Trop Med Public Health* 1997;28, 351–358.
9. Kuhn S, Manderson L. Health seeking and access to care for children with suspected dengue in Cambodia: an ethnographic study. *BMC Public Health* 2007;7(147):262.
10. Rosenstock IM. What research in motivation suggests for public health. *Am J Public Health* 1960;50:295–301.
11. Phuanukoonnon S, Brough M, Bryan JH. Fdk knowledge about dengue mosquitoes and contributions of health belief model in dengue control promotion in Northeast Thailand. *Acta Trop* 2006;99:6–14.
12. Rosenstock IM, Strecher VJ, Becker MH. Social learning theory and the health belief model. *Health Educ Q* 1988;15:175–183.
13. Chiaravalloti Neto F, de Moraes MS, Fernandes MA. Results of activities encouraging community participation in dengue control in an outlying neighborhood of Sao Jose do Rio Preto, Sao Paulo, and the relationship between the population's knowledge and habits. *Cad Saude Publica* 1998;14:101–109.
14. Degallier N, Vilarinhos PT, de Carvalho MS, Knox MB, Caetano J. People's knowledge and practice about dengue, its vectors, and control means in

Brasilia (DF), Brazil: its relevance with entomological factors. *J Am Mosq Control Assoc* 2000;16:114–123.

15. Gupta P, Kumar P, Aggarwal OP. Knowledge, attitude and practices related to dengue in rural and slum areas of Delhi after the dengue epidemic of 1996. *J Commun Dis* 1998;30:107–112.
16. Rosenbaum J, Nathan MB, Ragoonanansingh R, Rawlins S, Gayle C, Chadee DD, Lloyd LS. Community participation in dengue prevention and control: a survey of knowledge, attitudes, and practice in Trinidad and Tobago. *Am J Trop Med Hyg* 1995;53:111–117.
17. Swaddiwudhipong W, Lerdlukanavonge P, Khumklam P, Koonchote S, Nguntra P, Chaovakiratipong C. A survey of knowledge, attitude and practice of the prevention of dengue hemorrhagic fever in an urban community of Thailand. *Southeast Asian J Trop Med Public Health* 1992;23: 207–211.
18. Toledo ME, Baly A, Vanlerberghe V, Rodríguez M, Benitez JR, Duvergel J, Van der Stuyft P. The unbearable lightness of technocratic efforts at dengue control. *Trop Med Int Health* 2008;13:728–736.
19. Gordon AJ. Mixed strategies in health education and community participation: an evaluation of dengue control in the Dominican Republic. *Health Educ Res* 1988;3:399–419.
20. Winch P, Lloyd L, Godas MD, Kendall C. Beliefs about the prevention of dengue and other febrile illnesses in Merida, Mexico. *J Trop Med Hyg* 1991;94:377–387.
21. Reiter P. *Aedes albopictus* and the world trade in used tires, 1988–1995: the shape of things to come? *J Am Mosq Control Assoc* 1998;14:83–94.
22. Reiter P, Sprenger D. The used tire trade: a mechanism for the worldwide dispersal of container breeding mosquitoes. *J Am Mosq Control Assoc* 1987;3:494–501.
23. Mazine CA, Macoris ML, Andrighetti MT, Yasumaro S, Silva ME, Nelson MJ, Winch PJ. Disposable containers as larval habitats for *Aedes aegypti* in a city with regular refuse collection: a study in Marilia, Sao Paulo State, Brazil. *Acta Trop* 1996;62:1–13.
24. Hanson SM, Lampman RL, Novak RJ, Purseglove PM. *Aedes albopictus* and *Ae. triseriatus* eggs survive waste tire processing. *J Am Mosq Control Assoc* 1996;12:728–729.

25. Kendall C, Hudelson P, Leontsini E, Winch P, Lloyd L, Cruz F. Urbanization, dengue and the health transition: anthropological contributions to international health. *Med Anthropol Q* 1991;5:257–268.
26. Macoris ML, Mazine CA, Andrighetti MT, Yasumaro S, Silva ME, Nelson MJ, Winch PJ. Factors favoring houseplant container infestation with *Aedes aegypti* larvae in Marilia, Sao Paulo, Brazil. *Rev Panam Salud Publica* 1997;1:280–286.
27. Elder JP, Geller ES, Hovell MF, Mayer JA. *Motivating Health Behavior.* Albany, New York, 1994.
28. Focks DA, Brenner RJ, Hayes J, Daniels E. Transmission thresholds for dengue in terms of *Aedes aegypti* pupae per person with discussion of their utility in source reduction efforts. *Am J Trop Med Hyg* 2000;62: 11–18.
29. Tun-Lin W, Kay BH, Barnes A. Understanding productivity, a key to *Aedes aegypti* surveillance. *Am J Trop Med Hyg* 1995a;53:595–601.
30. Focks DA, Chadee DD. Pupal survey: an epidemiologically significant surveillance method for *Aedes aegypti*: an example using data from Trinidad. *Am J Trop Med Hyg* 1997;56:159–167.
31. Focks DA, Alexander N. Multicountry study of *Aedes aegypti* pupal productivity survey methodology: findings and recommendations. TDR/IRM/Den/06.1. WHO, Geneva.
32. Gómez-Dantés H, Rivas-Gutiérrez L, Canto-Celis S, Capetillo-Pasos M, Pacheco-Can G. *Waste Management, Domestic Hygiene and Community Participation in Dengue Control, Mérida, Yucatán, Mexico (1991–1994).* Final report submitted to The Rockefeller Foundation, New York, 1995.
33. Lloyd LS, Winch P, Ortega-Canto J, Kendall C. Results of a community-based *Aedes aegypti* control program in Merida, Yucatan, Mexico. *Am J Trop Med Hyg* 1992;46:635–642.
34. Méndez-Galván J, Rivas-Gutiérrez L, Nájera-Vázquez R, Inette-Burgos M, Canto-Celis S, Sabido-Montoya F. *Dengue Prevention and Control Project in Mérida, Yucatán, Mexico (1995–1996) — Workshop on Recent Advances in Community-Based Aedes aegypti Control: Honduras and Mexico.* Rockefeller Foundation, Merida, Mexico, 1997.
35. Lefebvre RC, Flora JA. Social marketing and public health intervention. *Health Educ Q* 1988;15:299–315.

36. Pai HH, Lu YL, Hong YJ, Hsu EL. The differences of dengue vectors and human behaviour between families with and without members having dengue fever/dengue hermorrhagic fever. *Int J Environ Health Res* 2005;15:263–269.
37. Dickin K, Griffiths M, Piwoz E. *Designing by Dialogue: A Program Planner's Guide to Consultative Research for Improving Young Child Feeding*. Academy for Educational Development, Washington DC, 1997, p. 325.
38. Sherman C, Fernandez EA, Chan AS, Lozano RC, Leontsini E, Winch PJ. La Untadita: a procedure for maintaining washbasins and drums free of *Aedes aegypti* based on modification of existing practices. *Am J Trop Med Hyg* 1998;58:257–262.
39. Fernandez EA, Leontsini E, Sherman C, Chan AS, Reyes CE, Lozano RC, Fuentes BA, Nichter M, Winch PJ. Trial of a community-based intervention to decrease infestation of *Aedes aegypti* mosquitoes in cement washbasins in El Progreso, Honduras. *Acta Trop* 1998;70:171–183.
40. Leontsini E, Rosenabaum J, Báez C, Solis A, Valera C, Gonzálvez G. NEgociación de PRÁcticas Mejoradas — NEPRAM (Negotiation of Improved Practices): the development of a national behaviour change strategy for community-based prevention of dengue fever in the Dominican Republic. *Dengue Bull* 2004;28(Suppl):22–25.
41. Barrera R, Navarro JC, Mora JD, Dominguez D, Gonzalez J. Public service deficiencies and *Aedes aegypti* breeding sites in Venezuela. *Bull Pan Am Health Organ* 1995;29:193–205.
42. Kittayapong P, Strickman D. Three simple devices for preventing development of *Aedes aegypti* larvae in water jars. *Am J Trop Med Hyg* 1993;49:158–165.
43. Kroeger A, Lenhart A, Ochoa M, Villegas E, Levy M, Alexander N, McCall PJ. Effective control of dengue vectors with curtains and water container covers treated with insecticide in Mexico and Venezuela: cluster randomised trials. *Br Med J* 2006;332(7552):1247–1252.
44. Tun-Lin W, Maung Maung M, Sein Maung T, Tin Maung M. Rapid and efficient removal of immature *Aedes aegypti* in metal drums by sweep net and modified sweeping method. *Southeast Asian J Trop Med Public Health* 1995b;26:754–759.
45. Kroeger A, Dehlinger U, Burkhardt G, Atehortua W, Anaya H, Becker N. Community-based dengue control in Columbia: people's knowledge and practice and the potential contribution of the biological larvicide

Bti (*Bacillus thuringiensis israelensis*). *Trop Med Parasitol* 1995;46: 241–246.

46. Marten GG, Borjas G, Cush M, Fernandez E, Reid JW. Control of larval *Aedes aegypti* (*Diptera*: *Culicidae*) by cyclopoid copepods in peridomestic breeding containers. *J Med Entomol* 1994;31:36–44.
47. Vu SN, Nguyen TY, Kay BH, Marten GG, Reid JW. Eradication of *Aedes aegypti* from a village in Vietnam, using copepods and community participation. *Am J Trop Med Hyg* 1998;59:657–660.
48. McQueen DV. A world behaving badly: the global challenge for behavioral surveillance. *Am J Public Health* 1999;89:1312–1314.
49. Chan AS, Sherman C, Lozano RC, Fernandez EA, Winch PJ, Leontsini E Development of an indicator to evaluate the impact, on a community-based *Aedes aegypti* control intervention, of improved cleaning of water-storage containers by householders. *Ann Trop Med Parasitol* 1998;92:317–329.
50. Sweat MD, Denison JA. Reducing HIV incidence in developing countries with structural and environmental interventions. *AIDS* 1995;9:S251–S257.
51. Tawil O, Verster A, O'Reilly KR. Enabling approaches for HIV/AIDS prevention: can we modify the environment and minimize the risk? *AIDS* 1995;9:1299–1306.
52. Winch P. Social and cultural responses to emerging vector-borne diseases. *J Vector Ecol* 1998;23:47–53.
53. Winch P, Kendall C, Gubler D. Effectiveness of community participation in vector-borne disease control. *Health Policy Plann* 1992;7:342–351.
54. Nathan MB. Critical review of *Aedes aegypti* control programs in the Caribbean and selected neighboring countries. *J Am Mosq Control Assoc* 1993;9:1–7.
55. Winch PJ, Leontsini E, Rigau-Pérez JG, Ruiz-Pérez M, Clark GG, Gubler DJ. Community-based prevention programs in Puerto-Rico: impact on knowledge, behavior, and residential mosquito infestation. *Am J Trop Med Hyg* 2002;67(4):363–370.
56. Jareg P, Kaseje DC. Growth of civil society in developing countries: implications for health. *Lancet* 1998;351:819–822.
57. Rifkin SB. The role of the public in the planning, management and evaluation of health activities and programmes, including self-care. *Soc Sci Med [A]* 1981;15:377–386.
58. Dias JC. Community participation and control of endemic diseases in Brazil: problems and possibilities. *Cad Saude Publica* 1998;14:19–37.

Controversies

As is appropriate to such a complicated group of diseases, the field of dengue diagnosis, epidemiology and research is filled with controversy.

Five important controversies have arisen which can be described by the following propositions: (1) the World Health Organization definition of dengue hemorrhagic fever is inadequate for clinical and epidemiological purposes; (2) the association of dengue hemorrhagic fever with second dengue infections is coincidental; (3) dengue hemorrhagic fever is caused by inherently virulent dengue viruses; (4) dengue hemorrhagic fever is caused by an accelerated or pathological T cell response; and (5) dengue hemorrhagic fever is caused by an autoimmune phenomenon triggered by dengue viral infection.

The fact that the editor of this book is embroiled on one side of many of these controversies has made it difficult to recruit articulate spokespersons to state and defend alternative points of view for the book. Dr. David Morens has graciously consented to prepare "pro" positions for four controversies and the editor has described and defended the opposing view. To give credence to each position, lengthy quotes from original proponents are provided. The reader will understand that greater insight into these complex arguments may require direct examination of the writings of proponents, many of which have been referenced.

14

Controversies

I. The World Health Organization Definition of Dengue Hemorrhagic Fever is Inadequate for Clinical and Epidemiological Purposes

Pro: *José G. Rigau-Pérez, MD, MPH*, Dengue Branch, Division of Vector-Borne Infectious Diseases, Centers for Disease Control and Prevention, San Juan, Puerto Rico, USA*

The WHO definition for DHF has been in use since 1975, with periodic revisions.[1,a] This case definition has come increasingly under scrutiny and

---

* Attribution refers to the position the author held at the time of original writing.

[a] *Case definition for dengue hemorrhagic fever*

The following four criteria must all be present:

(1) Fever, or history of acute fever, lasting 2–7 days
(2) Hemorrhagic tendencies, evidenced by at least one of the following:
 - a positive tourniquet test
 - petechiae, ecchhymosis or purpura
 - bleeding from the mucosa, gastrointestinal tract, injection sites or other locations
 - hematemesis or melena
(3) Thrombocytopenia (100,000 cells per mm^3 or less).
(4) Evidence of plasma leakage due to increased vascular permeability, manifested by at least one of the following:
 - a rise in the hematocrit equal to or greater than 20% above average for age, sex and population
 - a drop in the hematocrit following volume replacement treatment equal to or greater than 20% of the baseline
 - signs of plasma leakage, such as pleural effusion, ascites or hypoproteinemia

study, with many calls for revision.[2–5] The question addressed here is whether the case definition is applicable to the circumstances in which DHF is diagnosed, reported and studied in most areas of the world.

An important element in the case definition of DHF is the identification of cases that will be confirmed by the laboratory as being due to acute dengue infection. The "gold standard" evaluation of the definition was undertaken by an investigator who had strong virologic backup and an understanding of the timing of appearance of diagnostic signs. The strict application of the definition captured only 23 of 28 patients given the diagnosis of DHF by physicians at the Queen Sirikit National Center for Child Health (sensitivity of 82%, 95% confidence interval [CI] = 65–93%).[6]

For fatal cases, the negative predictive value (NPV) will be low when unusual manifestations of dengue account for a large proportion of deaths as occurred in Indonesia from 1975 to 1978. Sumarmo *et al.* observed that 9 of 30 virologically confirmed fatal cases did not fulfill DHF criteria.[7] The positive predictive value (PPV) will be lower when there is a high prevalence of other conditions (e.g. leptospirosis or bacterial sepsis) that may fulfill DHF criteria, especially if the timing of manifestations is not fully considered. A comparison of laboratory-positive fatal cases with laboratory-negative cases in Puerto Rico from 1992 to 1996 showed that the proportion of lab-positive deaths that fulfilled DHF criteria was 78% (95% CI = 56–93%), similar to lab-negatives (75%; 95% CI = 35–97%), suggesting the similarity of DHF to other diseases in the terminal stage.[8]

Paradoxically, the WHO case definition has been least useful for disease surveillance and monitoring the quality of patient care. In practice, the name of the disease often supplants the case definition, focusing the attention of physicians on hemorrhage and platelet concentrations instead of markers for excessive vascular permeability or impeding shock. The WHO case definition is complex, with many components requiring different and repeated tests (hematocrit, serum albumin or protein, X-ray, microscopic analysis of urine). As to hematocrit values, the WHO definition defines a diagnostic threshold achievable only after evolution of the disease. Early treatment with intravenous fluid makes it difficult to document significantly elevated hematocrit values; the definition may identify only severe or late, or poorly treated cases. Adopting a criterion of one hematocrit equal to or greater than 50% or hemoconcentration of equal to or greater than 10% almost doubled the number of DHF cases compared with using the WHO definition.[9] The difficulty in diagnosing DHF from hematocrit criteria has been recognized for some time in Southeast Asia and the Americas.[10,11] It is not rare to find that the WHO criteria are satisfied only when extensive clinical and laboratory observations are made.[6,11] The WHO definition was designed in the 1960's and, for example, does not define thrombocytopenia as a platelet count less than or equal to 150,000/mm^3, the current value, but as less than 100,000/mm^3, a value that corresponds to a blood smear with less than five platelets per high-powered field (100×) under oil immersion magnification.[1,12] There is some confusion about the end point of the tourniquet test. The 1997 WHO guidelines define a positive test as 20 petechiae per square inch, while the Guidelines for Treatment issued by the Southeast Asia Regional Office of WHO accept a more widely used cutoff of 10 petechiae or more per square inch.[1,13–16] Keying on the descriptive section of the "Clinical Diagnosis" chapter in the 1997 WHO Guidelines, many clinicians have added "hepatomegaly" to the DHF diagnostic criteria.

The WHO case definition requires validation in different and well-defined epidemiological and clinical settings to quantify its sensitivity, specificity and predictive positive and negative values. The reasons for lack of predictive power should be identified. A multicenter clinical study has been designed for this purpose and initiated by the Tropical Diseases Research Program of the World Health Organization.[17] This study may make it possible to issue a new case definition for DHF that can be applied with greater precision in diverse populations around the globe.

References

1. World Health Organization. *Dengue Hemorrhagic Fever: Diagnosis, Treatment, Prevention and Control*, 2nd. edn. World Health Organization, Geneva, 1997.
2. Balasubramanian S, Janakiraman L, Kumar SS, Muralinath S, Shivbalan S. A reappraisal of the criteria to diagnose plasma leakage in dengue hemorrhagic fever. *Indian Pediatr* 2006;43:334–339.

3. Deen JL, Harris E, Wills B, Balmaseda A, Hammond SN, Rocha C *et al.* The WHO dengue classification and case definitions: time for a reassessment. *Lancet* 2006;368:170–173.
4. Guzmán MG, García A, Kourí G. El dengue y el dengue hemorrágico: prioridades de investigacion. *Rev Panam Salud Publica* 2006;19:204–215.
5. Rigau-Pérez JG. Severe dengue: the need for new case definitions. *Lancet Infect Dis* 2006;6:297–302.
6. Kalayanarooj S, Vaughn DW, Nimmannitya S, Green S, Suntayakorn S, Kunentrasai N *et al.* Early clinical and laboratory indicators of acute dengue illness. *J Infect Dis* 1997;176(2):313–321.
7. Sumarmo, Wulur H, Jahja E, Gubler DJ, Suharyono W, Sorensen K. Clinical observations on virologically confirmed fatal dengue infections in Jakarta, Indonesia. *Bull World Health Organ* 1983;61(4):693–701.
8. Rigau-Pérez JG, Laufer MK. Dengue-related deaths in Puerto Rico, 1992–6; diagnosis and alarm signals. *Clin Infect Dis* 1999;42:1241–1246.
9. Rigau-Pérez JG, Bonilla GL, Puerto Rico Association of Epidemiologists. An evaluation of modified case definitions for the detection of dengue hemorrhagic fever. *Puerto Rico Health Sci J* 1999;18:347–352.
10. Sirinavin S, Hathirat P. Problems in severity grading of dengue hemorrhagic fever. *Southeast Asian J Trop Med Public Health* 1990;21(4):696.
11. Guzmán MG, Vázquez S, Martínez E, Álvarez M, Rodríguez R, Kourí G *et al.* Dengue in Nicaragua, 1994: reintroduction of serotype 3 in the Americas. *Rev Panam Salud Publica* 1997;1(3):193–199.
12. Lundberg G. JAMA instructions for authors. *J Am Med Assoc* 1997;271:74–82.
13. Teeraratkul A, Limpakanjanarat K, Nisalak A, Nimmannitya S. Predictive value of clinical and laboratory findings for diagnosis of dengue and dengue hemorrhagic fever. *Southeast Asian J Trop Med Public Health* 1990; 21(Suppl):696–697.
14. Lim LE, Quintos FN. Capillary fragility in children with Philippine hemorrhagic fever: the Rumpel–Leede test a screening test. *Philipp J Pediatr* 1954;3:1–9.
15. Tham VD, Hien HH, Long HD. [Value of the tourniquet sign in the diagnosis of hemorrhagic dengue (letter)]. *Med Trop (Mars)* 1996;56(1):99.
16. World Health Organization. *Guidelines for Treatment of Dengue Fever/Dengue Haemorrhagic Fever in Small Hospitals.* World Health Organizational Regional Office for Southeast Asia, New Delhi, 1999:2, note 1.

17. Bandyopadhyay S, Lum LCS, Kroeger A. Classifying dengue: a review of the difficulties in using the WHO case classification for dengue hemorrhagic fever. *Trop Med Int Health* 2006;11:1238–1255.

I. The World Health Organization Definition of Dengue Hemorrhagic Fever is Inadequate for Clinical and Epidemiological Purposes

Con: *Scott B. Halstead, MD*, Director, Supportive Research and Development, Pediatric Dengue Vaccine Initiative, International Vaccine Institute, Seoul, South Korea

The case definition for dengue hemorrhagic fever and dengue shock syndrome (DHF, DSS) was written by clinicians experienced in the diagnosis and treatment of dengue vasculopathy, initially in 1975 and in slightly revised versions in 1986 and 1997.[1–3] Case definitions were written primarily for the purpose of educating the physican and as a guide to life-saving treatment. In every edition the case definition was preceded by a carefully written narrative description of the presenting history, signs and symptoms of the dengue vascular permeability syndrome and followed by a detailed protocol describing fluid and colloid resuscitation. Strictly speaking, the Children's Hospital system of grading DHF/DSS is not a part of the case definition, although it is widely treated as such.[4]

Problems encountered in satisfying the WHO case definition seem mostly to result from the fact that this is a dynamic process with testing requirements that differ from case definitions for chronic disease syndromes such as AIDS. These problems are particularly acute to those who wish to establish a diagnosis of DHF/DSS retrospectively, using chart data available in the usual community hospital (see "pro" position, Chap. 14). Acute dengue vasculopathy generally lasts for 48 hours or less, presenting a physician with an array of rapidly changing pathophysiological conditions.[5] Because of this the assessment of patient status requires multiple observations, and a definitive diagnosis is a matter of timing. It is difficult to take seriously the complaint that the WHO case definition requires too many "repeated clinical tests."[4] It is also stated that early fluid repletion reduces hematocrit values, preventing a diagnosis of hypovolemia.[4] In carefully studied patients, however, third space effusions become prominent only once fluid resuscitation is begun. A diagnosis of dengue vasculopathy certainly is possible if the appropriate test is performed.[6]

Acute dengue vasculopathy is nearly unique in human medicine. The closest clinical analog is the vascular permeability that accompanies the Schwartzman phenomenon during meningococcal sepsis. Meningococcemia patients typically have an acute, sustained high fever and present with petechiae and purpura accompanying shock. In contrast, dengue vasculopathy is most often observed in afebrile patients, with few if any hemorrhagic signs. Important, but subtle, changes in blood pressure occur as dengue vascular permeability worsens. During the initial stage there is a compensatory increase in peripheral vascular resistance, with decreased cardiac output and normal or low central venous pressure. Uniquely, the diastolic pressure rises toward the systolic and the pulse pressure narrows. Patients in dengue shock remain conscious and lucid. The unprepared physician may measure a normal systolic pressure and misjudge the critical state of the patient. Peripheral blood pressures are helpful; use of invasive techniques to measure mean arterial blood pressure may subject patients to unnecessary risk, with little benefit.[7] Finally, there is decompensation and both pressures disappear abruptly. Prolonged deep shock and anoxia may lead to multiorgan failure and an extremely stormy clinical course. In children, severe bleeding occurs only rarely, and almost invariably in association with profound shock. Thrombotic complications are not seen.

A point of critical importance is that the lives of individuals with dengue vasculopathy can be saved through fluid resuscitation. But successful management relies on early recognition and on meticulous titration of parenteral fluid therapy during the period of increased vascular leakage. The physician must remember that all fluid administered will be reabsorbed and fluid overload can and often does result in cardiac failure and pulmonary edema.

Given the need to identify patients as early as possible in the course of a dengue infection, permitting parents to be warned or children hospitalized for observation, the introduction of such simple and inexpensive screening tests as the tourniquet test (or identifying visible hemorrhagic signs) and a platelet count can be regarded as defensible. Even if these tests lack specificity and sensitivity, a positive test, hemorrhagic sign or low platelet count can be detected prior to onset of shock, constituting a useful early warning signal. Considerable experience throughout the globe has called into question whether the tourniquet test, other hemorrhagic signs or low platelet counts are universally observed in cases of dengue vasculopathy. Certainly, in humans with dark skin, the tourniquet test is only rarely positive. The low sensitivity and specificity of these tests are described in detail in a review by Bandyopadhyay *et al.* including comments on the frequency of inappropriate test performance.[8] Nonetheless, this and another review document the fact that these tests are not essential diagnostic criteria for DHF/DSS.[9]

The single most important element in the WHO case definition is the requirement to develop objective evidence of clinically significant vascular leakage. As the review by Bandyopadhyay *et al.* makes clear, many physicians do not understand how to document vascular permeability and many case series have been reported in which such documentation is defective.[8] Some reports have described a clinical entity of shock without evidence of vascular permeability in patients with laboratory-confirmed dengue infections.[10] Such idiosyncratic reports suggest that there is a requirement for better and more standardized observations on clinical signs. Only massive gastrointestinal bleeding in patients with focal bleeding points is known to produce shock without vascular permeability in dengue patients.[11] Historically, the degree of hypovolemia has been estimated using repeated microhematocrit measurements. Because hematocrit values at peak hypovolemia can only be compared with a baseline value from the same individual (convalescent sample), the degree of hypovolemia cannot be determined until after the illness is over. This has led to a movement to establish normal population-based hematocrit values.[4] Unfortunately, emphasis on hematocrit determinations has blunted the requirement to look for the clinical signs of impending shock, such as cool extremities, a history of anuria and slow venous filling. These criticisms also miss the critical importance of obtaining repeated microhematocrit measurements during the preshock or shock stages for the purpose of estimating losses of fluid from the vascular compartment. They also provide the crucial data needed to manage the administration of intravenous fluids. In the Southeast Asian and Western Pacific WHO regions, it has been recommended that microhematocrit centrifuges be located on the nursing station directly on hemorrhagic fever or critical care wards. This innovation should be adopted universally.

It is now possible to visualize gall bladder wall thickening and serous effusions in the abdomen or thorax using noninvasive ultrasound scans. Unfortunately, and against the author's recommendation at the time, the current WHO case definition permits individual hematocrit values to be compared with community normals. Also, hypoproteinemia is listed as a laboratory test that signals vascular permeability. Of course, that is not true. Hypoproteinemia, specifically hypoalbuminemia, is observed in acute vascular permeability, but determining normal values and distinguishing chronic from acute hypoproteinemia seriously undermines the value of this test.

Many workers have called attention to encephalopathy, multiorgan failure, a prolonged course and severe bleeding in patients experiencing a dengue infection. Encephalopathy is probably a standard feature of DHF/DSS. Whether other clinical phenomena are the result of inadequate fluid repletion, are symptoms of fluid overload or are the direct result of dengue infection requires careful study, which should include postmortem examination. The occurrence of relatively rare syndromes should not distract physicians from focusing their efforts on identifying vascular permeability or on treating it promptly and accurately.

Finally, the most important deficiency in the existing WHO case definition is the differences observed between children and adults. Careful studies in Vietnam have established the fact that the intrinsic strength of the capillary bed differs with age with the youngest children having the greatest capillary fragility and adults the least.[12] Undoubtedly, this intrinsic difference in capillary fragility is reflected in the observed susceptibility of the youngest children to vasculopathy during second dengue infections.[13,14] By way of understanding the severe bleeding observed in dengue-infected adults who are not in shock, it is important to recall that severe bleeding in children is coupled with prolonged hypovolemia. It is possible that damaged endothelial cells that leak fluid also mediate thrombocytopenia and altered hemostasis. The same disease response mediators in adults and children may produce hemorrhaging but this occurs in adults without a clinically significant loss of fluid from the vascular compartment.

In sum, eliminating the tourniquet test and platelet count from the WHO case definition may accede to existing realities. This change will not improve the lives of patients unless the nature of dengue vasculopathy is more widely understood and all physicians in dengue-endemic areas participate in an informed and diligent effort to detect and treat capillary permeability at an early stage of presumptive dengue illnesses.

References

1. World Health Organization. *Technical Guides for Diagnosis, Treatment, Surveillance, Prevention and Control of Dengue Hemorrhagic Fever*. World Health Organization, Geneva, 1975.
2. World Health Organization. *Dengue Hemorrhagic Fever: Diagnosis, Treatment and Control*. World Health Organization, Geneva, 1987.
3. World Health Organization. *Dengue Hemorrhagic Fever: Diagnosis, Treatment, Prevention and Control*, 2nd ed., Technical Report. WHO, Geneva, 1997.
4. Rigau-Perez JG. Severe dengue: the need for new case definitions. *Lancet Infect Dis* 2006;6:297–302.
5. Wills BA, Oragui EE, Dung NM, Loan HT, Chau NV, Farrar JJ *et al*. Size and charge characteristics of the protein leak in dengue shock syndrome. *J Infect Dis* 2004;190:810–818.
6. Srikiatkhachorn A, Krautrachue A, Ratanaprakarn W, Wongtapradit L, Nithipanya N, Kalayanrooj S *et al*. Natural history of plasma leakage in dengue hemorrhagic fever: a serial ultrasonographic study. *Pediatr Infect Dis J* 2007;26:283–290.
7. Alfaro A, Pizzaro D, Navas L, Kivers G, Penniecook T, Perez E. La organizacion ye efectividad de una unidad especial de atencion de dengue del area de salud de Limon, Costa Rica. *Mem Acad Nacl Cienc* (*Costa Rica*) 1999;7:11–21.
8. Bandyopadhyay S, Lum LCS, Kroeger A. Classifying dengue: a review of the difficulties in using the WHO case classification for dengue haemorrhagic fever. *Trop Med Int Health* 2006;11:1238–1255.

9. Deen JL, Harris E, Wills B, Balmaseda A, Hammond SN, Rocha C *et al.* The WHO dengue classification and case definitions: time for a reassessment. *Lancet* 2006;368:170–173.
10. Balmaseda A, Hammond SN, Perez MA, Cuadra R, Solano S, Rocha J *et al.* Short report: assessment of the World Health Organization scheme for classification of dengue severity in Nicaragua. *Am J Trop Med Hyg* 2005;73:1059–1062.
11. Wang JY, Tseng CC, Lee CS, Cheng KP. Clinical and upper gastroendoscopic features of patients with dengue virus infection [see comments]. *J Gastroenterol Hepatol* 1990;5:664–668.
12. Gamble J, Bethell D, Day NP, Loc PP, Phu NH, Gartside IB *et al.* Age-related changes in microvascular permeability: a significant factor in the susceptibility of children to shock? *Clin Sci (Colch)* 2000;98(2):211–216.
13. Bethell DB, Gamble J, Pham PL, Nguyen MD, Tran TH, Ha TH *et al.* Noninvasive measurement of microvascular leakage in patients with dengue hemorrhagic fever. *Clin Infect Dis* 2001;32(2):243–253.
14. Guzman MG, Kouri G, Bravo J, Valdes L, Vazquez S, Halstead SB. Effect of age on outcome of secondary dengue 2 infections. *Int J Infect Dis* 2002;6:118–124.

II. The Association Between Dengue Hemorrhagic Fever and Second Dengue Infections is Simply Coincidental

Pro: *David M. Morens, MD*, Office of the Director, National Institute of Allergy and Infectious Disease, NIH, Bethesda, MD, USA

Review of the literature

The first report that severe dengue disease (DHF/DSS) was related to a second or third heterologous dengue infection was from a retrospective analysis of a large series of virologically defined patients admitted to Bangkok Children's Hospital in 1962–64. A statistically significant difference in the incidence of secondary-type immune responses was observed in patients with severe compared with less severe dengue illnesses.[1] This was called the "the two-infection phenomenon."[2]

In his 1977 American Society of Tropical Medicine and Hygiene Presidential Address, Rosen questioned the two-infection hypothesis of DHF/DSS, asking whether the associations between type of antibody response and DHF/DSS were statistically significant.[3] He suggested that high rates of secondary infections associated with dengue shock syndrome might simply reflect the high prevalence of dengue antibodies in the population and gave an example, a 1975 study from Rangoon that described the prevalence of dengue HI antibodies in children under 5 years to be 97%.[4] He concluded that if "a particularly virulent...type of dengue virus...infects everyone...it can be seen that...more than 97% of all shock cases observed would show secondary type antibody responses." Rosen also argued that conclusions about the high incidence of secondary compared with primary antibody responses in dengue shock syndrome were "potentially flawed" because of "(1) differences in patient age, (2) different dengue serotypes and strains and (3) errors in classification of the nature of the antibody response." He questioned whether it was possible "that persons with primary dengue infections do not have a lower incidence of shock syndrome as compared with those with secondary infections but, rather, a higher incidence of undifferentiated febrile syndrome." Rosen believed that severe dengue disease was simply a low frequency outcome of an infection with a virulent dengue virus regardless of whether it occurred during a primary or secondary infection.

Some of these issues were described in greater detail by other investigators. The problem of possible age differences was addressed by Deparis *et al.* after studying the 1996–97 DEN-2 epidemic in Tahiti, which followed seven years of DEN-3 circulation.[5] A population of 174 persons under the age of 21 years presenting with suspected dengue was followed for clinical outcomes and tested for the development of IgM and IgG dengue ELISA antibodies. Since this paper has been widely cited in the literature, an extensive quote from the Discussion is provided:

> "Our results show that dengue incidence or dengue [period] prevalence depend on [length of] time of exposure to the risk of receiving infected mosquitoes (*sic*) bites...and thus strongly on age in endemic

> and in epidemic countries. In our study, before the DEN-2 epidemic, analysis of dengue prevalence revealed a strong relationship with age because the [length of] time of exposure to the risk of receiving infected mosquitoes (*sic*) bites was a function of the age of the individuals. By contrast, in the same study population, the DEN-2 prevalence was not related to age because the time of exposure was identical for the entire study population. To control for the effect of the [length of] time of exposure, adjusted age rates must be used for comparison of dengue incidence or dengue prevalence between populations."

> "But most of the numerous epidemiological studies in endemic regions were conducted without adjustment for age, in cohorts of hospitalized cases not representative of the whole population of dengue cases, leading to the conclusion that DHF is more frequent during secondary than primary dengue infection.... With adjustment for age, Glaziou *et al.*[6] concluded that during the DEN-3 epidemic in 1989 in Tahiti, DHF was as frequent in primary as it was in secondary dengue cases."

Rosen felt that the difficulty of making clinical classifications impeded the analysis of data. For example, "It is difficult to make quantitative comparisons between different geographic areas because of problems in defining severe dengue.... While it might seem logical to label a dengue infection with hemorrhage as dengue hemorrhagic fever, strange as it may seem the World Health Organization (WHO) in the latest edition of guidelines on the subject distinguishes between 'dengue hemorrhagic fever' on one hand and 'dengue with hemorrhagic manifestations' on the other!.... Since thrombocytopenia is a common manifestation in all types of dengue infection and since children with fever, vomiting, or diarrhea from any cause can also have hemoconcentration it is not hard to understand why most clinicians have difficulty in distinguishing between the two putative entities. I am not aware of any data, published or unpublished, which justify the distinction made by WHO guidelines on either pathogenic or prognostic grounds."[7] Many of these arguments were restated expressing the view that it was not logical to require hemoconcentration and not hemorrhage to satisfy the WHO case definition for DHF.[8–11]

Finally, it remains to be explained how the two-infection hypothesis can account for documented instances in which primary infections cause a high rate of severe dengue disease,[12] in which epidemics producing high rates of secondary infections produce few or no cases of severe dengue disease,[13] and instances of clear associations between imported virulent dengue strains and severe disease.[14]

Summation

Alleged proofs of a significant association between secondary dengue infections and severe dengue syndromes are not convincing. Age and diagnostic categories are confounding variables. Dengue strain-associated virulence is a clear and apparently independent risk factor for disease severity, and there remains no proof that host factors increase this risk.

References

1. Halstead SB, Nimmannitya S, Yamarat C, Russell PK. Hemorrhagic fever in Thailand; recent knowledge regarding etiology. *Jpn J Med Sci Biol* 1967; 20s:96–103.
2. Russell PK, Yuill TM, Nisalak A, Udomsakdi S, Gould D, Winter PE. An insular outbreak of dengue hemorrhagic fever. II. Virologic and serologic studies. *Am J Trop Med Hyg* 1968;17(4):600–608.
3. Rosen L. The emperor's new clothes revisited, or reflections on the pathogenesis of dengue hemorrhagic fever. *Am J Trop Med Hyg* 1977;26: 337–343.
4. Thaung U, Ming CK, Swe T, Thein S. Epidemiological features of dengue and chikungunya infections in Burma. *Southeast Asian J Trop Med Public Health* 1975;6:276–283.
5. Deparis X, Roche C, Murgue B, Chungue E. Possible dengue sequential infection: dengue spread in a neighbourhood during the 1996/97 dengue-2 epidemic in French Polynesia. *Trop Med Int Health* 1998;3(11):866–871.
6. Glaziou P, Chungue E, Gestas P, Soulignac O, Couter JP, Plichart R *et al.* Dengue fever and dengue shock syndrome in French Polynesia. *Southeast Asian J Trop Med Public Health* 1992;23:531–532.
7. Rosen L. Dengue hemorrhagic fever. *Bull Soc Pathol Exot* 1996;89:91–94.

8. Rosen L. The pathogenesis of dengue haemorrhagic fever: a critical appraisal of current hypotheses. *S Afr Med J* 1986;Suppl:40–42.
9. Rosen L. [Pathogenesis of hemorrhagic dengue: critical discussion of current hypotheses] La pathogenese de la dengue hemorragique: discussion critique des hypotheses actuelles. *Bull Soc Pathol Exot Filiales* 1986;79:342–349.
10. Rosen L. Disease exacerbation caused by sequential dengue infections: myth or reality? *Rev Infect Dis* 1989;11(Suppl 4):S840–S842.
11. Rosen L. Comments on the epidemiology, pathogenesis, and control of dengue. *Med Trop* 1999;59(4):495–498.
12. Barnes WJS, Rosen L. Fatal hemorrhagic disease and shock associated with primary dengue infection on a Pacific island. *Am J Trop Med Hyg* 1974; 23(3):495–506.
13. Rico-Hesse R, Harrison LM, Salas RA, Tovar D, Nisalak A, Ramos C *et al.* Origins of dengue type 2 viruses associated with increased pathogenicity in the Americas. *Virology* 1997;230:244–251.
14. Kochel TJ, Watts DM, Halstead SB, Hayes CG, Espinosa A, Felices V, Caceda R, Bautista T, Montoya Y, Douglas S, Russell KL. Effect of dengue-1 antibodies on American dengue-2 viral infection and dengue haemorrhagic fever. *Lancet* 2002;360:310–312.

II. The Association Between Dengue Hemorrhagic Fever and Second Dengue Infections is Simply Coincidental

Con: *Scott B. Halstead, MD*, Director, Supportive Research and Development, Pediatric Dengue Vaccine Initiative, International Vaccine Institute, Seoul, South Korea

Review of the literature

From the perspective of dengue literature in the 21st century, the arguments set forth in the "pro" position seem inappropriate and dated, but nonetheless they merit criticism because many of these arguments and papers continue to be cited. The association between dengue hemorrhagic fever (dengue vasculopathy) syndrome and second dengue infections has clearly been established in numerous prospective cohort studies and epidemiological observations performed over many years in different countries and in different hemispheres.[1–10] Inasmuch as five of these prospective cohort studies were published prior to 1989, it is curious to find Rosen's contention that "A well designed prospective epidemiological study would resolve the question — if exacerbation of dengue were demonstrated. However, the low and unpredictable incidence of the shock syndrome, even in highly endemic areas, poses formidable problems for the design of such a study. There may be a greater chance of obtaining unequivocally positive data from retrospective investigations in particularly favorable environments."[11] None of the published prospective studies reported encountered these "formidable" difficulties in identifying cohort members with severe dengue illnesses.

Indeed, as predicted by Rosen,[12] retrospective studies in the "favorable environment" of Cuba provide unequivocal evidence that individuals circulating dengue 1 antibodies were at risk to DHF during dengue 2 or dengue 3 infections.[9,10,13] There is not much left of the contention that severe dengue disease is only randomly associated with secondary dengue infections.

But the relationship between second infections and dengue vasculopathy is complex. Not all sequential dengue infections result in DHF.[14] Known constraints are imposed by both host and virus. In the host, blacks are at lower risk of developing DHF during second dengue infections than are whites or Asians[15]; at times this protection is nearly complete.[16] Among genetically susceptible humans, adults are at lower risk of dengue vasculopathy than children during second dengue infections,[17] and adults infected at short intervals are at lower risk than adults infected at longer intervals.[18] Some degree of heterotypic cross-protection follows infection with the dengue viruses. Cross-protection between dengue viruses was discovered by Sabin, who attempted to infect volunteers with dengue 1 and then 2 viruses at intervals of a few weeks. When a dengue 1 infection was followed by dengue 2 in less than three months, no illness was observed.[19] An analogous phenomenon may have occurred with dengue 2 American genotype (AM) virus infections in dengue-1-immune humans.[14] The mild secondary dengue 2 infections that did occur might plausibly have been the outcome of down-regulation of disease by the strong cross-neutralization of DENV-2 AM viruses by dengue 1 antibodies.[20] A similar down-regulation of disease severity had been observed in schoolchildren who, prior to a secondary Asian dengue 2 infection, circulated heterotypic antibodies that neutralized dengue 2 viruses at low dilutions of serum when assayed in human monocyte cultures.[21] Preillness sera from children who were hospitalized during a secondary dengue 2 infection did not neutralize and actually enhanced dengue 2 infections.

The natural history of dengue antibody responses to primary dengue infections has not been subjected to thorough study. As a result the phenotypic attributes of heterotypic dengue antibodies are not described. Are heterotypic dengue antibodies raised in response to epitopes on dengue viral envelope proteins or do host genetics provide for variety in epitope recognition and resultant immune responses? Heterotypic antibodies are likely to be important mechanisms controlling the outcome of the various sequences of two infections (of which 12 variations are possible). The exact contribution of each sequence to severe outcome is as yet not known. Dengue 1 followed by dengue 2 and dengue 1 followed by dengue 3 do cause severe dengue disease.[13,22] In addition, there is evidence that DHF accompanies a third dengue infection, but rarely.

Critics have pointed to age as a confounding variable that affects disease response to primary or secondary dengue antibody responses.[12,23] Rosen illustrated his criticism with an anecdote, supposing that virtually all dengue illnesses would be accompanied by secondary-type antibody responses in children whose preoutbreak dengue HI antibody prevalence had been 97%.[12] This anecdote was composed in 1977, eight years after dengue HI antibody prevalence data from a large age-stratified serum collection obtained from residents of Bangkok in 1962 were published.[24] In these only 50% of five-year-old Thai children had experienced a single dengue infection. If severe disease occurred by chance, then 50% of DSS cases in five-year-olds should have been accompanied by a primary dengue antibody response — not the 3% observed — a percentage that remained stable across all childhood age groups.[24] Contradictory and nearly unintelligible statements of the confounding effect of age are quoted at length in the "pro" paper.[23]

A second complaint by critics has been that dengue nomenclature contributes to misclassification. For example, DHF is not always accompanied by severe hemorrhaging while dengue fever syndrome may be.[25] This seeming idiosyncrasy has been resolved. DHF, an inaptly named syndrome, is characterized notably by capillary leakage, which if uncorrected may lead to hemorrhage as a late complication.[26,27] Bleeding may accompany dengue fever due to a common hemorrhagic diathesis. Gastrointestinal hemorrhages were observed in the 1987–88 epidemic in Taiwan, during primary dengue 1 infections in adults who had underlying peptic ulcer disease.[28] It should no longer be possible to say "I am not aware of any data."[12]

Some criticisms seem to miss the mark or set up straw men. Referring to published prospective studies, Rosen observed, "I do not find these studies convincing because their conclusions are based on certain explicit or implied assumptions that either are not supported by data or are actually contrary to available information."[11] The "main points of contention" include:

(1) *There is no serological test that can determine whether a person who has had more than one previous dengue infection is susceptible to further infection.* In cohort studies, the clinical outcome of the infection of susceptible children is compared with outcomes in children who had dengue antibodies to one previous infection.
(2) *In the absence of virus isolation, it is impossible to identify the infecting serotype.* Viruses are routinely isolated from cohort children experiencing dengue illnesses at high rates while silent infections can be reliably identified using the dengue neutralization test.
(3) *It cannot be assumed that an entire population will be uniformly exposed to infection with circulating serotypes.* Seroconversions have been measured in samples or entire cohorts of 2000–3000 by detecting antibodies in sera collected before and after the dengue virus transmission season.
(4) *Age susceptibilities may differ.* Susceptibility to vascular permeability in different age groups during second dengue infections has been measured, but in cohort studies comparisons of disease incidence are made in individuals in the same age group, differing only by their infection experience.

(5) *Fatal cases, if not virologically confirmed, might be excluded from the numerator.* No fatal cases have been reported in prospective cohort studies.
(6) *Dengue illnesses in cohort members may be incorrectly classified as primary or secondary.* Rosen cites his own report of high antibody titers in noncohort islanders assumed to have had primary dengue infections,[29] but in cohort studies primary versus secondary responses are classified by the status of the preillness serum and also observing the antibody response in paired sera.

Finally, some critics contend that the two-infection phenomenon has been responsible for delaying public health solutions to the dengue problem. In their view, ADE is viewed as a "politically correct" position leading to a widespread belief in the dangers of partial dengue immunity. These views have impeded progress leading to the abandonment of monovalent candidate dengue vaccines begun in the 1970's.[30] This writer is unaware of any effort to develop monovalent dengue vaccines, and dengue vaccine development certainly has not been abandoned.

Summation

The allegation that severe dengue syndromes are only coincidentally related to secondary dengue infections is without merit.

References

1. Winter PE, Yuill TM, Udomsakdi S, Gould D, Nantapanich S, Russell PK. An insular outbreak of dengue hemorrhagic fever. I. Epidemiologic observations. *Am J Trop Med Hyg* 1968;17(4):590–599.
2. Winter PE, Nantapanich S, Nisalak A, Udomsakdi S, Dewey RW, Russell PK. Recurrence of epidemic dengue hemorrhagic fever in an insular setting. *Am J Trop Med Hyg* 1969;18(4):573–579.
3. Russell PK, Yuill TM, Nisalak A, Udomsakdi S, Gould D, Winter PE. An insular outbreak of dengue hemorrhagic fever. II. Virologic and serologic studies. *Am J Trop Med Hyg* 1968;17(4):600–608.

4. Sangkawibha N, Rojanasuphot S, Ahandrik S, Viriyapongse S, Jatanasen S, Salitul V *et al.* Risk factors in dengue shock syndrome: a prospective epidemiologic study in Rayong, Thailand. I. The 1980 outbreak. *Am J Epidemiol* 1984;120:653–669.
5. Thein S, Aung MM, Shwe TN, Aye M, Zaw A, Aye K *et al.* Risk factors in dengue shock syndrome. *Am J Trop Med Hyg* 1997;56(5):566–572.
6. Burke DS, Nisalak A, Johnson DE, Scott RM. A prospective study of dengue infections in Bangkok. *Am J Trop Med Hyg* 1988;38(1):172–180.
7. Graham RR, Juffrie M, Tan R, Hayes CG, Laksono I, Ma'roef C *et al.* A prospective seroepidemiologic study on dengue in children four to nine years of age in Yogyakarta, Indonesia I. Studies in 1995–1996. *Am J Trop Med Hyg* 1999;61(3):412–419.
8. Endy TP, Chunsittiwat S, Nisalak A, Libraty D, Green S, Rothman A *et al.* Epidemiology of inapparent and symptomatic acute dengue virus infection: a prospective study of primary school children in Kamphaeng Phet, Thailand. *Am J Epidemiol* 2002;156:40–51.
9. Guzman MG, Kouri GP, Bravo J, Soler M, Vazquez S, Morier L. Dengue hemorrhagic fever in Cuba, 1981: a retrospective seroepidemiologic study. *Am J Trop Med Hyg* 1990;42:179–184.
10. Guzman MG, Kouri G, Valdes L, Bravo J, Alvarez M, Vazquez S *et al.* Epidemiologic studies on dengue in Santiago de Cuba, 1997. *Am J Epidemiol* 2000;152(9):793–799.
11. Rosen L. Disease exacerbation caused by sequential dengue infections: myth or reality? *Rev Infect Dis* 1989;11(Suppl 4):S840–S842.
12. Rosen L. The emperor's new clothes revisited, or reflections on the pathogenesis of dengue hemorrhagic fever. *Am J Trop Med Hyg* 1977;26: 337–343.
13. Alvarez M, Rodriguez R, Bernardo L, Vasquez S, Morier L, Gonzalez D *et al.* Dengue hemorrhagic fever caused by sequential dengue 1–3 infections at a long interval: Havana epidemic, 2001–2002. *Am J Trop Med Hyg* 2006;75:1113–1117.
14. Watts DM, Porter KR, Putvatana P, Vasquez B, Calampa C, Hayes CG *et al.* Failure of secondary infection with American genotype dengue 2 to cause dengue haemorrhagic fever [see comments]. *Lancet* 1999;354(9188): 1431–1434.

15. Kouri GP, Guzman MG, Bravo JR, Triana C. Dengue haemorrhagic fever / dengue shock syndrome: lessons from the Cuban epidemic, 1981. *Bull World Health Organ* 1989;67:375–380.
16. Halstead SB, Streit TG, Lafontant JG, Putvatana R, Russell K, Sun W *et al.* Haiti: absence of dengue hemorrhagic fever despite hyperendemic dengue virus transmission. *Am J Trop Med Hyg* 2001;65:180–183.
17. Guzman MG, Kouri G, Bravo J, Valdes L, Vazquez S, Halstead SB. Effect of age on outcome of secondary dengue 2 infections. *Int J Infect Dis* 2002;6:118–124.
18. Guzman MG, Kouri G, Valdes L, Bravo J, Vazquez S, Halstead SB. Enhanced severity of secondary dengue 2 infections occurring at an interval of 20 compared with 4 years after dengue 1 infection. *PAHO J Epidemiol* 2002;81:223–227.
19. Sabin AB. Research on dengue during World War II. *Am J Trop Med Hyg* 1952;1:30–50.
20. Kochel TJ, Watts DM, Halstead SB, Hayes CG, Espinosa A, Felices V, Caceda R, Bautista T, Montoya Y, Douglas S, Russell KL. Effect of dengue-1 antibodies on American dengue-2 viral infection and dengue haemorrhagic fever. *Lancet* 2002;360:310–312.
21. Kliks SC, Nisalak A, Brandt WE, Wahl L, Burke DS. Antibody-dependent enhancement of dengue virus growth in human monocytes as a risk factor for dengue hemorrhagic fever. *Am J Trop Med Hyg* 1989;40(4):444–451.
22. Guzman MG, Kouri G, Bravo J, Soler M, Martinez E. Sequential infection as risk factor for dengue hemorrhagic fever/dengue shock syndrome (DHF/DSS) during the 1981 dengue hemorrhagic Cuban epidemic. *Mem Inst Oswaldo Cruz* (*Rio de Janeiro*) 1991;86(3):367.
23. Deparis X, Roche C, Murgue B, Chungue E. Possible dengue sequential infection: dengue spread in a neighbourhood during the 1996/97 dengue-2 epidemic in French Polynesia. *Trop Med Int Health* 1998;3(11):866–871.
24. Halstead SB, Scanlon J, Umpaivit P, Udomsakdi S. Dengue and chikungunya virus infection in man in Thailand, 1962–1964. IV. Epidemiologic studies in the Bangkok metropolitan area. *Am J Trop Med Hyg* 1969;18(6):997–1021.
25. Rosen L. Dengue hemorrhagic fever. *Bull Soc Pathol Exot* 1996;89:91–94.
26. Wills BA, Oragui EE, Dung NM, Loan HT, Chau NV, Farrar JJ *et al.* Size and charge characteristics of the protein leak in dengue shock syndrome. *J Infect Dis* 2004;190:810–818.

27. Wills BA, Oragui EE, Stephens AC, Daramola OA, Dung NM, Loan HT *et al.* Coagulation abnormalities in dengue hemorrhagic fever: serial investigations in 167 Vietnamese children with dengue shock syndrome. *Clin Infect Dis* 2002;35:277–285.
28. Tsai CJ, Kuo CH, Chen PC, Changcheng CS. Upper gastrointestinal bleeding in dengue fever. *Am J Gastroenterol* 1991;86:33–35.
29. Barnes WJS, Rosen L. Fatal hemorrhagic disease and shock associated with primary dengue infection on a Pacific island. *Am J Trop Med Hyg* 1974; 23(3):495–506.
30. Murgue B, Cassar O, Roche C, Deparis X. [Pathogenesis of dengue: the emperor is still naked!]. *Med Mal Infect* 2004;34(Suppl 1):S31–S33.

III. Dengue Hemorrhagic Fever is Caused by Virulent Dengue Viruses

Pro: *David M. Morens, MD*, Office of the Director, NIAID, NIH, Bethesda, MD, USA

Brief review of the early literature

The idea of dengue virus virulence arose during sequential epidemics in Athens/Piraeus in 1927 and 1928. Observers at the time were influenced by the memory of the "Spanish influenza" pandemic a decade earlier, which had appeared in three successive waves in 1918–1919. Whereas the first wave in spring/summer 1918 had been universally mild, the second and third, in fall 1918 and winter 1918–1919, respectively, were both deadly. A prevailing theory was that the presumed microbial agent of influenza had increased in pathogenicity upon passage through human populations, leading to the emergence of a more deadly strain to cause the second wave. The same argument was invoked in 1928, that the severe dengue hemorrhagic fever cases appearing that year were attributable to increased virulence of the agent after months of circulation in humans. (Subsequently, retrospective data were published that supported this view, while other data suggested that the 1928 phenomenon had been caused by sequential DEN-1/DEN-2 epidemics.[1,2])

In the era of modern virology, virulent dengue viruses were first suggested by Hammon and colleagues, who studied patients with a "hemorrhagic fever" in the Philippines in 1956, recovering two viruses new to science — dengue types 3 and 4. In addition, viruses initially called dengue types 5 and 6 were isolated from patients with severe dengue infections in the Bangkok outbreaks of 1958 and 1960.[3,4] These new viruses were thought to be responsible for producing the hemorrhagic fever syndrome.

In 1973, Hammon explained his thinking on the issue of viral virulence as the cause of severe dengue disease as follows: "At the moment, it appears reasonable to continue to examine the hypotheses which point to the agent or agents producing infection as responsible for the change which has led to this new epidemic of hemorrhagic disease, rather than to discard such and place all emphasis on the inviting and in many respects plausible immunologic disease hypothesis for the total disease spectrum."[5] The motivation for this opinion was stated repeatedly over the years: "Several groups of workers are now engaged in dengue vaccine studies, but if hemorrhagic fever is entirely an immunologic disease, immunization could be the worst possible weapon to employ in its control."

In 1972, an outbreak of dengue 2 on Niue, a small, isolated Pacific island with a population of 4600, occurred during the months of March to June, with 790 cases reported. A survey performed in August estimated that 44.8% of the total population had experienced a febrile disease during the epidemic and around a quarter of these illnesses were accompanied by hemorrhages varying from epistaxis to gastrointestinal bleeding. A total of 12 individuals died during the outbreak, 7 of whom were under the age of 20 and therefore could not have been infected in the dengue 1 epidemic reported during World War II. Illness in fatal cases was of relatively short duration and terminated with death, often preceded by generalized ecchymoses or gastrointestinal bleeding, and in one instance, in a two-year-old, death was attributed to shock. Monotypic dengue 2 neutralizing antibody titers were detected in sera obtained in either August 1972 or early 1973 from five children who survived a febrile disease accompanied by ecchymoses and other hemorrhagic signs. Sera from four lifelong residents of Niue who were sick during the epidemic contained neutralizing antibodies to both dengue type 1 and type 2 viruses. It was concluded that the Niue outbreak supported a contention that not all DHF cases are associated with a "second dengue infection in an individual older than one year of age." Alternative hypotheses were proposed, "(1) that some strains of dengue possess unusual potential to cause hemorrhagic disease; or, (2) that some of the residents of Niue were especially susceptible to the hemorrhagic potential of an ordinary dengue virus."[6] The Niue outbreak is now widely cited as evidence that severe dengue disease accompanies primary dengue infections (i.e. dengue viruses are inherently virulent).

Scott *et al.*[7] described illnesses in three children older than one year who were admitted to Bangkok Children's Hospital in 1973. Each had an elevated hematocrit, thrombocytopenia and a narrow pulse pressure together with a primary-type antibody response. None of these cases was in overt shock. Antibody patterns suggested that two of these cases were caused by dengue 1 while a dengue 2 virus was isolated from the third patient. Earlier, severe illnesses had been observed in an outbreak in the rural town of Ubol, Thailand, in 1964, in which some cases had primary dengue 1 antibody responses.[8]

In his 1977 presidential address before the American Society of Tropical Medicine and Hygiene, Rosen summarized two lines of evidence supporting dengue virus virulence: (1) he doubted the statistical associations between type of antibody response and DHF/DSS[9] (this argument will be presented in greater detail in the "pro" position on controversy II); and (2) he supported the hypothesis that severe disease was simply a low frequency outcome of infection with a virulent dengue virus regardless whether in primary or secondary infection order, the present epidemic DHF/DSS in Southeast Asia being attributed to the high endemicity of dengue viruses. And, "the apparent absence of dengue hemorrhagic fever and dengue shock syndrome in such dengue-infected areas as the Caribbean can be explained as the result of the circulation of only a few dengue strains which do not happen to be particularly virulent and the failure to recognize the rare severe clinical forms of dengue when they do occur."

An *in vitro* study of virus growth properties supports the importance of virus virulence to disease severity. A panel of 13 low passage DEN-2 isolates obtained during the 1980 Bangkok epidemic showed that virus growth in nonimmune human macrophages was correlated with disease severity.[10]

American genotype dengue 2 viruses

This latter statement seemed prescient when it was reported in 1999 that DHF/DSS was not observed when humans were infected by the American (AM) genotype dengue 2 as either a primary or a secondary infection. Clinical and serological observations on AM dengue 2 infections were made in 1995 during a prospective cohort study in Iquitos, in Amazonian Peru.[10] Dengue 1 virus had been continuously endemic in Iquitos since 1990. In 1995, an estimated 65% of the population experienced a secondary dengue 2 infection in the sequence dengue 1 followed by dengue 2. Infections in this same sequence had resulted in a large DHF/DSS outbreak in Cuba in 1981[11] caused by a dengue 2 virus of Southeast Asian origin (SEA).[12] An editorial in *The Lancet* suggested that there might be virulence differences between the Asian and American dengue 2 viruses, calling for a comparison of their gene sequences.[13] This request was fulfilled by Rebecca Rico-Hesse and colleagues, who found structural differences between the SEA and AM dengue 2 viruses in the prM gene and amino acid 390 in the E protein and in nucleotides 68–80 in the 5′ nontranslated region (NTR) and in the upstream 300 nucleotides of the 3′ NTR.[14]

In addition to predicting that these changes were responsible for reduced virulence in human beings, Rico-Hesse attributed the apparent displacement of AM by SEA dengue 2 viruses to the possibility that AM viruses, are less transmissible by *Aedes aegypti* mosquitoes than SEA viruses and further that AM replicative constraints produce lower titered viremias in humans than do SEA dengue 2 viruses.[15] To support these contentions, the research group found differing infection rates of Texan and Mexican *A. aegypti* following oral feeding of several low passage strains of AM and SEA genotype dengue 2 viruses.[16,17] Some but not all SEA genotype dengue 2 strains grew to higher titers in infected mosquitoes than in AM genotype strains. In a follow-up study, virus replicated in the midgut to significantly higher in SEA- compared with AM-infected mosquitoes, and virus-specific proteins could be detected in salivary glands seven days earlier in SEA- than in AM-infected mosquitoes. This was thought to be due either to differences in replicative efficiency or to differences in translation ability.[18] The net predicted effect would be a shorter extrinsic incubation period in mosquitoes infected with SEA than with AM dengue 2 viruses, resulting in a larger portion of the SEA-dengue-2-infected female mosquito population surviving long enough to transmit the virus to humans.

A further prediction is that AM dengue 2 viruses would demonstrate reduced ability to replicate in putative human dengue target cells.[14] In an *in vitro* study it was observed that AM viruses replicated less well than SEA viruses in human-monocyte-derived macrophages, while substitution of the SEA amino acid 390 into the E protein of an AM construct reduced viral replication in macrophages.[19] In a separate study, SEA dengue 2 chimeras containing the AM 390 amino acid exhibited reduced virus replication in human monocytes and dendritic cells. Introducing AM changes in the 3′ and 5′ NTR's also reduced infectivity. But all three AM changes were needed to produce a virus with the replicative properties of wild-type American genotype dengue 2.[20]

Summation

Dengue disease severity is related to documented genetic differences in dengue strains. A strong case has been made that the consistent association of American genotype dengue 2 viruses with mild dengue disease during primary or secondary dengue infections is explained by low viral virulence.

References

1. Rosen L. The pathogenesis of dengue haemorrhagic fever: a critical appraisal of current hypotheses. *S Afr Med J* 1986;Suppl:40–42.
2. Halstead SB, Papaevangelou G. Transmission of dengue 1 and 2 viruses in Greece in 1928. *Am J Trop Med Hyg* 1980;29(4):635–637.
3. Hammon WM, Rudnick A, Sather GE. Viruses associated with epidemic hemorrhagic fevers of the Philippines and Thailand. *Science* 1960;131:1102–1103.
4. Hammon WM, Sather, GE. Virological findings in the 1960 hemorrhagic fever epidemic (dengue) in Thailand. *Am J Trop Med Hyg* 1964;13:629–641.

5. Hammon WM. Dengue hemorrhagic fever — do we know its cause? *Am J Trop Med Hyg* 1973;22:82–91.
6. Barnes WJS, Rosen L. Fatal hemorrhagic disease and shock associated with primary dengue infection on a Pacific island. *Am J Trop Med Hyg* 1974; 23(3):495–506.
7. Scott RM, Nimmannitya S, Bancroft WH, Mansuwan P. Shock syndrome in primary dengue infections. *Am J Trop Med Hyg* 1976;25(6):866–874.
8. Halstead SB, Yamarat C. Recent epidemics of hemorrhagic fever in Thailand: observations related to pathogenesis of a "new" dengue disease. *J Am Public Health Assoc* 1965;55(9):1386–1395.
9. Rosen L. The emperor's new clothes revisited, or reflections on the pathogenesis of dengue hemorrhagic fever. *Am J Trop Med Hyg* 1977;26: 337–343.
10. Watts DM, Porter KR, Putvatana P, Vasquez B, Calampa C, Hayes CG *et al.* Failure of secondary infection with American genotype dengue 2 to cause dengue haemorrhagic fever [see comments]. *Lancet* 1999;354(9188): 1431–1434.
11. Guzman MG, Kouri GP, Bravo J, Soler M, Vazquez S, Morier L. Dengue hemorrhagic fever in Cuba, 1981: a retrospective seroepidemiologic study. *Am J Trop Med Hyg* 1990;42:179–184.
12. Guzman MG, Deubel V, Pelegrino JL, Rosario D, Marrero M, Sariol C *et al.* Partial nucleotide and amino acid sequences of the envelope and the envelope/nonstructural protein-1 gene junction of four dengue-2 virus strains isolated during the 1981 Cuban epidemic. *Am J Trop Med Hyg* 1995;52: 241–246.
13. White NJ. Variation in virulence of dengue virus [comment]. *Lancet* 1999;354(9188):1401–1402.
14. Leitmeyer KC, Vaughn DW, Watts DM, Salas R, Villalobos I, de C *et al.* Dengue virus structural differences that correlate with pathogenesis. *J Virol* 1999;73(6):4738–4747.
15. Rico-Hesse R. Microevolution and virulence of dengue viruses. *Adv Virus Res* 2003;59:315–341.
16. Armstrong PM, Rico-Hesse R. Differential susceptibility of *Aedes aegypti* to infection by the American and Southeast Asian genotypes of dengue type 2 virus. *Vector Borne Zoonotic Dis* 2001;1:159–168.

17. Armstrong PM, Rico-Hesse R. Efficiency of dengue serotype 2 virus strains to infect and disseminate in *Aedes aegypti*. *Am J Trop Med Hyg* 2003; 68:539–544.
18. Anderson JR, Rico-Hesse R. *Aedes aegypti* vectorial capacity is determined by the infecting genotype of dengue virus. *Am J Trop Med Hyg* 2006;75: 886–892.
19. Pryor MJ, Carr JM, Hocking H, Davidson AD, Li PX, Wright PJ. Replication of dengue virus type 2 in human monocyte-derived macrophages: comparisons of isolates and recombinant viruses with substitutions at amino acid 390 in the envelope glycoprotein. *Am J Trop Med Hyg* 2001;65:427–434.
20. Cologna R, Rico-Hesse R. American genotype structures decrease dengue virus output from human monocytes and dendritic cells. *J Virol* 2003;77:3929–3938.

III. Dengue Hemorrhagic Fever is Caused by Virulent Dengue Viruses

Con: *Scott B. Halstead, MD*, Director, Supportive Research and Development, Pediatric Dengue Vaccine Initiative, International Vaccine Institute, Seoul, South Korea

The role that dengue antibodies play in modulating dengue disease expression up or down has best been documented in prospective seroepidemiological studies and island outbreaks. These are reviewed in Chap. 7. Other than in infants born to dengue-immune mothers, there are authentic instances of severe dengue illnesses accompanying primary dengue virus infections in children and adults. The contribution that idiosyncratic host responses make to these responses is poorly studied. It is known that peptic ulcer disease may result in severe gastrointestinal hemorrhages during primary dengue infections. So far, this observation has been limited to dengue 1.[1,2]

To understand conflicting views on dengue biology it should be recognized that *heterogeneities in dengue viruses themselves are responsible for many diverse clinical responses.* Dengue viruses have the properties of pathogenicity and virulence. *Pathogenicity* describes the spectrum of disease syndromes associated with dengue infection. Within this spectrum, the frequency of specific syndromes may vary. The ratio of severe syndromes to total dengue infections can be measured and is referred to as *virulence*.

Observations on specific dengue viruses

Island epidemics and human volunteer studies provide strong evidence that different strains within genotypes of dengue viruses vary greatly in intrinsic *pathogenicity*. Here, "intrinsic" is measured by the behavior of dengue virus infections in flavivirus naïve hosts, principally adults.

Dengue 1

There has been a prolonged human experience with primary dengue 1 infections. For example, it is now known that Simmons *et al.* induced primary dengue 1 infections in 80 susceptible young adult volunteers in 1928–30 in the Philippines.[3–5] Dengue 1 was transmitted widely to combatants and to civilian populations during World War II[6,7] and viruses isolated from these epidemics were used to produce clinical disease in susceptible adult Americans and Japanese.[8,9] In the modern era, dengue 1 was introduced into the Western Hemisphere in 1977[10] and as it spread the virus encountered many populations that were essentially flavivirus-naïve.[11] Dengue 1 was introduced into Taiwan in 1987, encountering a population that had been free of dengue since the end of World War II,[12] and into Hawaii in 2001–02 in a population composed almost completely of flavivirus susceptibles.[13] Most of what we have learned about classical dengue fever derives from this extensive experience. In general, during primary dengue 1 outbreaks morbidity/infection ratios were high, especially in combat troops. However, mortality rates were low.[14] With only a few exceptions in children, disease has been mild during virgin soil dengue 1 outbreaks.[15] Nonetheless, cases of nonshock dengue hemorrhagic fever syndrome have been documented in Thai children who have a primary-type serological response to dengue 1,[16,17] and occasionally these infections have progressed to narrowed pulse pressure.[18] It is important to note that dengue 1 infections exhibit an important hemorrhagic diathesis. Some individuals with pre-existing peptic ulcer disease and dengue 1 infections suffered gastrointestinal hemorrhaging, in some cases severe.[1]

Dengue 2

The greatest variability in pathogenicity has been described for dengue 2 viruses. Some dengue 2 strains appear to produce little or no disease in susceptibles. For example, genotype III SEA dengue 2 virus infections in the 1997 Santiago de Cuba outbreak resulted in almost no overt disease in flavivirus-naïve children or adults.[19] But the same virus that infected adults who were circulating dengue 1 antibodies from infections acquired in 1977–79 had an overt disease ratio of nearly 1, with cases occurring across a spectrum from DF to DHF/DSS and deaths.[19,20] The same is true for Asian children. Extensive studies at Bangkok Children's Hospital over many years have established that very few children with dengue illnesses are seen during primary-type antibody response to dengue 2 viruses. By contrast, when dengue 2, New Guinea C strain, genotype III, was inoculated into susceptible adult volunteers most developed classical dengue fever.[8,21] A SEA strain of dengue 2, when inoculated into flavivirus-naïve adults with advanced neurosyphilis, produced overt dengue fever.[22] Unfortunately, the overt disease to infection ratios were not reported but, as described, can be assumed to be fairly high.[8,22] In the modern era, a high ratio of overt disease to primary dengue 2 infections (Cosmopolitan genotype) was observed in expatriate Chinese construction workers in Singapore.[23]

Gubler described variations in the severity of dengue 2 illnesses in island outbreaks in the Pacific during 1971–74. These occurred in Tahiti, New Caledonia, American Samoa and Tonga, all caused by American genotype V viruses.[6,24] Although they were often described as "virgin soil" epidemics, in fact, many residents of these islands over the age of 25 had antibodies to dengue 1. For both children and young adults, outbreaks in Tahiti and New Caledonia were explosive and accompanied by high overt disease attack rates, while dengue in American Samoa smoldered, originally being diagnosed as rubella. The outbreak in Tonga was especially mild.[24] The rates of virus isolation from acute phase sera in these outbreaks varied directly with outbreak "severity."[6] The dengue 2 epidemic on Niue in 1972 did not yield a virus isolate but is considered to be part of the Pacific-wide dissemination of the AM virus. A very large outbreak of what was probably AM genotype dengue 2 occurred in the southern US in 1922–23, with an estimated 1–2 million cases.[14] The accompanying disease was classical dengue fever. Observers noted moderate to severe menorrhagia in menstruating women.[25,26] With the exception of Singapore, a population that is infected by multiple dengue 2 genotypes, island outbreaks and infections in human volunteers can reasonably be expected to have been caused by a single virus strain. The disease variation observed after the clonal introductions of dengue viruses points to a marked phenotypic diversity among human infections caused by different dengue 2 strains within at least two different genotypes.

Dengue 3

This virus resembles dengue 1 in its pathogenicity spectrum. Consistently, a small percentage of children admitted to hospitals in Southeast Asia with the diagnosis of DHF have primary antibody responses.[17,27]

Dengue 4

This virus resembles dengue 2. In Southeast Asia, primary dengue 4 infections in children are nearly all silent. Overt disease is nearly always accompanied by a secondary-type antibody response.[17,28] Data on dengue 4 are relatively meager compared with dengue 2, as this virus cycles at longer intervals than the other three dengue viruses.[29] Overt secondary dengue infection disease has been observed in the sequences dengue 1–4 and 2–4 (R. Gibbons, personal communication).

Genetic correlates of virulence

Significant efforts have been directed toward finding genetically distinct viruses that cause severe or mild dengue disease. Many attempts failed. In Southeast Asia, essentially no differences were found in dengue 2 envelope proteins from viruses recovered from severe and mild disease,[30–32] in the 240 nucleotides of the E/NS1 junction of 73 dengue 2 viruses isolated from Thai children with mild or severe dengue disease,[33] or in the full length sequences of 7 dengue 3 strains isolated between 1974 and 1998 from Indonesia and Thailand.[34] In some studies dengue 2 viruses have been classified into "subtypes" that appear to be associated with either mild or severe disease.[32,35,36] These changes have not been reported by other workers. The same laboratory identified changes in the 3′ nontranslated region (NTR) of 4 dengue 2 viruses that were associated with disease severity.[37] However, a much larger study of all 4 dengue viruses isolated from patients hospitalized in Bangkok over a 30-year period found no changes in the 3′ NTR that significantly correlated with disease severity.[38]

Special case of American genotype dengue 2 viruses

Many of the predictions about the "low virulence" of American genotype dengue 2 viruses are based upon the reduced replicative efficiency observed *in vitro* or in experimental systems, such as laboratory-reared *Aedes aegypti*. But are these predictions borne out in real life? The conclusion that AM dengue 2 viruses are being replaced by SEA viruses is not consistent with the fact that in 1995 AM viruses arrived well before SEA viruses to the furthest reaches of the Amazon.[39] Because in the American AM virus has always produced mild human illness, it may not be possible to monitor the spread of this virus using passive surveillance of clinical disease. It seems to be true that AM virus has successfully circulated in the Western Hemisphere for 100 or more years. Despite laboratory markers this virus has exhibited successful transmission via diverse populations of *Aedes aegypti* and humans belonging to many racial groups.

The single most important evidence against the hypothesis that AM viruses have inherently low virulence is the existence of a plausible alternative hypothesis. This hypothesis specifically explains the failure of secondary dengue infections with AM viruses to result in severe disease. In dengue-1-immunes infection outcomes with AM viruses were completely different than with SEA viruses. As opposed to the Cuban SEA virus, the Iquitos AM dengue 2 viruses have dengue-1-like epitopes that result in neutralization by dengue-1-immune sera.[40] In Aotus monkeys, AM dengue 2 infections were down-regulated in dengue-1-immune individuals.[41] It can be predicted that in humans immune to dengue 1, AM virus infection will be down-regulated. This should moderate accompanying disease expression.

Summation

Severe dengue (vascular permeability syndrome) is not caused by inherently virulent dengue viruses. It occurs consistently during dengue infections in individuals circulating heterotypic dengue antibodies at enhancing concentrations. Diverse clinical responses to primary infections with different dengue virus strains do occur. Disease expression during primary dengue infections includes subclinical vascular permeability, which occasionally is observed as overt hypovolemia. The fact that AM dengue 2 viruses produce only mild disease during secondary dengue infections is best explained by the down-regulation of disease by dengue 1 antibodies directed at a unique antigenic structure expressed on AM dengue 2 viruses.

References

1. Tsai CJ, Kuo CH, Chen PC, Changcheng CS. Upper gastrointestinal bleeding in dengue fever. *Am J Gastroenterol* 1991;86:33–35.
2. Wang JY, Tseng CC, Lee CS, Cheng KP. Clinical and upper gastroendoscopic features of patients with dengue virus infection [see comments]. *J Gastroenterol Hepatol* 1990;5:664–668.
3. Simmons JS, St John JH, Reynolds FHK. Experimental studies of dengue. *Philipp J Sci* 1931;44:1–252.
4. Halstead SB. Etiologies of the experimental dengues of Siler and Simmons. *Am J Trop Med Hyg* 1974;23(5):974–982.
5. Nishiura H, Halstead SB. Natural history of dengue virus (DEN)-1 and DEN-4 infections: reanalysis of classical studies. *J Infect Dis* 2007;195: 1007–1013.

6. Gubler DJ. Dengue and dengue hemorrhagic fever: its history and resurgence as a global public health problem. In: Gubler DJ, Kuno G (eds.) *Dengue and Dengue Hemorrhagic Fever*. CAB, New York, 1997, pp. 1–22.
7. Hotta S. Dengue epidemics in Japan, 1942–1945. *J Trop Med Hyg* 1953;56: 83–89.
8. Sabin AB. Research on dengue during World War II. *Am J Trop Med Hyg* 1952;1:30–50.
9. Hotta S. Experimental studies in dengue I. Isolation, identification and modification of the virus. *J Infect Dis* 1952;90(1):1–9.
10. Pan American Health Organization. *Dengue in the Caribbean, 1977. Proceedings of the Workshop in Montego Bay, Jamaica 8–11 May 1978,* PAHO Scientific Publication No. 375. PAHO, Washington, D.C., 1979.
11. Mas P. Dengue fever in Cuba in 1977: some laboratory aspects. In: *Proceedings of Dengue in the Caribbean, 1977. 8–11 May 1979, Montego Bay, Jamaica.* PAHO, Washington, D.C., 1979, pp. 40–42.
12. Liu HW, Ho TL, Hwang CS, Liao YH. Clinical observations of virologically confirmed dengue fever in the 1987 outbreak in southern Taiwan. *Kao-Hsiung I Hsueh Ko Hsueh Tsa Chih* (*Kaohsiung J Med Sci*) 1989; 5:42–49.
13. Effler PV, Pang L, Kitsutani P, Vorndam V, Nakata M, Ayers T *et al.* Dengue fever, Hawaii, 2001–2002. *Emerg Infect Dis* 2005;11:742–749.
14. Sabin AB. Dengue. In: Rivers TM (ed.) *Viral and Rickettsial Infections of Man*, 2nd ed. J.B. Lippincott, Philadelphia, 1952, pp. 556–568.
15. Kuberski T, Rosen L, Reed D, Mataika J. Clinical and laboratory observations on patients with primary and secondary dengue type 1 infections with hemorrhagic manifestations in Fiji. *Am J Trop Med Hyg* 1977;26(4):775–783.
16. Nimmannitya S, Halstead SB, Cohen S, Margiotta MR. Dengue and chikungunya virus infection in man in Thailand, 1962–1964. I. Observations on hospitalized patients with hemorrhagic fever. *Am J Trop Med Hyg* 1969; 18(6):954–971.
17. Vaughn DW, Green S, Kalayanarooj S, Innis BL, Nimmannitya S, Suntayakorn S *et al.* Dengue viremia titer, antibody response pattern, and virus serotype correlate with disease severity. *J Infect Dis* 2000; 181(1):2–9.
18. Scott RM, Nimmannitya S, Bancroft WH, Mansuwan P. Shock syndrome in primary dengue infections. *Am J Trop Med Hyg* 1976;25(6):866–874.

19. Guzman MG, Kouri G, Valdes L, Bravo J, Alvarez M, Vazquez S *et al.* Epidemiologic studies on dengue in Santiago de Cuba, 1997. *Am J Epidemiol* 2000;152(9):793–799.
20. Kouri GP, Guzman MG, Bravo JR, Triana C. Dengue haemorrhagic fever / dengue shock syndrome: lessons from the Cuban epidemic, 1981. *Bull World Health Organ* 1989;67:375–380.
21. Schlesinger RW, Gordon I, Frankel JW, Winter JW, Patterson PR, Dorrance WR. Clinical and serologic response of man to immunization with attenuated dengue and yellow fever viruses. *J Immunol* 1956;77(5):352–364.
22. Hammon WM. Observations on dengue fever, benign protector and killer: a Dr Jekyll and Mr Hyde. *Am J Trop Med Hyg* 1969;18(2):159–165.
23. Seet RCS, Ooi E, Wong HB, Paton NI. An outbreak of primary dengue infection among migrant Chinese workers in Singapore characterized by prominent gastrointestinal symptoms and a high proportion of symptomatic cases. *J Clin Virol* 2005;33:336–340.
24. Gubler DJ, Reed D, Rosen L, Hitchcock JC, Jr. Epidemiological, clinical, and virologic observations on dengue in the kingdom of Tonga. *Am J Trop Med Hyg* 1978;27(3):581–589.
25. Chandler AG, Rice L. Observations on the etiology of dengue fever. *Am J Trop Med* 1923;3:233–262.
26. Rice L. A clinical report of the Galveston epidemic of 1922. *Am J Trop Med* 1923;3:73–90.
27. Gubler DJ, Suharyono W, Lubis I, Eram S, Suliantisaroso J. Epidemic dengue hemorrhagic fever in rural Indonesia. I. Virological and epidemiological studies. *Am J Trop Med Hyg* 1979;28(4):701–710.
28. Buchy P, Vo VL, Trinh TX, Glaziou P, Le TT, Le VL *et al.* Secondary dengue virus type 4 infections in Vietnam. *Southeast Asian J Trop Med Public Health* 2005;36:178–185.
29. Nisalak A, Endy TP, Nimmannitya S, Kalayanarooj K, Thisyakorn U, Scott RM *et al.* Serotype-specific dengue virus circulation and dengue disease in Bangkok, Thailand from 1973 to 1999. *Am J Trop Med Hyg* 2003;68:191–202.
30. Duangchanda S, Tanaka M, Morita K, Rojanasuphot S, Igarashi A. Comparative nucleotide and deduced amino acid sequence of the envelope glycoprotein gene among three dengue virus type 2 strains isolated from patients with different disease severities in Maha Sarakham, northeast Thailand. *Southeast Asian J Trop Med Public Health* 1994;25:243–251.

31. Thant KZ, Morita K, Igarashi A. Sequences of E/NS1 gene junction from four dengue-2 viruses of northeastern Thailand and their evolutionary relationships with other dengue-2 viruses. *Microbiol Immunol* 1995;39:581–590.
32. Thant KZ, Morita K, Igarashi A. Detection of the disease severity-related molecular differences among new Thai dengue-2 isolates in 1993, based on their structural proteins and major non-structural protein NS1 sequences. *Microbiol Immunol* 1996;40:205–216.
33. Rico-Hesse R, Harrison LM, Nisalak A, Vaughn DW, Kalayanarooj S, Green S *et al.* Molecular evolution of dengue type 2 virus in Thailand. *Am J Trop Med Hyg* 1998;58(1):96–101.
34. Raekiansyah M, Pramesyanti A, Bela B, Kosasih H, Ma'roef CN, Tobing SY *et al.* Genetic variations and relationship among dengue virus type 3 strains isolated from patients with mild or severe form of dengue disease in Indonesia and Thailand. *Southeast Asian J Trop Med Public Health* 2005;36: 1187–1197.
35. Pandey BD, Igarashi A. Severity-related molecular differences among nineteen strains of dengue type 2 viruses. *Microbiol Immunol* 2000;44(3): 179–188.
36. Pandey BD, Morita K, Hasebe F, Parquet MC, Igarashi A. Molecular evolution, distribution and genetic relationship among the dengue 2 viruses isolated from different clinical severity. *Southeast Asian J Trop Med Public Health* 2000;31:266–272.
37. Mangada MN, Igarashi A. Sequences of terminal non-coding regions from four dengue-2 viruses isolated from patients exhibiting different disease severities. *Virus Genes* 1997;14(1):5–12.
38. Zhou Y, Mammen MP Jr, Klungthong C, Chinnawirotpisan P, Vaughn DW, Nimmannitya S *et al.* Comparative analysis reveals no consistent association between the secondary structure of the 3′-untranslated region of dengue viruses and disease syndrome. *J Gen Virol* 2006;87(Pt 9): 2595–2603.
39. Watts DM, Porter KR, Putvatana P, Vasquez B, Calampa C, Hayes CG *et al.* Failure of secondary infection with American genotype dengue 2 to cause dengue haemorrhagic fever [see comments]. *Lancet* 1999;354(9188): 1431–1434.
40. Kochel TJ, Watts DM, Halstead SB, Hayes CG, Espinosa A, Felices V, Caceda R, Bautista T, Montoya Y, Douglas S, Russell KL. Effect of dengue-1

antibodies on American dengue-2 viral infection and dengue haemorrhagic fever. *Lancet* 2002;360:310–312.

41. Kochel TJ, Watts DM, Gozalo AS, Ewing DF, Porter KH, Russell KL. Cross-serotype neutralization of dengue virus in *Aotus nancymae* monkeys. *J Infect Dis* 2005;191:1000–1004.

IV. Dengue Hemorrhagic Fever is Caused by an Abnormal or Accelerated T Cell Response to Infection

Pro: *David M. Morens, MD*, Office of the Director, NIAID, NIH, Bethesda, MD, USA

Abnormal T cell responses

Dengue is thought to be an example of a general immunological phenomenon in which prior immune experience affects immune response.[4,5] Dengue virus type-specific and cross-reactive T cells have been detected in the blood of individuals during the acute phase of dengue virus infections. Studies of CD4 and CD8 T-cell responses after primary dengue virus infections find that CD4 T cells make greater amounts of IFN-γ when stimulated by homotypic viral antigens while the production of TNF-α was higher than that of IFN-γ stimulation by heterotypic viral CD4 epitopes.[1] CD8 T cells have been found to exhibit partial agonist responses.

It has been hypothesized that serotype cross-reactive T cells raised to the first infection dengue serotype are of low avidity and predominate during a secondary dengue infection — a phenomenon categorized by its authors as an example of "original antigenic sin."[2] In children with severe dengue disease there were lower numbers of dengue peptide-specific CD4+ T cells directed to infecting viruses than to heterologous viruses in PBMC's stained with human leukocyte antigen (HLA) A11 tetramers.[2] More importantly, the numbers of cytotoxic T lymphocytes were lower in DHF than in less severe dengue disease or healthy controls, with the majority of CD8+ T cells undergoing apoptosis. Studies of HLA A2-restricted CD 8 T cells responses in individuals immune to a single dengue virus showed both quantitative and qualitative differences in their cytokine responses to variant dengue epitopes, which suggests that the previous infection history as well as the sequence of heterologous dengue virus infection may alter subsequent clinical outcome.[3] For some epitopes, a single dengue variant was able to elicit the highest response in all donors, regardless of the infecting serotype. Moreover, an HLA A11-restricted NS3 T cell epitope has been found to be presented by HLA A24, an allele belonging to a different HLA superfamily.[6] These findings suggest that variant immunodominant dengue epitopes may control the outcome of heterotypic dengue infections.

During primary dengue infections high avidity T cells competently terminate cellular infection, while during heterotypic dengue infections low avidity T cells may fail to eliminate dengue-infected cells, resulting in "enhanced" infection and severe disease.[7] It has been suggested that "dysfunction of DENV-specific CD8+ lymphocytes due to apoptosis and inappropriate stimulation of lymphocytes contributes to the high number of viral infected cells which may lead to the production of various mediators that regulate vascular permeability and damage of the blood clotting system."[8]

Accelerated immune response

The secondary immune response has been identified by many workers as an important contributor to enhanced severity of dengue infections. Representative statements are briefly abstracted below:

> "One of the most striking epidemiological observations is the association of DHF with secondary DENV infections. DHF can clearly occur during a primary DENV infection. This has been observed in infants during the first year of life who have residual anti-DENV antibody from transfer across the placenta…. However, observational studies as well as several well-designed prospective cohort studies have demonstrated convincingly that the risk of DHF is 15–80 times higher during secondary DENV infections."[9]

"This [secondary infection] stimulates the efferent response phase of the disease involving the activation of DENV-reactive memory T lymphocytes to proliferate and produce proinflammatory type 1 cytokines, such as IFN-γ and TNF-α, which directly affect vascular endothelial cells to produce plasma leakage."[9]

"The increased incidence of DHF in secondary DENV infections is explained in this model by the greater magnitude and more rapid kinetics of activation of memory T lymphocytes already present at the time of secondary DENV infections."[9]

"Analysis of T cell responses to DENV prior to secondary DENV infections showed an association between *in vitro* TNF-α responses to DENV antigens and more severe disease during the subsequent infection."[10]

"In the middle stages of infection, the level of T lymphocyte activation is markedly increased, reflecting the increased antigen presentation, the increased frequency of dengue virus-specific T lymphocytes in secondary infection and the more rapid activation and proliferation of memory T lymphocytes. Positive feedback effects of activated dengue virus-specific T lymphocytes on monocytes, thought to be the action of IFN-γ, further contribute to the dysregulation of cytokine production."[11]

Summation

Evidence suggests that secondary infection T cell responses are inefficient in killing dengue-infected cells and at the same time generate high quantities of damaging cytokines. These attributes of secondary T cell responses may be responsible for DHF.

References

1. Mangada MM, Rothman AL. Altered cytokine responses of dengue-specific CD4+ T cells to heterologous serotypes. *J Immunol* 2005;175:2676–2683.

2. Mongkolsapaya J, Dejnirattisai W, Xu XN. Original antigenic sin and apoptosis in the pathogenesis of dengue hemorrhagic fever. *Nat Med* 2003;9: 921–927.
3. Bashyam HS, Green S, Rothman AL. Dengue virus-reactive CD8+ T cells display quantitative and qualitative differences in their response to variant epitopes of heterologous viral serotypes. *J Immunol* 2006;176:2817–2824.
4. Welsh RM, Selin LK. No one is naive: the significance of heterologus T-cell immunity. *Nat Rev Immunol* 2002;2002:417–426.
5. Budd RC. Activation-induced cell death. *Curr Opin Immunol* 2001;13: 356–362.
6. Mongkolsapaya J, Dkuaangchinda T, Dejnirattisai W, Vasanawathana S, Avirutnan P, Jairungsri A *et al.* T cell responses in dengue hemorrhagic fever: are cross-reactive T cells suboptimal? *J Immunol* 2006;176:3821–3829.
7. Rothman A. Dengue: defining protective versus pathologic immunity. *J Clin Invest* 2004;113:946–951.
8. Ubol S, Masrinoul P, Jaijarnwanich J, Kalayanarooj SM, Chareonsirisuthikul T, Kasisith J. Differences in PBMC-global gene expression indicate a significant role in the innate responses in DF but not DHF progression. *J Infect Dis* 2007, in press.
9. Rothman AL. Immunology and immunopathogenesis of dengue disease. *Adv Virus Res* 2003;60:397–419.
10. Mangada MM, Endy TP, Nisalak A, Chunsuttiwat S, Vaughn DW, Libraty DH *et al.* Dengue-specific T cell responses in peripheral blood mononuclear cells obtained prior to secondary dengue virus infections in Thai schoolchildren. *J Infect Dis* 2002;185(12):1697–1703.
11. Rothman AL, Ennis FA. Immunopathogenesis of dengue hemorrhagic fever. *Virology* 1999;257(1):1–6.

IV. Dengue Hemorrhagic Fever is Caused by an Abnormal or Accelerated T Cell Response to Infection

Con: *Scott B. Halstead, MD*, Director, Supportive Research and Development, Pediatric Dengue Vaccine Initiative, International Vaccine Institute, Seoul, South Korea

The hypotheses and conclusions that inefficiency in the elimination of heterologous dengue-infected cells results in an enhanced number of dengue-infected cells or that accelerated secondary T cell responses contribute to enhancing the severity of the immune elimination response in dengue-infected individuals are based on a crucial misunderstanding of dengue epidemiology. Classic dengue vascular permeability syndrome in countries with endemic transmission of multiple dengue viruses occurs regularly in two groups: individuals experiencing heterologous dengue infections and infants born to dengue-immune mothers during a primary infection.[1–3] These two groups constitute 95% and 5% of hospitalized cases in children, respectively.[2] Published descriptions of the clinical course in each group show the disease to be virtually identical, although somewhat more severe in infants than in children.[4–6] Circulating cytokine levels in both groups are similar.[5] Crucially, both groups are hospitalized on the same median day after onset of fever and the duration of hospitalization is similar.[6]

Mothers of infants with DHF/DSS circulate antibodies against three or four dengue viruses[3] (Simmons C, personal communication). Rothman has concluded that the risk of acquiring DHF during a second dengue infection may be "15–80-fold higher" than in infants with a primary dengue infection.[7] However, he has misquoted the source. Children with second dengue infections have a 15–80-fold increased relative risk of acquiring DHF as compared with children in the same population with primary dengue infections.[8] Passively acquired antibodies are very efficient in producing DHF during primary infections of infants. Infant cases, if distributed over the full 12 months, would constitute 10% of all cases.[9] Assuming that infants are at risk of enhanced infections for only a single month, then 10% of dengue-infected infants develop DHF/DSS, a rate that is nearly five times higher than hospitalization rates for second dengue infections at any age.[3,9,10]

Because primary dengue infections result in authentic DHF, a secondary immune response cannot be required to produce this syndrome. T cell researchers need to study this group of cases and search for immunological mechanisms that unify primary and secondary infection DHF. Clearly, T cells responding to a first infection are equally efficient, as is the T cell response to heterologous infection as the kinetics of the elimination of virus-infected cells during primary immune responses in infants is the same as during secondary immune responses in older children. If an observed smaller population of CD8+ T cells targeted at heterologous antigens is inefficient in achieving effective elimination, the onset of vascular permeability during secondary dengue infections should be delayed. This phenomenon has not been observed. A misunderstanding of the differing kinetics of the B and T cell responses may contribute to the problem. As summarized by Green, "According to the theory of antibody-dependent enhancement, plasma leakage in DHF is a direct consequence of increased viral burden. The onset of plasma leakage, however, occurs up to several days after viremia has been significantly reduced or cleared, which suggests an immune-mediated mechanism for DHF."[11] Green mistakes the removal of extracellular virus, a function of the early appearance of antibodies during both primary and secondary infections, for the final elimination of virus-infected cells, a T cell function. As summarized in Chap. 9, studies on rhesus monkeys observed peak cellular infection to occur a day or two after viremia terminated.

The speculations about aberrant T cell responses are totally unnecessary. It has long been noted that individuals with DHF are unusually healthy, with normal immune responses. In the future, if time and effort are invested in studying the immunology of infant DHF/DSS, it is to be expected that a unitary explanation will link all DHF cases — when dengue infections occur in the presence of enhancing antibodies, an expanded infected cell mass results. Quite simply, primary or secondary inflammatory T cell responses are proportional to the authentic antigenic load.

Summation

Hypotheses of T cell responses that consider only secondary dengue infections ignore classical DHF cases that occur during late infancy in dengue-endemic countries. The authentic role of T cells must be established by comprehensive studies on infant DHF. Passive antibodies enhance disease efficiently. When corrected for the size of the age group,

DHF occurs in infants at a higher rate than DHF in older children during second dengue infections.

References

1. Halstead SB. Observations related to pathogenesis of dengue hemorrhagic fever. VI. Hypotheses and discussion. *Yale J Biol Med* 1970;42:350–362.
2. Halstead SB, Lan NT, Myint TT, Shwe TN, Nisalak A, Soegijanto S *et al.* Infant dengue hemorrhagic fever: research opportunities ignored. *Emerg Infect Dis* 2002;12:1474–1479.
3. Kliks SC, Nimmannitya S, Nisalak A, Burke DS. Evidence that maternal dengue antibodies are important in the development of dengue hemorrhagic fever in infants. *Am J Trop Med Hyg* 1988;38(2):411–419.
4. Hung NT, Lan NT, Lei HY, Lien LB, Huang KJ, Lin CF *et al.* Association between sex, nutritional status, severity of dengue hemorrhagic fever and immune status in infants with dengue hemorrhagic fever. *Am J Trop Med Hyg* 2005;72:370–374.
5. Hung NT, Lei HY, Lan NT, Lin YS, Huang KJ, Lien LB *et al.* Dengue hemorrhagic fever in infants: a study of clinical and cytokine profiles. *J Infect Dis* 2004;189:221–232.
6. Kalayanarooj S, Nimmannitya S. Clinical presentations of dengue hemorrhagic fever in infants compared to children. *J Med Assoc Thai* 2003;86(Suppl 3):S673–680.
7. Rothman AL. Immunology and immunopathogenesis of dengue disease. *Adv Virus Res* 2003;60:397–419.
8. Sangkawibha N, Rojanasuphot S, Ahandrik S, Viriyapongse S, Jatanasen S, Salitul V *et al.* Risk factors in dengue shock syndrome: a prospective epidemiologic study in Rayong, Thailand. I. The 1980 outbreak. *Am J Epidemiol* 1984;120:653–669.
9. Halstead SB. Immunological parameters of *Togavirus* disease syndromes. In: Schlesinger RW (ed.) *The Togaviruses: Biology, Structure, Replication.* Academic, New York, 1980, pp. 107–173.
10. Halstead SB. Neutralization and antibody-dependent enhancement of dengue viruses. In: Chambers TJ, Monath TP (eds.) *The Flaviviruses: Pathogenesis and Immunity*. Elsevier, New York, 2003, pp. 422–467.
11. Green S, Rothman A. Immunopathological mechanisms in dengue and dengue hemorrhagic fever. *Curr Opin Infect Dis* 2006;19:429–436.

V. Dengue Hemorrhagic Fever is Caused by Autoimmune Phenomena Triggered by a Dengue Viral Infection

Pro: *David M. Morens, MD*, Office of the Director, NIAID, NIH, Bethesda, MD, USA

There are currently several proposed mechanisms for explaining autoimmune responses to viral infections, including molecular mimicry, bystander activation and viral persistence.[1–3] Markoff *et al.* and Falconar introduced the concept of viral–human mimicry in dengue.[4,5] A 20-amino-acid sequence in the envelope region of dengue 4 was found to be similar to sequences in a family of clotting proteins including plasminogen. Antibodies to plasminogen peptides were detected in 70% of acute phase sera from Thai dengue patients; in Tahiti such antibodies were correlated with secondary infections and with hemorrhage.[4,6] In a Taiwan study, 1–4-month convalescent sera from 16 patients with dengue fever reacted with peptides 759–779 of human plasminogen and produced modest reductions in serum plasmin activity.[7] Falconar observed that DENV NS1 antibodies generated in mice cross-reacted with human blood clotting proteins, integrin/adhesion proteins, platelets and endothelial cells.[5] Furthermore, a cross-reactive DENV NS1 monoclonal antibody injected in mice produced hemorrhages.

Other studies have shown that antibodies in sera from dengue patients cross-react with endothelial cells.[8] Levels of antiplatelet and antiendothelial cell autoantibodies in sera from DHF/DSS patients were higher than those in sera from DF patients. Absorption experiments revealed that anti-DENV NS1 antibodies partly accounted for the cross-reactivity and endothelial cell apoptosis. Dengue patient sera resulted in complement-mediated platelet lysis and inhibited ADP-induced platelet aggregation. Further evidence of a role for autoantibodies was the observation in DENV-infected mice of transient thrombocytopenia associated with the generation of antiplatelet antibodies.[9] *In vitro*, the binding of DENV NS1 induced apoptosis of endothelial cells that was mediated by nitric oxide. Production of NO caused upregulation of p53 and Bax, as well as downregulation of Bcl-2 and Bcl-xL, leading to cytochrome c release and caspase-3 activation.[10,11]

An important point is the correlation between the production of antibodies and damage to platelets. This has been described as follows: "The kinetics of anti-platelet autoantibodies showed a gradual decrease along acute, convalescent and later stages.... The platelet counts reached normal range in both dengue fever and DHF/DSS patients in convalescent phase. However, the anti-platelet IgM autoantibodies were still present at significant levels during this time period. It is speculated that in the acute phase those platelets expressing specific surface molecules are recognized by the autoantibodies and subsequently lysed. The newly replaced platelets from hematopoetic progenitors in the convalescent phase do not or slightly express these surface molecules, the platelet levels therefore maintain normal...only some population of (normal) platelets could react with patient IgM when platelets from healthy volunteers were used...platelet surface molecules recognized by autoantibodies present in patient sera need to be identified."[12]

In addition to platelet autoantibodies, anti-DENV NS1 may induce inflammatory activation of endothelial cells. The expression of IL-6, IL-8 and MCP-1 in endothelial cells increased after incubation with anti-DENV NS1.[13] Increases in both adhesion molecule ICAM-1 expression and the adhesion ability of peripheral blood mononuclear cells to endothelial cells were also observed. Unpublished data from these authors showed that mice treated with anti-DENV NS1 leaked dye from vessels into skin tissue.

Summation

Molecular mimicry occurs between components of dengue viruses and elements of the platelet and endothelial systems. The presence of autoantibodies raises troubling questions about consequent damage to tissues. These questions should be answered.

References

1. Fujinami RS, von Herrath MG, Christen U. Molecular mimicry, bystander activation, or viral persistence: infections and autoimmune disease. *Clin Microbiol Rev* 2006;19:80–94.

2. Kim B, Kaistha SD, Rouse BT. Viruses and autoimmunity. *Autoimmunity* 2006;39:71–77.
3. Rouse BT, Deshpande S. Viruses and autoimmunity: an affair but not a marriage contract. *Rev Med Virol* 2002;12:107–113.
4. Markoff LJ, Innis BL, Houghten R, Henchal LS. Development of cross-reactive antibodies to plasminogen during the immune response to dengue virus infection. *J Infect Dis* 1991;164:294–301.
5. Falconar AK. The dengue virus nonstructural-1 protein (NS1) generates antibodies to common epitopes on human blood clotting, integrin/adhesin proteins and binds to human endothelial cells: potential implications in haemorrhagic fever pathogenesis. *Arch Virol* 1997;142(5):897–916.
6. Chungue E, Poli L, Roche C, Gestas P, Glaziou P, Markoff LJ. Correlation between detection of plasminogen cross-reactive antibodies and hemorrhage in dengue virus infection. *J Infect Dis* 1994;170:1304–1307.
7. Huang YH, Chang BJ, Lei YH, Liu HS, Liu CC, Wu HL *et al.* Antibodies against dengue virus E protein pepide bind to human plasminogen and inhibit plasmin activity. *Clin Exp Immunol* 1997;110(3):35–40.
8. Lin CF, Lei HY, Shiau AL, Liu CC, Liu HS, Yeh TM *et al.* Antibodies from dengue patient sera cross-react with endothelial cells and induce damage. *J Med Virol* 2003;69:82–90.
9. Huang KJ, Li SY, Chen SC, Liu HS, Lin YS, Yeh TM *et al.* Manifestation of thrombocytopenia in dengue-2-virus-infected mice. *J Gen Virol* 2000;81(Pt 9): 2177–2182.
10. Lin CF, Lei HY, Shiau AL, Liu HS, Yeh TM, Chen SH *et al.* Endothelial cell apoptosis induced by antibodies against dengue virus nonstructural protein 1 via production of nitric oxide. *J Immunol* 2002;169:657–664.
11. Lin YS, Lin CF, Lei HY, Liu HS, Yeh TM, Chen SH *et al.* Antibody-mediated endothelial cell damage via nitric oxide. *Curr Pharm Des* 2004;10:213–221.
12. Lin CF, Lei HY, Liu CC, Liu HS, Yeh TM, Wang ST *et al.* Generation of IgM anti-platelet autoantibody in dengue patients. *J Med Virol* 2001; 63(2):143–149.
13. Lin CF, Chiu SC, Hsiao YL, Wan SW, Lei HY, Shiau AL *et al.* Expression of cytokine, chemokine and adhesion molecules during endothelial cell activation induced by antibodies against dengue virus nonstructural protein 1. *J Immunol* 2005;174:395–403.

V. Dengue Hemorrhagic Fever is Caused by Autoimmune Phenomena Triggered by a Dengue Viral Infection

Con: *Scott B. Halstead, MD*, Director, Supportive Research and Development, Pediatric Dengue Vaccine Initiative, International Vaccine Institute, Seoul, South Korea

The proponents of hypotheses about the autoimmune mechanism causing the dengue hemorrhagic fever syndrome center many arguments upon the observation that NS1 protein exhibits molecular structural mimicry with human platelets, endothelial cells and blood clotting proteins. Antibodies raised in mice to DENV NS1 react with these mimetic epitopes on human platelets and endothelial cells, resulting in short-lived thrombocytopenia in mice and vascular permeability in *in vitro* systems.

Heterophile antibodies occur frequently in normal vertebrates and, in some instances, accompany viral infections. The most notable illustration of heterophile antibodies accompanying a viral infection is the occurrence of such antibodies accompanying EB virus infections.[1] Nonspecific inhibitors present in normal human sera are removed by incubation with guinea pig kidney and beef erythrocytes. Following EB virus infections, early convalescent phase adsorbed sera react with sheep or horse erythrocytes. The literature on autoantibodies following viral infections is filled with suggested immunopathologic mechanisms, many of which explain the etiology of chronic diseases.[2]

Putting the complex issue of the biological relevance of heterophile or autoantibodies aside, problems not satisfactorily solved by a role for autoimmunity in dengue pathophysiology are their inconsistencies with the epidemiology of dengue hemorrhagic fever and the matter of timing. DHF regularly occurs during first infections in less than 12-month-old infants who had been born to dengue-immune mothers. The kinetics of the production of IgM DENV NS1 antibodies in infants with primary dengue infections has not been studied carefully. But it is likely that such antibodies do not appear earlier than 5 days after the onset of fever. However, thrombocytopenia in these infants regularly starts on day 2 or 3 and vascular permeability occurs around 5 days after the onset of viremia (fever). These clinical phenomena occur before or just as IgM antibodies to dengue antigens become detectable. Furthermore, thrombocytopenia and vascular permeability both end just as quickly as they begin — a return to normal values starting around one day after the end of fever. One cannot imagine how NS1 antibodies can produce thrombocytopenia and endothelial damage leading to vascular permeability so early in the evolution of the primary antibody response. What is more difficult to explain is the rapid reversion of thrombocytopenia and vascular permeability to normal at the time when large amounts of anti-NS1 are being produced. Since the anti-NS1 response persists, why then are infants not left with chronic vascular permeability and thrombocytopenia? What is the precedent for the concept that antigens on the surface of platelets are removed and not present on replacement platelets?[3]

Similar problems attend the timing of thrombocytopenia and vascular permeability in patients with the much-more-studied cases of dengue diseases accompanied by secondary responses. It is feasible to posit that anamnestic anti-NS1 IgG antibodies appear in the blood early during the viremic period. But the highest titers of anti-NS1 antibodies circulate many days after the end of viremia, long after platelet counts have returned to normal and vascular leaks have ceased. In one report, antibodies reactive with a 20-amino-acid segment of the plasminogen molecule have been observed to appear in the acute stage and disappear in convalescence[4] and in another to be present many months after the end of infections.[5] Recently, interest in the antiplasminogen phenomena seems to have waned, but, as with other autoimmune hypotheses, this one also must resolve the timing problem and provide explanatory mechanisms for the clinical phenomena observed in the two immunological forms of DHF.

Summation

Hypotheses concerning autoimmune contributions of dengue viral infection to thrombocytopenia and vascular permeability are at variance with the kinetics of the pathophysiological and immune responses.

References

1. Paul JR, Bunnell WW. The presence of heterophile antibodies in infectious mononucleosis. *Am J Med Sci* 1932;183:90–104.
2. Fairweather D, Kaya Z, Shellam GR *et al.* From infection to autoimmunity. *J Autoimmun* 2001;16(3):175–186.
3. Lin CF, Lei HY, Liu CC *et al.* Generation of IgM anti-platelet autoantibody in dengue patients. *J Med Virol* 2001;63(2):143–149.
4. Markoff LJ, Innis BL, Houghten R, Henchal LS. Development of cross-reactive antibodies to plasminogen during the immune response to dengue virus infection. *J Infect Dis* 1991;164:294–301.
5. Huang YH, Chang BJ, Lei YH *et al.* Antibodies against dengue virus E protein peptide bind to human plasminogen and inhibit plasmin activity. *Clin Exp Immunol* 1997;110(3):35–40.

Index